JAMES K. CLARKE

MATHEMATICS APPLICATIONS
UNIT 2

Copyright © 2015 JAMES K. CLARKE

Published by Vivid Publishing
P.O. Box 948, Fremantle
Western Australia 6959
www.vividpublishing.com.au

Mathematics Applications Unit 2 - 1st edition

Cover designed by Nicholas Miles-Gray
Answers checked by Mitchell Ford, Travis Williams
Project edited by Isaac Maldonado, Deepesh Gendah

National Library of Australia Cataloguing-in-Publication data:
Creator: Clarke, James K., author.
Title: Mathematics applications Unit 2. Book 1 / James K. Clarke.
ISBN: 9781925209754 (paperback)
Target Audience: For secondary school age.
Subjects: Mathematics--Study and teaching (Secondary)--Australia.
            Mathematics--Problems, exercises, etc.
Dewey Number: 510.76

Updated edition. All rights reserved. No part of this publication may be reproduced, stored in a retrieval system or transmitted in any form or by any means, electronic, mechanical, photocopying, recording or otherwise, without the prior written permission of the copyright holder. The information, views, opinions and visuals expressed in this publication are solely those of the author(s) and do not necessarily reflect those of the publisher. The publisher disclaims any liabilities or responsibilities whatsoever for any damages, libel or liabilities arising directly or indirectly from the contents of this publication.

PREFACE

Mathematics Applications Unit 2 is a course-book based designed for students who want to prepare for the Western Australian Certificate of Education (WACE) examination. The scope of the book is to extend the mathematical skills of students beyond Year 10 level.

Mathematics Applications **Unit 2** is a course-book which focuses on the use of mathematics to solve problems in contexts that involve mathematical modelling. The book covers three main topics which are: **Univariate data analysis and the statistical process**, **Linear equations and their graphs** and **Applications of Trigonometry**. The latter have been subdivided into thirteen chapters as shown in the table of contents.

Univariate data analysis and the statistical process develops students' ability to organise and summarise univariate data in the context of conducting a statistical investigation.

Linear equations and their graphs, on the other hand, uses linear equations and straight-line graphs, as well as linear-piece-wise and step graphs to model and analyse practical situations.

Applications of trigonometry extends students' knowledge of trigonometry to solve practical problems involving non-right- angled triangles in both two and three dimensions, including problems involving the use of angles of elevation and depression and bearings in navigation.

Classroom access to the technology necessary to support the graphical, computational and statistical aspects of this unit is assumed.

Mathematics Applications Unit 2 is a book that includes notes, formulae, guidance of how to make use of technology to solve problems and worked examples that follow a simple, logical and easy-to-understand approach. The exercises have been carefully constructed to reinforce the students' learning of concepts and skills in Mathematics taught in the classroom. The questions are interesting, stimulating and challenging, and will help students develop skills and understanding necessary for them to excel in Mathematics.

Worked solutions have been included for each topic, at the end of the book, to assist in the teaching and learning of mathematical skills.

TABLE OF CONTENTS

1. **Univariate Data: Classify, Organise and Display** page 7
   - 1A Types of data page 7
   - 1B Displaying categorical data 10
   - 1C Displaying discrete variables 13
   - 1D Displaying continuous random variables 17

2. **Summarising Data** page 23
   - 2A Measures of central tendency 23
   - 2B Calculating mean, mode , median and range for raw data 25
   - 2C Finding averages for a frequency table 26
   - 2D Using technology to find averages for raw numbers 31
   - 2E Using technology to find averages for frequency tables 34
   - 2F Using CAS to find averages for grouped frequency tables 37
   - 2G Applications 40

3. **Measures of dispersion or spread** page 43
   - 3A Calculating standard deviation by long hand 43
   - 3B Using technology to calculate standard deviation of raw numbers 47
   - 3C Using technology to calculate standard deviation for frequency tables 49
   - 3D Using technology to calculate standard deviation for grouped frequency tables numbers 53
   - 3E Standard deviation from mean 58

4. **Box Plots and describing distributions** page 61
   - 4A Box and whisker plots 61
   - 4B Identifying outliers 65
   - 4C Skewness 67
   - 4D Comparing box plots 68
   - 4E Describing other distributions 70

5. **Statistical Investigation process** page 73

6. **Solving Equations** page 75
   - 6A Solving equations using technology 75
   - 6B Simple Linear equations 77
   - 6C Equations having unknown on both sides 79
   - 6D Solving equations using cross multiplication 80
   - 6E Using LCM to solve equations 81
   - 6F Applications 82

7. Using equations to solve problems — page 85
- 7A Algebraic equations and expressions — 85
- 7B Math Pyramids — 87
- 7C Applications — 90
- 7D Solving equations using ratios — 93

8. Linear Relationships — page 97
- 8A Gradient — 97
- 8B Determining gradient of a line from a graph — 98
- 8C Determining gradient given two points — 101
- 8D Finding gradient and y-intercept from an equation — 102
- 8E Given a rule how to tabulate — 104
- 8F Linear or not? — 105
- 8G Equation of a line: given gradient and y-intercept — 108
- 8H Equation of a line: gradient given and a point — 109
- 8I Equation of a line passing through two points — 110
- 8J Does the point lie on the line? — 112
- 8K How to sketch a line without a calculator? — 113
- 8L Horizontal and vertical lines — 116
- 8M Sketching lines using technology — 118
- 8N How to find equation of a line given the sketch? — 119
- 8O Applications — 121

9. Piecewise functions — page 127
- 9A Graphing piecewise functions — 127
- 9B Determining equations given graphs of piecewise functions — 129
- 9C Using technology to graph piecewise functions — 131
- 9D Applications — 132

10. Trigonometry for right triangles — page 137
- 10A Hypotenuse, Adjacent and Opposite — 137
- 10B Trigonometric ratios — 138
- 10C Using the calculator — 139
- 10D Finding side lengths — 140
- 10E Finding missing angles — 143
- 10F Applications : Angle of elevation and depression — 146
- 10G Bearings — 149

11. Trigonometry for non-right angled triangles — 153
- 11A Area of triangle: base and height known — 153
- 11B Area of triangle : height unknown — 154
- 11C Area of triangle : Heron's formula — 156
- 11D The Sine Rule : Calculator Assumed — 158
- 11E Sine Rule : Non Calculator — 161

11F	The Cosine Rule (I) finding sides		163
11G	The Cosine Rule (II) finding angles		165
11H	The Cosine Rule (III) finding smallest and largest angle		166
11I	Sine Rule, Cosine Rule and Area Applications		167

12. Simultaneous Linear equations — **173**

12A	Elimination method	173
12B	Substitution method	177
12C	Simultaneous equations : using solve capacity	180
12D	Using graphical facility on calculator	181
12E	The graphical method step by step approach	183
12F	Applications	187

13. The Normal Distribution — **191**

13A	Using Technology	192
13B	The 68%, 95% and 99.7% rule	194
13C	The Inverse Normal Distribution	197
13D	Quantiles or percentiles	200
13E	The z-score or standard score	201

WORKED SOLUTIONS — **203+**

CHAPTER 1

UNIVARIATE DATA : CLASSIFY, ORGANISE AND DISPLAY

1A   TYPES OF DATA

Data can be defined as 'a series of observation, measurement or facts.' Thus to obtain data we need to observe, measure and collect facts about a variable. This variable can take the form of swimming level, country of birth, favourite sport, height, number of rooms in a house, temperature in December and so on.

Group A	Group B
• swimming level • country of birth • favourite sport	• height • number of rooms in a house • Temperature in December

If we study the above table carefully, we can notice important differences between the variables and the way they are measured or observed. In Group A, for instance, we can see that the data are mostly non-numerical and cannot be measured. They are usually termed as **categorical data**. For Group B, on the other hand, all the variables have numbers assigned to them and can be measured. These types of variables are termed as **numerical data**.

Categorical variables classified into two types data : nominal and ordinal. Each of these will be visited in detail with examples below. Numerical data likewise can be subdivided into two types: discrete data and continuous data. These two types of data will be explained with examples as well.

CATEGORICAL VARIABLES

Categorical variables are values or observations that can be sorted into groups or categories. Examples of categorical variables are eye colour, sex, age group, and educational level. Categorical variables can further be divided into two groups: **nominal data** and **ordinal data**.

NOMINAL VARIABLES

Nominal data are values or observations that can be assigned a code in the form of a number where the numbers are simply labels. Nominal data can be counted but not ordered or measured. For example, in a data set of students boys could be coded as 0, girls as 1. The following are examples of nominal variables.

What is your gender?	Which state were you born?	What is the colour of your car?
➢ Male (0) ➢ Female (1)	➢ New South Wales (1) ➢ Western Australia (2) ➢ South Australia (3) ➢ Queensland (4) ➢ Victoria (5)	➢ Red ➢ White ➢ Brown ➢ Green ➢ Grey

ORDINAL VARIABLES

Ordinal variables are values or observations that can have numerical values attached to them; they can be counted and ordered **but not measured**. In the table below, there are some examples of ordinal variables.

Your income level	How do rate the new shops?	Karate level belts
➢ high ➢ medium ➢ low	➢ Unsatisfactory ➢ Satisfactory ➢ Good ➢ Very good ➢ Excellent	➢ White ➢ Yellow ➢ Orange ➢ Green ➢ Blue

NUMERICAL VARIABLES

Numerical data are values or observations that involve measurement or count. We can make measurements such as a student's weight, a tree's height, the temperature in different suburbs at 8 am on Sunday. As a count, we can have examples such as number of classrooms in a school, shoe size at the show store or number of pages of homework done.

Numerical data can further be split into two groups: discrete and variable.

DISCRETE AND CONTINUOUS VARIABLES

Discrete variables represent data that can be counted; it takes particular values and changes in steps. Shoe size is a concrete example of a discrete variable. When we ask our friends their shoe size we expect answers such as 7 or 8.5. The sizes are either a whole number or half a size. Similarly, the number of people attending the Sunday prayer during the month of January is a perfect example of discrete data. A teacher can help 8, 9 or 10 students but not 10.2 students. Continuous variables are measurements that can take any value within a certain range. Continuous data can be counted, ordered and measured. For example, height (1.35m), weight (75.45 kg), temperature (21.45°C), the amount of vitamin C in an orange (35 cl), time to do one lap of the oval (52.546 s). When we measure someone's weight we do not obtain exact values such as 75 kg. The person's weight can be close to 75 kg. As a result we might say that the person's weight lies between 75 kg and 76 kg. Similarly, the time to finish a race is not an exact measurement as it might involve fractions of seconds as well.

The diagram below is a summary of the different types of data.

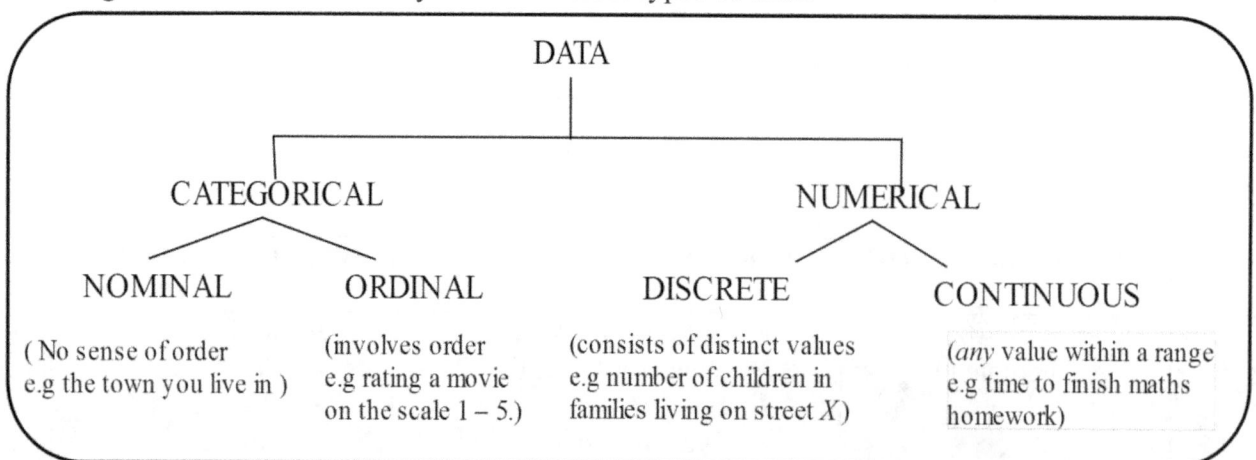

CHAPTER 1 : UNIVARIATE DATA – CLASSIFY, ORGANISE & DISPLAY

EXERCISE 1A

For each of the following categorical data sets, tick whether it is nominal or ordinal.

	Nominal	Ordinal
1. The eye colour of members of your family	✓	
2. Your favourite sport		
3. Language spoken at home		
4. Capital cities in Europe		
5. Grade descriptors in Mathematics		
6. Eyesight test results of tennis players at US Open		

For each of the following numerical data sets, tick whether it is discrete or continuous.

	Discrete	Continuous
7. The number of oranges in each 1 kg bag	✓	
8. The height of students in Year 11		
9. The average time to run a 100m race		
10. The number of children in each family in your street		
11. Number of admissions in Year 1		
12. Shoe size of players in Man Utd team		
13. Size of a car's gas tank		

For each of the following data sets, tick the appropriate box.

	CATEGORICAL		NUMERICAL	
	Nominal	Ordinal	Discrete	Continuous
14. Year level at school		✓		
15. Travel time to work				
16. Colour of cars in a parking lot				
17. Blood type (A, B, O, AB)				
18. Number of pets in a family				
19. Age of patients in a nursing home				
20. Blood pressure of patients				
21. Stage of Cancer (I, II, III or IV)				
22. The number of staff in schools				
23. The cholesterol level in teenagers				
24. The number of lollies in 50g packets				
25. Number of goals scored in a match				

1B DISPLAYING CATEGORICAL DATA

Categorical data are usually displayed with tables, bar charts and pie graphs. An example of each has been given below.

EXAMPLE

The table below shows the subject selection of 80 students in Year 11 Mathematics ATAR Courses at Fluency Senior High.

Courses	Number enrolled
Mathematics Essentials	30
Mathematics Applications	40
Mathematics Methods	10

Display the information as a

(a) Bar chart
(b) Pie chart

SOLUTION

(a)

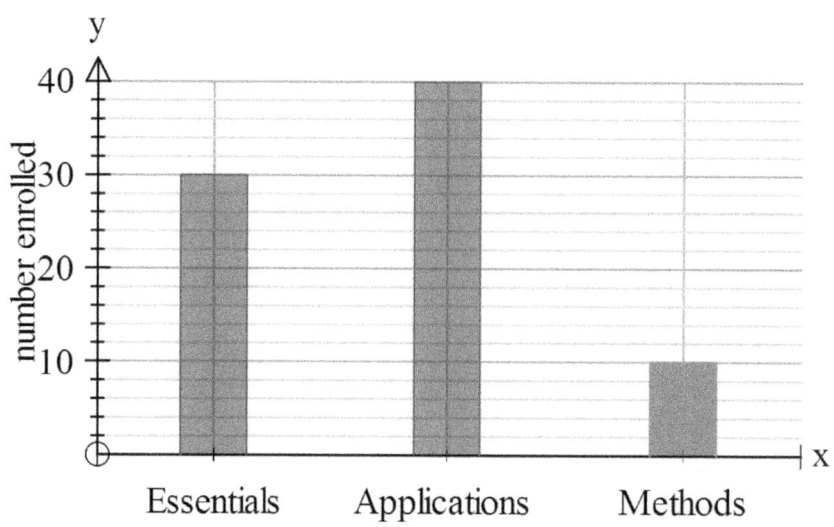

(b)

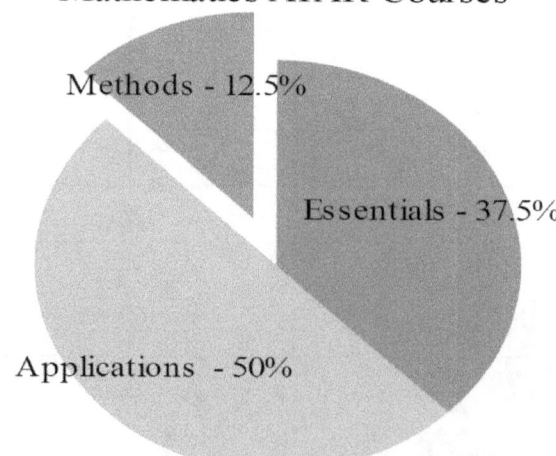

CHAPTER 1 : UNIVARIATE DATA – CLASSIFY, ORGANISE & DISPLAY

EXERCISE 1B

1. Little Johnny has a huge collection of toy cars in his playroom. The table below shows the colours and the number of each car.

Colour	Number of cars
Red	8
Yellow	2
Green	4
Blue	6

Display the above information as a bar chart and pie chart. Use the axes and circle given below.

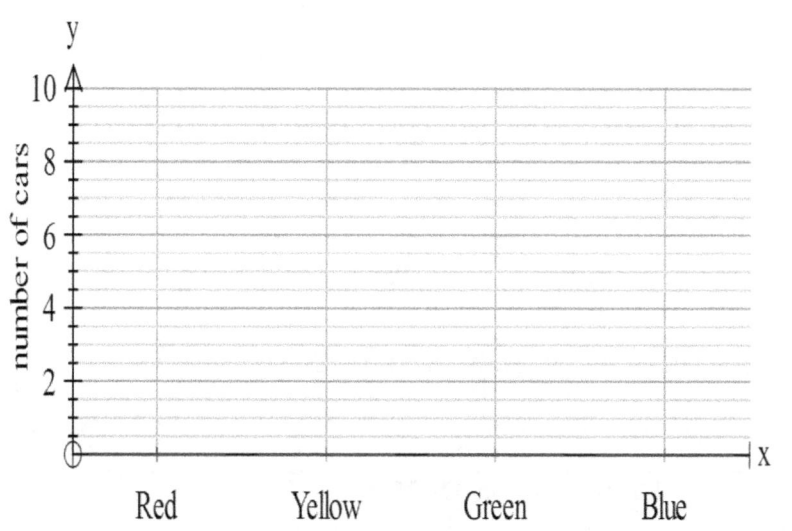
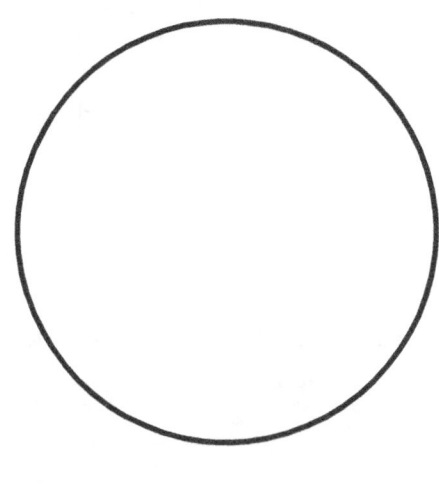

2. Mr Bates did a survey about the eye colour of his students in Year 11 Application Mathematics. He recorded his results in the table below.

Eye Colour	Number of students
Brown	12
Blue	8
Other	5

Display the information on a bar chart below.

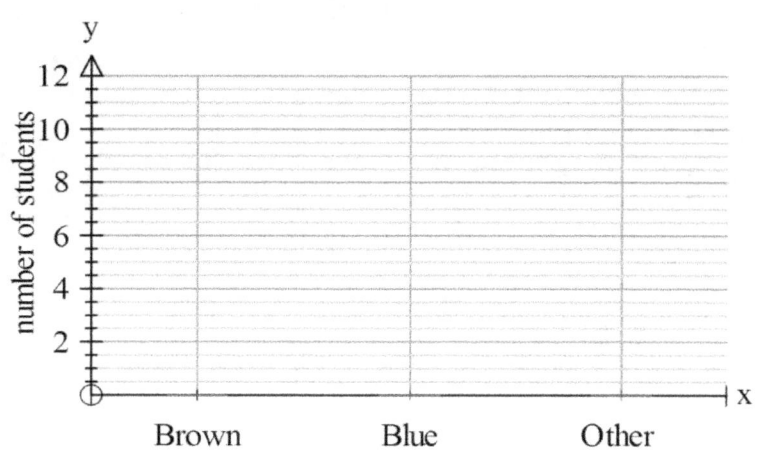

3. Are the recreational facilities at the new sport complex adequate? The responses produced the following results.

	Males	Females
Agree	8	5
Neutral	10	15
Disagree	12	10

Display the above table on a side-by-side bar graph below.

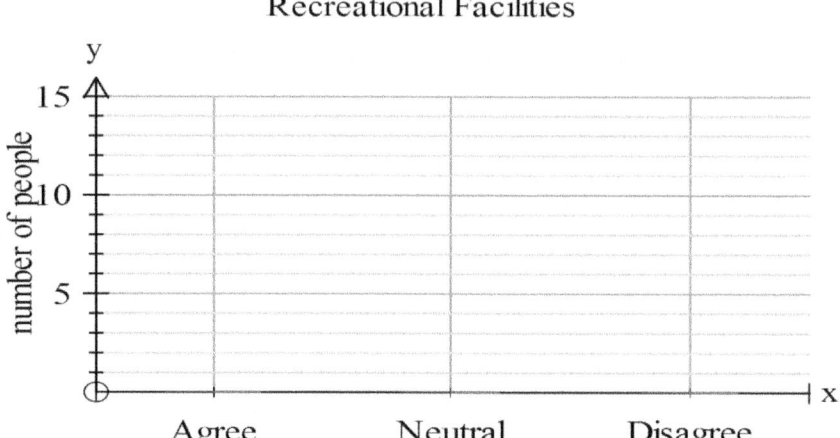

Recreational Facilities

4. The table below shows the results of a survey relating to the BMI of 120 staff members of a company.

		BODY MASS INDEX			
		Underweight	About Right	Overweight	Total
Gender	Male	10	50	20	80
	Female	18	20	2	40
	Total	28	70	22	120

Display the above table on a side-by-side percentage bar graph for each gender below.

Body Mass Index

CHAPTER 1 : UNIVARIATE DATA – CLASSIFY, ORGANISE & DISPLAY

1C   DISPLAYING DISCRETE VARIABLES

The Natural Valley Pre-School Playgroup has 30 students in its kindergarten class. In an attempt to find out how many books each child has read, the following data was collected.

Number of books	23	24	25	26	27	28	29	30
Frequency	2	0	4	2	5	1	6	2

Clearly, the variable here which is number of books can be counted and as a result is a discrete variable. We can display the above table on a **dot frequency diagram** as shown.

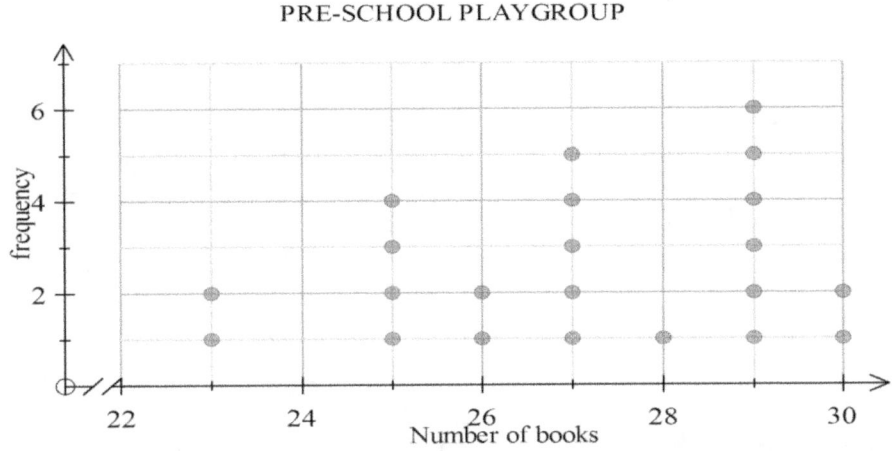

The above table can also be represented by a frequency histogram as shown below.

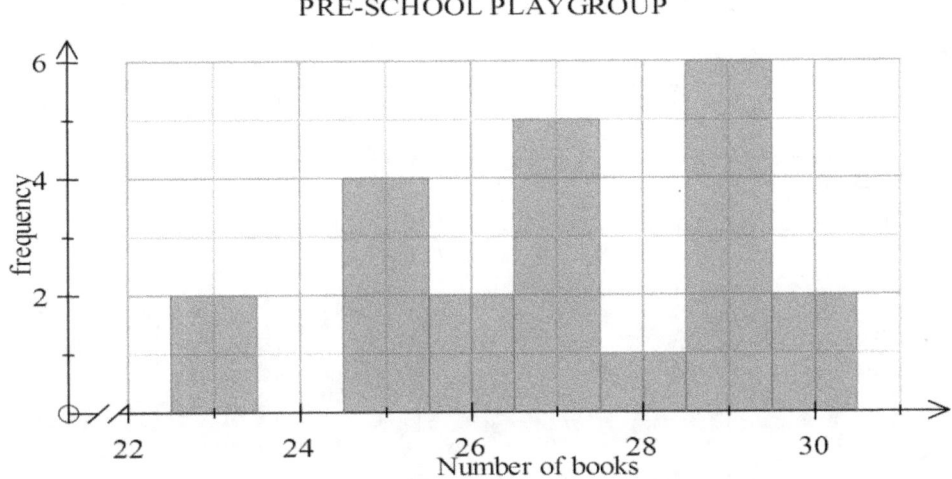

Note that the scores must be written half-way in between the rectangles. The score of 23 lies between 22.5. and 23.5. The number 22.5 is called the **lower boundary** whereas 23.5 is termed as the **upper boundary**. Similarly the score 30 lies between a lower boundary of 29.5 and an upper boundary of 30.5.

As we can see a histogram can be a popular graphing tool in statistics. It can be used to summarise discrete data as well as continuous data. The latter will be dealt in the next part of the chapter. A histogram has an appearance similar to a vertical bar graph as seen earlier, but there are no gaps between the bars. (Unless we have a frequency of zero). In the above example no child read 24 books accounting for the gap between 23.5 and 24.5 on the horizontal axis.

EXERCISE 1C

Draw a frequency histogram for the following:

1.

Score	5	6	7	8	9
frequency	2	7	5	1	4

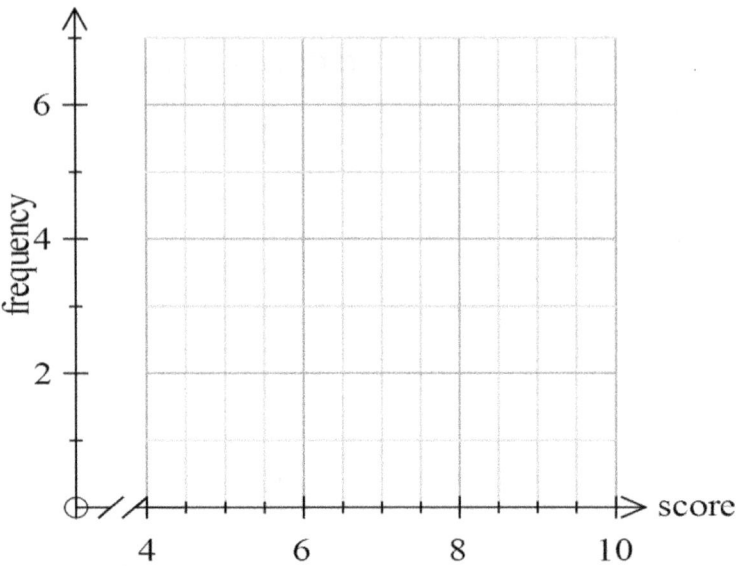

2.

Score	10	11	12	13	14
frequency	3	5	8	2	1

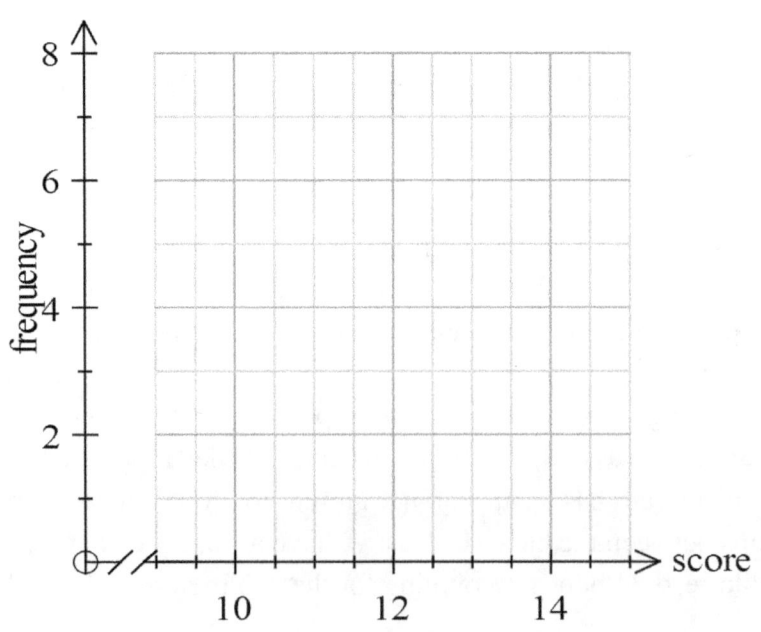

3. An ordinary die was rolled 40 times and the results listed below.

2	5	4	6	1	1	3	3
3	4	1	6	2	2	1	3
6	1	1	2	1	5	6	6
3	1	5	3	4	3	3	6
5	1	1	2	5	6	2	3

Complete the table below and hence draw a dot frequency diagram on the given set of axes.

score	Frequency	Relative frequency $\left(\dfrac{frequency}{total\ frequency}\right)$	Percentage frequency
1	10		
2	6		
3	9		
4	3		
5	5		
6	7		
Total			

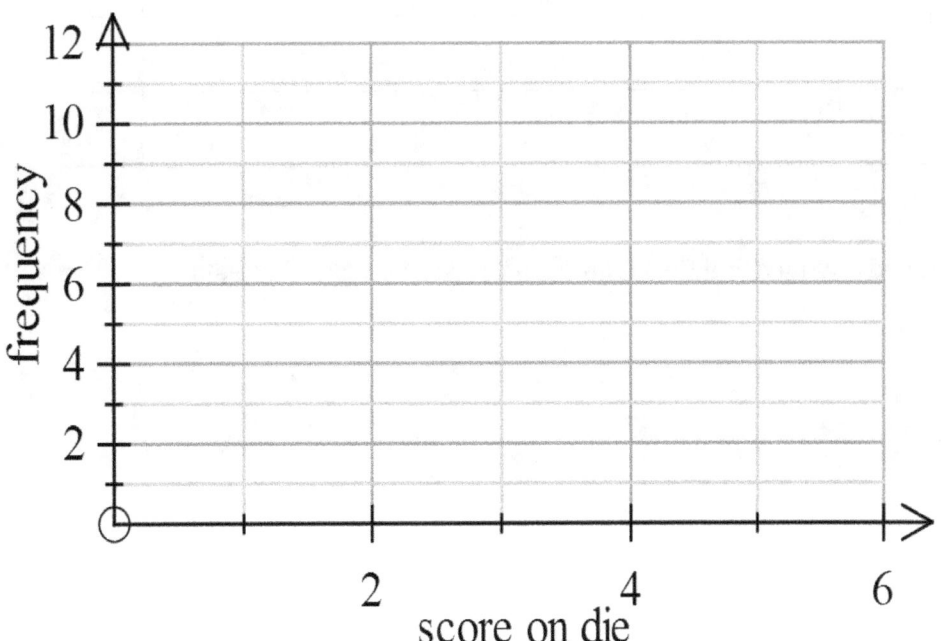

ROLL A DIE

4. The dot frequency diagram shows the amount of pocket money received on a daily basis by a group of 25 students while attending school.

DAILY POCKET MONEY

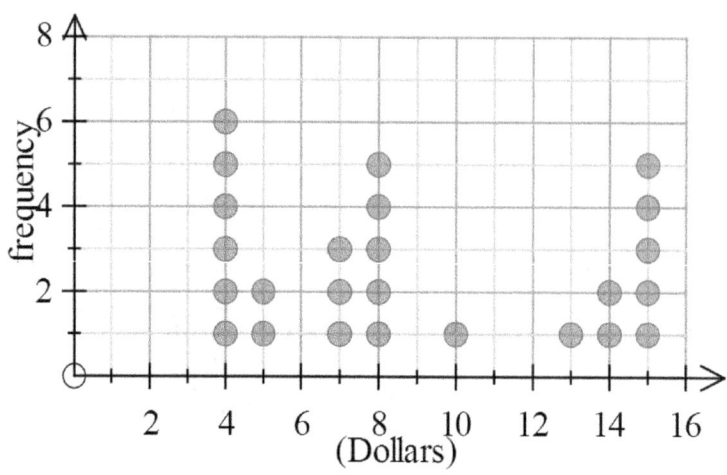

Use the diagram to complete the following table.

Pocket money ($)	Frequency	Relative frequency	Percentage frequency
4			
5			
7			
8			
10			
13			
14			
15			

5. List some advantages and disadvantages of using categorical data.

CHAPTER 1 : UNIVARIATE DATA – CLASSIFY, ORGANISE & DISPLAY

1D   DISPLAYING CONTINUOUS VARIABLES

Imagine there is a large data spread for the amount of pocket money received by a group of students. Then to display the information on a dot frequency diagram might be very time consuming and tiresome. In such a case we can resort to grouping the data into classes and display the information as a frequency histogram as illustrated in the example below.

EXAMPLE 1
25 students travel the following distances in km to attend school every week days.

3	5	4	8	7
4	11	15	2	23
13	10	9	8	29
20	32	37	11	6
7	8	19	52	25

(a) Tabulate the above results in a grouped frequency table.

Distance travelled (km)	frequency
$0 \leq d < 10$	12
$10 \leq d < 20$	6
$20 \leq d < 30$	4
$30 \leq d < 40$	2
$40 \leq d < 50$	0
$50 \leq d < 60$	1

Note that the class $0 \leq d < 10$ includes numbers ranging from 0 to 9 and the class $10 \leq d < 20$ contains numbers from 10 to 19 inclusive.

(b) Hence display the information as a frequency histogram...

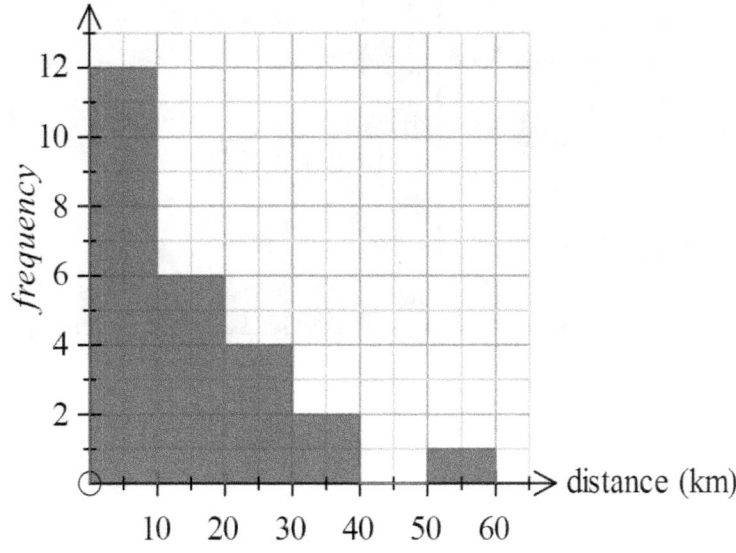

EXAMPLE 2

A group of people carried out a survey to find out how far workers in a particular factory travel to get to work on a particular day. The travelling distance were recorded, to the nearest kilometre, for each of the 30 workers. The results are shown in the table.

Distance to work (minutes)	Number of workers
$5 < x \leq 9$	7
$10 < x \leq 14$	2
$15 < x \leq 19$	10
$20 < x \leq 24$	6
$25 < x \leq 29$	5

Display the above table in a frequency histogram and include the frequency histogram.

SOLUTION

As we can observe, the classes are all of equal width but are discontinuous. The task of the reader is to make the classes continuous implying that each rectangle starts where the previous one leaves off. This can be done as shown in the table.

Distance to work (km)	CONTINUITY (subtract 0.5 from the lower class boundary and add 0.5 to the upper class boundary)	Number of workers
$5 < x \leq 9$	$4.5 < x \leq 9.5$	7
$10 < x \leq 14$	$9.5 < x \leq 14.5$	2
$15 < x \leq 19$	$14.5 < x \leq 19.5$	10
$20 < x \leq 24$	$19.5 < x \leq 24.5$	6
$25 < x \leq 29$	$24.5 < x \leq 29.5$	5

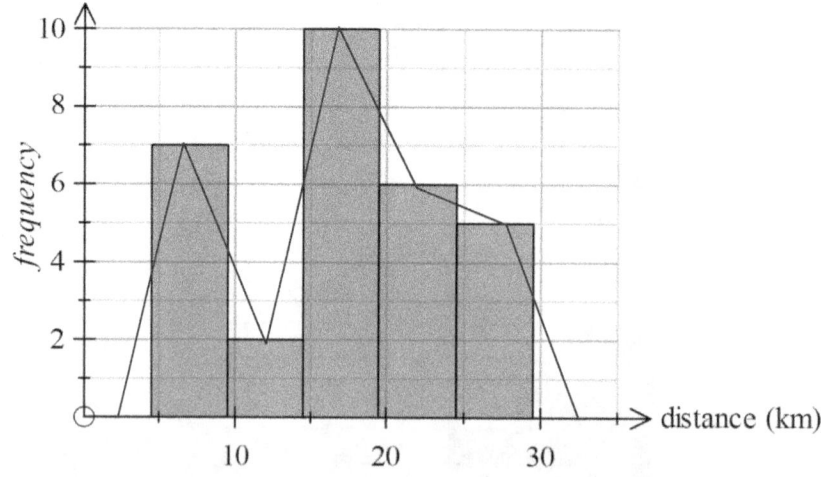

The frequency histogram can be drawn by joining the midpoint of each rectangle by means of straight lines a shown.

CHAPTER 1 : UNIVARIATE DATA – CLASSIFY, ORGANISE & DISPLAY

EXERCISE 1D

1. The table below shows the age distribution of 16 people in a holiday home during Christmas. Display the information as a frequency histogram.

Age (x years)	$0 < x \leq 10$	$10 < x \leq 20$	$20 < x \leq 30$	$30 < x \leq 40$
Number of people	8	3	4	1

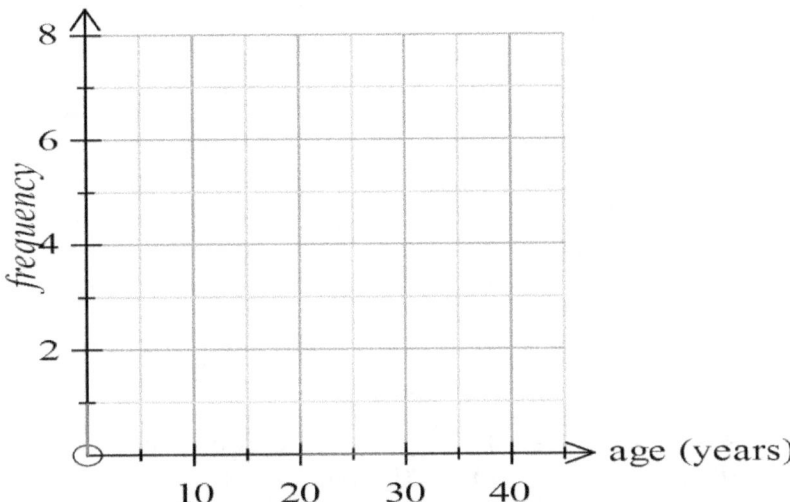

2. Auto Repairs Ltd advertises the company as one of the quickest and most efficient. The table below shows the time taken, in minutes, to get a car repair fixed.

Time in minutes	$0 < x \leq 15$	$15 < x \leq 30$	$30 < x \leq 45$	$45 < x \leq 60$
Number of repairs	2	5	3	6

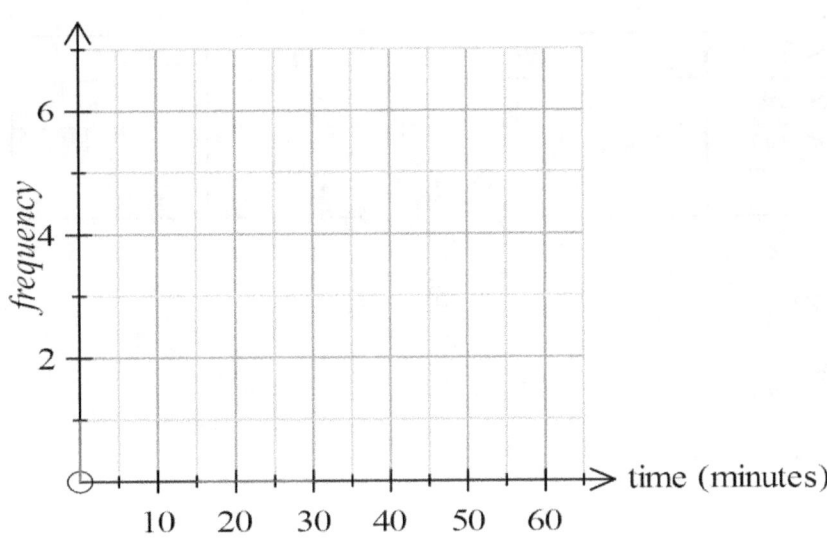

3. The table below shows the speed of cars recorded from 7.30 am to 90 am in a 40 km/h school zone in a suburb. Display the information as a frequency histogram.

Speed km/h	CONTINUITY	Number of cars
$30 < x \leq 39$		6
$40 < x \leq 49$		10
$50 < x \leq 59$		7
$60 < x \leq 69$		2
$70 < x \leq 79$		1

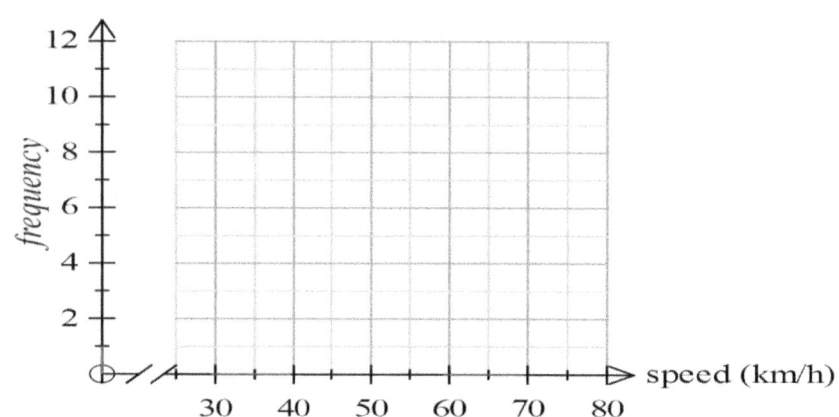

4. The table below shows the marks obtained by a group of 25 Year 11 Mathematics Applications students in a recently held assessment. Draw a percentage frequency histogram.

Marks	CONTINUITY	Number of students (f)	Percentage frequency
$10 < x \leq 14$		3	
$15 < x \leq 19$		7	
$20 < x \leq 24$		11	
$25 < x \leq 29$		2	
$30 < x \leq 34$		0	
$35 < x \leq 39$		2	

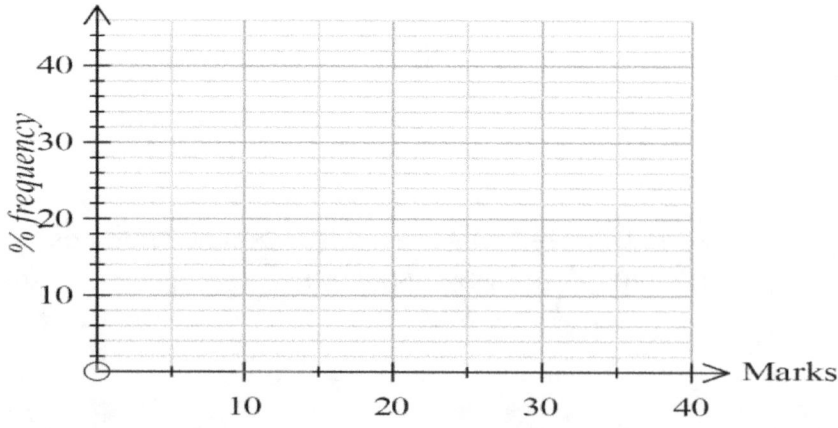

5. Thomas is fond of his dice. He has loads of them and likes to experiment. During his first experiment he rolled a tetrahedral die 40 times having sides labelled 1, 2, 3 and 4. Which of the following histogram is most likely to be obtained from his results?

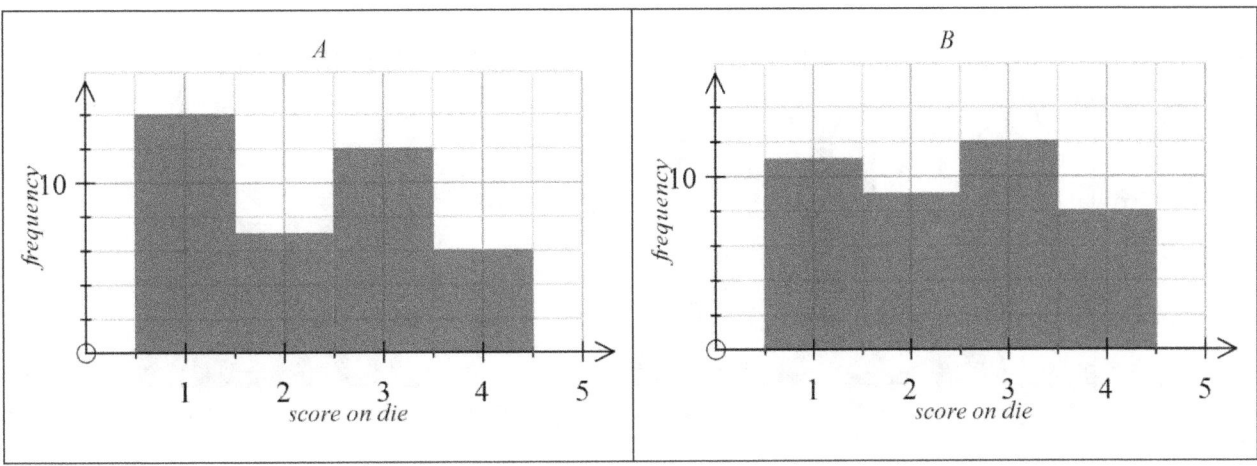

6. In his next experiment he took 5 ordinary dice and rolled them 12 times. Which of the following histogram is more likely to be displayed?

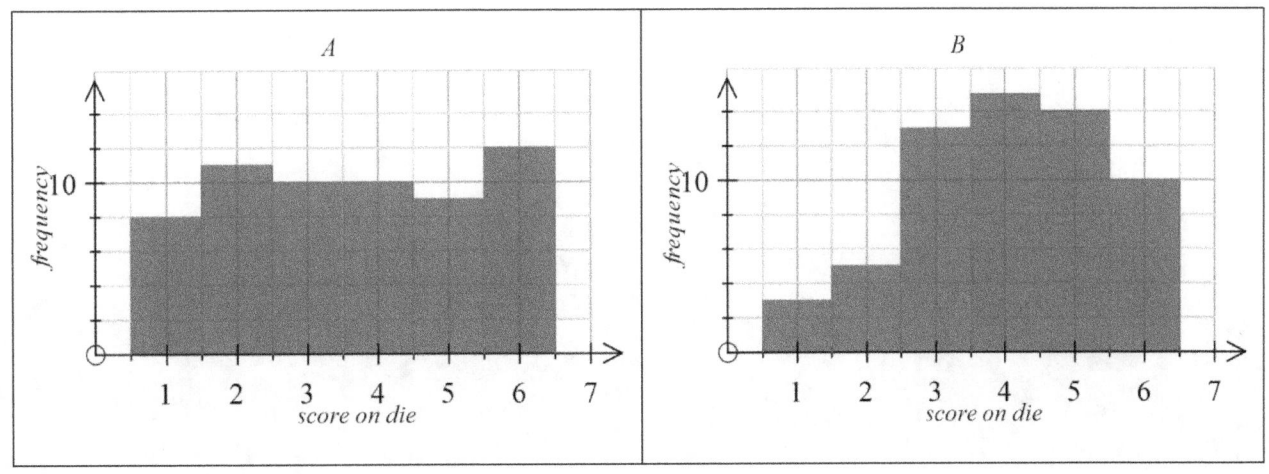

7. In his last experiment, Thomas recorded the marks obtained by all his classmates in their recent semester one exams in mathematics. He grouped the scores and came up with a histogram. Which of the following can be his class performance given that his teacher is lenient and set easy exams?

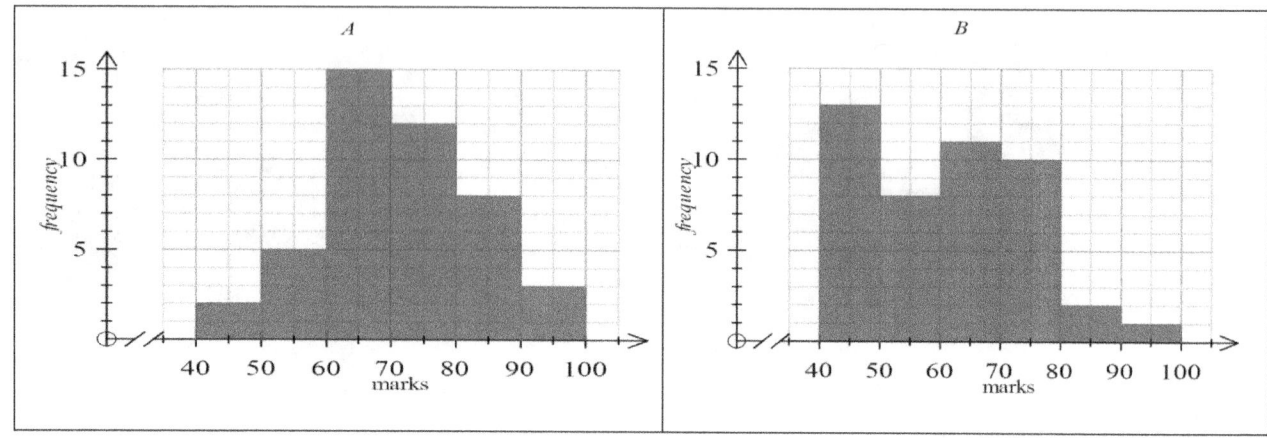

CHAPTER 2

SUMMARISING DATA

2A MEASURES OF CENTRAL TENDENCY

Also known as measures of central location, a measure of central tendency is a single value that attempts to describe a set of data by identifying the central position within that set of data. The mean, median and mode are all very useful statistical measures, but under different situations, some measures of central tendency become more appropriate to use than others. In this part of the chapter, we will look at the mean, mode and median, and learn how to calculate them and under what conditions they are most appropriate to be used.

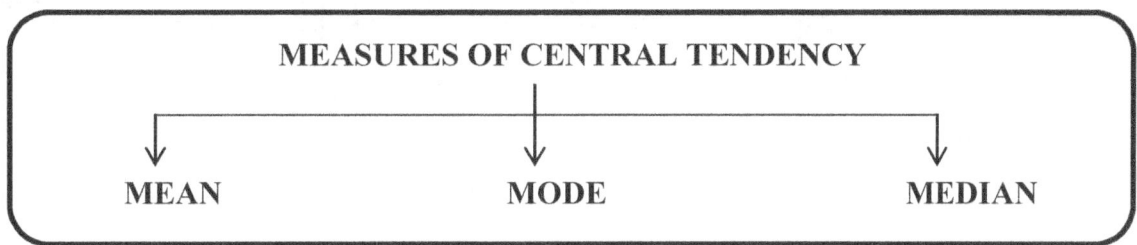

MEAN
The mean or simply known as average is the most popular measure of central tendency. To calculate the mean we divide the sum of all the values in the data set by the number of values in the data set. Mathematically, we denote the mean by the symbol $\bar{x}$ (pronounced x bar), and calculated as

$$\bar{x} = \frac{\Sigma x}{n} \text{ or } \bar{x} = \frac{\Sigma xf}{\Sigma f}$$

where Σ, the Greek letter, pronounced "sigma", which means "the sum of...":

MODE
The mode is the most frequent score in a data set. As such, it is the number with the highest frequency.

MEDIAN
The median is the middle score for a set of data that has been arranged in order. To calculate the median we first need to rearrange that data into ascending order and find the middle score. The position of the median value can be calculated using the formula

median location $= \frac{n+1}{2}$, where n is the number of values in the data set.

RANGE
Range is the difference between the highest and the lowest data.
Range is not a measure of central tendency as it measures a spread of a set of data. However, since it is very easy to compute and it will be seen quite a few times during the course of this chapter, it has been included in some of the exercises to familiarise students with this measure of dispersion or spread.

In the table an attempt has been made to compare the three measures of central tendency by looking at each other's advantages and disadvantages.

	Advantages	Disadvantages
Mean	➤ Most popular in fields such as engineering and business ➤ all the data used to find the answer ➤ useful for comparing data sets ➤ It is unique as it has only one answer	➤ Affected by extreme values (outliers) ➤ Cannot be calculated for nominal data
Mode	➤ Is an actual value of the data ➤ Quick and easy to determine ➤ Not affected by outliers ➤ Only average that can be found for nominal data	➤ Often more than answer or none ➤ Not used for anything else ➤ Can change from sample to sample
Median	➤ Works on ordered data as well as numerical data ➤ Not affected by extreme scores ➤ Useful in calculating percentiles ➤ It is unique as there is only one answer	➤ May not be representative of a sample ➤ Not appropriate for nominal data

EXERCISE 2A

Tick the most appropriate measure of central tendency in each case.

	mean	mode	median
1. Wages of staff at a high school			
2. Most common form of transport			
3. Performance in a recent assessment			
4. House prices in a suburb			
5. Most frequent burger purchased			
6. Yummy chocolate bars sold			
7. Daily high temperature during a week in December			
8. Hours spent playing online games			
9. Allowance received by retired politicians			
10. Favourite cereal for breakfast			
11. Age of Year 11 students at a school			
12. Distance travelled to school			
13. Favourite movie of college students			
14. Preferred sports channel			
15. Age of people at a family get-together			

2B CALCULATING MEAN, MODE, MEDIAN AND RANGE FOR RAW DATA

EXAMPLES

1. Find the mean, mode and median of 2, 3, 5, 8, 3, 4, 10 Also state the range. **SOLUTION** mean, $\bar{x} = \frac{2+3+5+8+3+4+10}{7} = \frac{35}{7} = 5$ Since 3 is the most frequent score, mode = 3 To find median, first, we re-arrange the scores in ascending order 2, 3, 3, 4, 5, 8, 10 Clearly the median is 4. Range = 10 – 2 = 8	2. Find the mean, mode and median of 1, 2, 2, 6, 7, 8, 9, 13 **SOLUTION** mean, $\bar{x} = \frac{1+2+2+6+7+8+9+13}{8} = \frac{48}{8} = 6$ Since 2 is the most frequent score, mode = 2 1, 2, 2, 6, 7, 8, 9, 13 Median location = $\frac{n+1}{2} = \frac{8+1}{2} = 4.5th\ number$ To find the median, we find the average of the 4th and the 5th number. Hence median = $\frac{6+7}{2} = 6.5$

EXERCISE 2B

Without using a calculator, determine the three measures of central tendency: mean, mode and median. Determine the range in each case as well.

1. 4, 5, 10, 7, 4	2. 10, -7, 7, 8, 2, 10
3. 2, 5, 7, 11, 13, -11, 3, 2	4. 4, -4, 5, -5, 8, 12, 8

2C FINDING AVERAGES FOR A FREQUENCY TABLE

EXAMPLE 1
Without using your calculator find the mean, mode and median of the following frequency table.

SOLUTION

Mean, $\bar{x} = \dfrac{\sum xf}{\sum f} = \dfrac{28}{28} = 1$

Scores (x)	Frequency (f)	x f
0	10	0
1	10	10
2	6	12
3	2	6
	$\sum f = 28$	$\sum fx = 28$

The score 0 and 1 have the highest frequencies of 10,
∴ modes = 0 and 1 (bi-modal)

Median location = $\dfrac{n+1}{2} = \dfrac{28+1}{2}$ th score = 14.5th score

To find the median score, keep adding up the frequencies until you go past 14.5

∴ $median = 2$

EXAMPLE 2

The dot frequency shows the daily pocket money received by a group of high school students.

Determine the mean, mode and median amount of pocket money. Which measure of central tendency would be most appropriate in this situation?

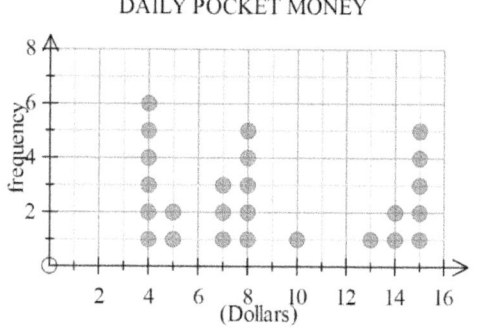

DAILY POCKET MONEY

SOLUTION

We can first express the dot frequency as a frequency table and then calculate the averages.

Pocket money ($) x	4	5	7	8	10	13	14	15	
Frequency f	6	2	3	5	1	1	2	5	$\sum f = 25$
xf	24	10	21	40	10	13	28	75	$\sum fx = 221$

Mean, $\bar{x} = \dfrac{\sum xf}{\sum f} = \dfrac{221}{25} = 8.84$

The score 4 has the highest frequencies of 6, ∴ modes = $4

Median location = $\dfrac{n+1}{2} = \dfrac{25+1}{2}$ th score = 13th score

To find the median score, keep adding up the frequencies until you go past 13

∴ $median = \$8$

EXERCISE 2C

Without using your calculator find the mean, mode and median for the frequency tables and dot frequency diagrams below.

1.

Scores (x)	Frequency (f)	x f
1	6	
2	4	
3	7	
4	8	
	$\sum f =$	$\sum fx =$

2.

Scores (x)	Frequency (f)	x f
5	1	
10	2	
15	5	
20	7	
	$\sum f =$	$\sum fx =$

3.

Scores (x)	Frequency (f)	x f
1	5	
2	10	
3	6	
4	9	
	$\sum f =$	$\sum fx =$

4.

Scores (x)	Frequency (f)	x f
2	6	
4	7	
6	1	
8	6	
	$\sum f =$	$\sum fx =$

5.

Scores (x)	Frequency (f)	x f
0	5	
2	11	
4	4	
7	15	
	$\sum f =$	$\sum fx =$

6.

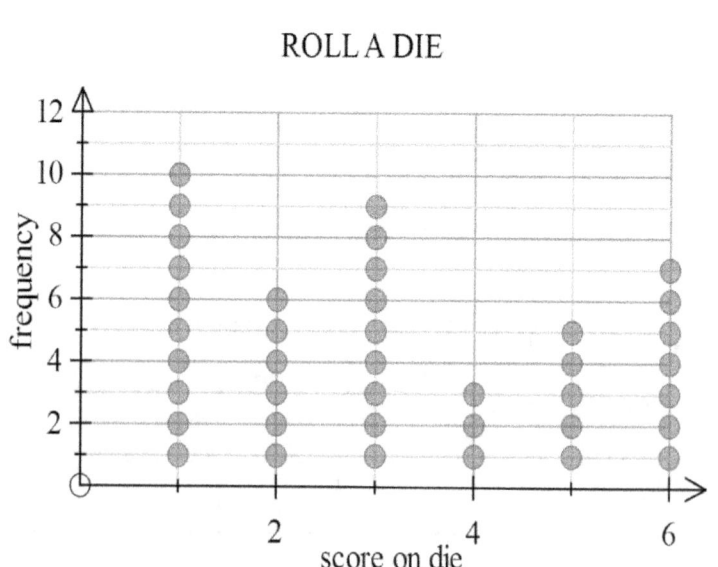

ROLL A DIE

7. The dot frequency diagram shows the number of books read by the pre-school playgroup. Calculate the mean, mode and median.

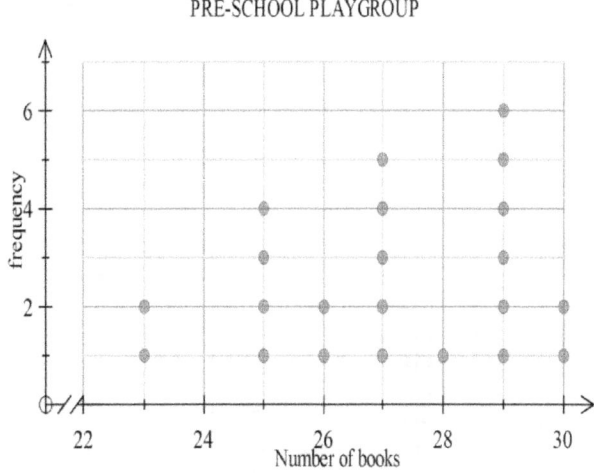

PRE-SCHOOL PLAYGROUP

8. The stem and leaf on the right shows the marks of a group of Year 11 students in an Algebra assessment.
   (a) state the lowest score,

   (b) state the highest score,

   (c) the mean, mode and median.

   ```
   1 | 5 8
   2 | 5 8 9
   3 | 0 2 6 6 6
   4 | 0 0 2 5 8
   ```

9. The stem and leaf diagram on the right shows the distance run, in kilometres, by 22 members of a golf club to commemorate the 50th anniversary of the club.
   (a) State the smallest distance run,

   (b) State the longest distance run,

   (c) The mean, mode and median.

   Distance Run (km)
   ```
   0 | 8 9
   1 | 1 3 4 5 5 5 6
   2 | 0 0 3 8
   3 | 0 3 8 9
   4 | 2 9
   5 | 1 3 4
   ```

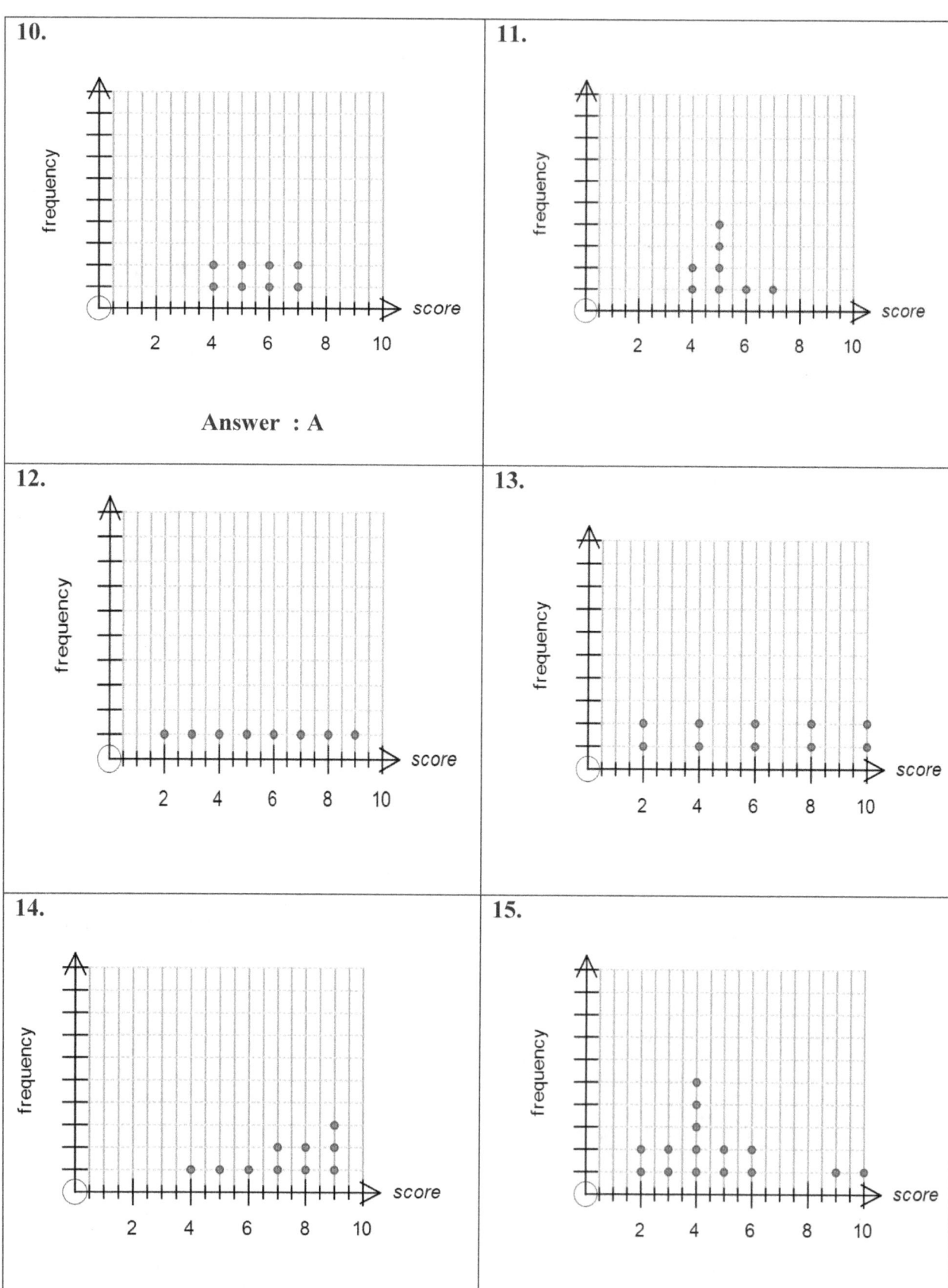

2D USING TECHNOLOGY TO FIND AVERAGES FOR RAW NUMBERS

So far we have tried as far as possible to limit the use of calculators. However, we can make use of technology to find the measures of central tendency without doing any calculations. The examples and steps are shown below for raw numbers. A calculator display has also been shown so as to give the reader an idea how the final results look like. The results have been displayed in two different boxes as we have to scroll down to see the results of the bottom box.

EXAMPLE

Use the statistical capability of your calculator to find the mean, mode and median of the following set of raw data.

$$11 \quad 13 \quad 15 \quad 14 \quad 21 \quad 28 \quad 29 \quad 35 \quad 14 \quad 14 \quad 28$$

SOLUTION

Use the following steps on your CAS to find the averages of raw numbers.

step 1 : Menu
step 2 : Statistics
step 3 : Enter the scores in list 1, do not order them.
step 4 : Calc
step 5 : One-variable
step 6 : OK

All the answers will be displayed as shown on the right.
At this stage of the chapter, we are only going to take the bold ones into account.

$\bar{x} = 20.181818$
$\Sigma x = 222$
$\Sigma x^2 = 5178$
$\sigma_x = 7.9637609$
$S_x = 8.3524629$
$n = 11$
$minX = 11$
$Q_1 = 14$
$Med = 15$

$\bar{x} = 20.181818$ is the mean

$\Sigma x = 222$ is the sum of all the numbers

$n = 11$ implies there are 11 scores in the data set.

$minX = 11$ means that the smallest number is 11

$Med = 15$ is the short form for median

$MaxX = 35$ shows the highest score

$Q_3 = 28$
$MaxX = 35$
$Mode = 14$
$Mode\ N = 1$
$Mode\ F = 3$

$Mode = 14$
$Mode\ N = 1$ means that there is a unique mode.
$Mode\ F = 3$ implies that the mode appears three times in the data set. That is, 14 is the mode and has a frequency of 3.

We could also find the mean by using the formula $\bar{x} = \frac{\Sigma x}{n} = \frac{222}{11} = 20.18$ as above.
Hence mean = 20.18, mode = 14 and median = 15.

EXERCISE 2D

Use the statistical capability of your calculator to complete the following table.

		mean	mode	median
1.	4,5,4,7,8,9,7,11,13,4			
2.	10,12,13,14,21,15,17,19,18,14,14			
3.	14,14,15,14,17,15,16,14,15,16			
4.	11,13,11,15,11,20,25,28			
5.	-20,-25,-36,-78,-89,-91,-105,-25			

6. For the dot frequency on the right, determine the mean, mode and median.

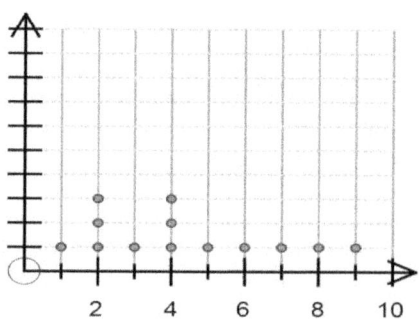

7. Determine the three measures of central tendency for the dot frequency diagram.

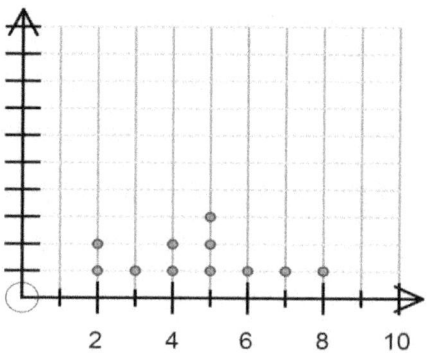

8. On a farm the number of workers picking oranges on eleven days last April were
   3, 8, 7, 11, 9, 2, 11, 11, 9, 9, 5
   Find the mean, mode and median.

9. A study was made of the BMI of 12 athletes at the Olympic Games. The results, to the nearest whole number, were recorded for as shown in the table.

BODY MASS INDEX		
28	16	19
17	32	22
25	31	30
24	29	26

For the group of 12 athletes, calculate

(i) mean (ii) median

A 13th athlete recorded a BMI Index of 40 and added to the existing 12 scores. What effect if any (do no calculate) would it have on the

(iii) mean (iv) median

10. Calculate the mean of each of the four dot frequency diagrams and hence rank them in ascending order.

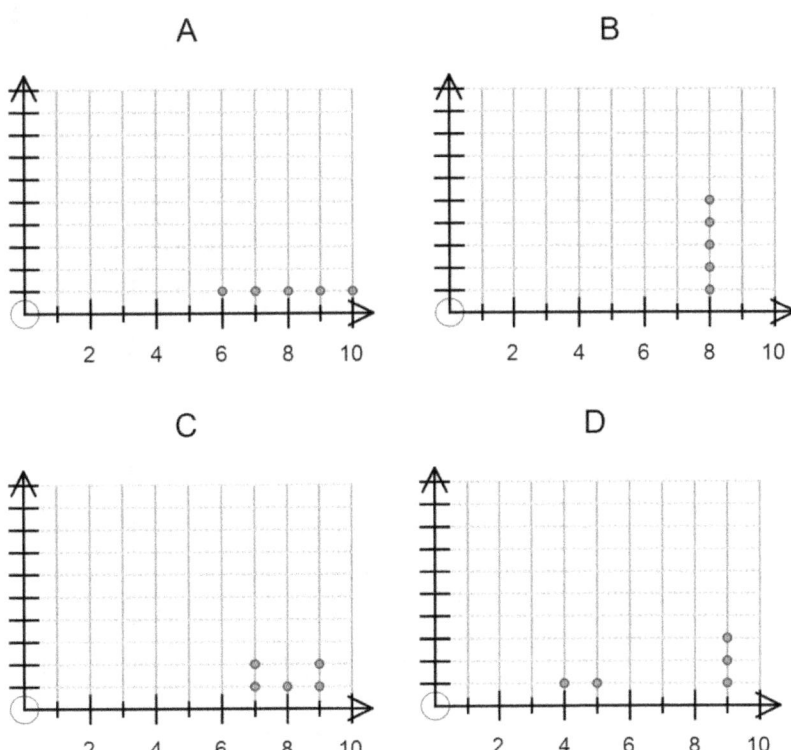

2E USING TECHNOLOGY TO FIND AVERAGES FOR FREQUENCY TABLES

This section is very similar to the previous one. The only slight change is that we use more columns to display the data sets. A sample has been illustrated below to facilitate the teaching and learning of the capabilities of the statistical tool in our calculators.

EXAMPLE

Use the statistical capability of your calculator to find the mean, mode and median of the frequency table given below.

Score on die	1	2	3	4	5	6
Frequency	7	11	13	18	5	2

SOLUTION

Use the following steps on your CAS to find the averages.

step 1 : Menu
step 2 : Statistics
step 3 : Enter the score on die in list 1
step 4 : Enter frequency in list 2
step 5 : Calc
step 5 : One-variable   XList : list1
                                      Freq : list2
step 6 : OK

Hence mean $\bar{x} = 3.16$,
        mode = 4 and
        median = 3

	list1	list2	list3
1	1	7	
2	2	11	
3	3	13	
4	4	18	
5	5	5	
6	6	2	
7			
8			
9			

$\bar{x} = 3.1607143$
$\Sigma x = 177$
$\Sigma x^2 = 653$
$\sigma_x = 1.2925167$
$S_x = 1.3042139$
$n = 56$
$minX = 1$
$Q_1 = 2$
$Med = 3$

$Q_3 = 4$
$MaxX = 6$
$Mode = 4$
$Mode\ N = 1$
$Mode\ F = 18$

EXERCISE 2E

Use the statistical capability of your calculator to find the mean, mode and median for each of the following.

1.

Scores (x)	Frequency (f)
0	10
1	15
2	18
3	11
4	14

2.

Scores (x)	Frequency (f)
2	7
4	11
6	13
8	5
10	10

3.

Scores (x)	Frequency (f)
8	7
9	12
10	20
12	6
15	10

4.

Scores (x)	Frequency (f)
5	2
10	3
15	8
20	7
25	15

5. Tickets for a show cost $5, $8, $10, $15 or $25. The number of tickets sold at each price is shown in the table.

Price ($)	5	8	10	15	25
Number of tickets sold	80	50	120	60	12

(a) Write down the modal price.

(b) Use your calculator to find the median and mean prices.

6. A survey was carried out to find the number of children in each of 400 families living in an affluent suburb. The results are shown in the table below.

Number of children in family	0	1	2	3	4
Number of families	46	92	98	104	60

(a) Write down the modal number of children.

(b) Use your calculator to find the mean number of children per family.

7. The number of goals scored by a soccer team in each of 30 matches was as follows:

2 3 1 3 0 0 4 2 1 1 3 5 0 1 2
4 1 0 0 2 1 2 3 6 1 4 2 0 0 1

(a) Complete the table in the answer space.

Number of goals scored	0	1	2	3	4	5	6
Number of matches	7					1	1

(b) For this distribution, find the mean and state the mode.

CHAPTER 2 : SUMMARISING DATA AND DESCRIBING DISTRIBUTIONS SOLUTIONS

2F USING CAS TO FIND AVERAGES FOR GROUPED FREQUENCY TABLES

For a grouped frequency table, we use two columns same like we did for frequency table. However, the only difference is that we have to find the midpoints of each class and insert in list 1.

EXAMPLE

Use the statistical capability of your calculator to find the mean, modal class and the median class of the grouped frequency distribution.

Score	1-5	6-10	11-15	16-20	21-25	26-30
Frequency	8	12	19	3	8	10

SOLUTION

Use the following steps on your CAS to find the averages.

step 1 : Menu
step 2 : Statistics
step 3 : Enter midpoint of each class in list 1
step 4 : Enter frequency in list 2
step 5 : Calc
step 5 : One-variable   XList : list1
                          Freq : list2
step 6 : OK

$\bar{x} = 14.75$
$\sum x = 885$
$\sum x^2 = 17095$
$\sigma_x = 8.2069584$
$S_x = 8.27621679$
$n = 60$
$minX = 3$
$Q_1 = 8$
$Med = 13$

$Q_3 = 23$
$MaxX = 28$
$Mode = 13$
$Mode\ N = 1$
$Mode\ F = 19$

Hence mean $\bar{x} = 3.16$,
modal class = 11 - 15 and
median class = 11-15.

EXERCISE 2F

1. Use the statistical capability of your calculator to find the mean, modal class and the median class of the following grouped frequency tables.

Scores	Midpoint (x)	Frequency (f)
0-10		7
10-20		15
20-30		19
30-40		12
40-50		9

2.

Scores	Midpoint (x)	Frequency (f)
20-30		7
30-50		11
50-70		13
70-100		5
100-120		10

3.

Scores	Frequency (f)
10-19	7
20-29	12
30-39	25
40-49	6
50-59	10

4.

Scores	Frequency (f)
20-24	2
25-29	3
30-34	10
35-39	7
40-44	11

5. **The heights of 40 children are shown in the table below.**

Height (x cm)	$60 < x \leq 70$	$70 < x \leq 80$	$80 < x \leq 90$	$90 < x \leq 100$
Number of children	12	18	7	3

Calculate an estimate of the mean height of the children.

6. The length of time taken by all of the 200 pupils of a school to complete a task is given in the table below.

Time in minutes	$35 < x \leq 45$	$45 < x \leq 55$	$55 < x \leq 65$	$65 < x \leq 75$
Number of pupils	55	a	25	68

State the value of a and hence calculate an estimate of the mean time taken to complete the task.

CHAPTER 2 : SUMMARISING DATA AND DESCRIBING DISTRIBUTIONS SOLUTIONS

7. The daily high temperature in degrees Fahrenheit registered at a Metrological station during a particular month is given below:

    61 70 62 64 64 70 74 70 62 67 65 68 71 66 59 78
    58 60 59 56 53 51 55 56 50 53 57 55 50 46

Display the above information in the frequency table.

Interval	frequency
45-49	
50-54	
55-59	
60-64	
65-69	
70-74	
75-79	

(a) The true boundary of the class 50-54 is 49.5-54.5. State the true boundary of the median class.

(b) Calculate an estimate of the mean temperature.

8. A test was set to group of students and marked out of 70. The minimum mark for a grade A is 58, and a minimum mark for a grade B is 40. Marks of 30 or more score a C grade. The results has been represented on a stem and leaf plot as under.

(a) How many students sat the test?

Algebra Test

0	6
1	1 5
2	2 8
3	2 5 7 9 9
4	0 3 3 5 5 6 6 8
5	1 3 3 9
6	4 9

(b) State a suitable diagram that can used to illustrate the percentage of students in the three different grades. Determine the percentage of students in each grade.

(c) Use your calculator to find the mean mark and state the median mark.

(d) Is the mean or median a better indicator of how the students performed in the test?

2G APPLICATIONS

EXAMPLES

1. Four numbers have a mean 50. If three of the numbers are 20, 60 and 80, find the 4th number. **Solution** Total for the 4 numbers = 50 × 4 = 200 Total of the three given numbers = 20 + 60 + 80 = 160 The 4th number = 200 − 160 = 40.	2. In a test the 10 boys in a class score a mean mark of 60 and the 15 girls score a mean mark of 50. Calculate the mean mark for the whole group of 25. **Solution** Marks scored by boys = 10 × 60 = 600 Marks scored by the girls = 15 × 50 = 750 Total marks scored = 600 + 750 = 1350 Mean mark = $\dfrac{1350}{25}$ = 54

EXERCISE 2G

1. Five numbers have a mean 30. If four of the numbers are 10, 20, 30 and 25, find the 5th number.	2. The mean of a set of 20 numbers is 65. When a 21st number is added, the mean increases to 70. Calculate the 21st number.
3. The mean of a set of 9 numbers is 40. When a 10th number is added, the mean increases by 3. Calculate the 10th number.	4. The mean of a set of 14 numbers is 63. When a 15th number is added, the mean decreases by 2. Calculate the 15th number.
5. Paul scored 48%, 61% and 53% in his last three tests. He needs an average of 60% to pass his course. Find how much he needs to score in his fourth test to pass the course.	6. Six numbers have a mean 40. If four of the numbers are 10, 20, 30 and 40, and the remaining two numbers are equal, find one of the remaining number.
7. The mean of a set of 10 numbers is 76. When an 11th number is added, the new mean is 80. Calculate the 11th number.	8. In a test the 15 boys in a class score a mean mark of 62 and the 12 girls score a mean mark of 54. Calculate the mean mark for the whole group.

CHAPTER 2 : SUMMARISING DATA AND DESCRIBING DISTRIBUTIONS SOLUTIONS

9. The mean weight of 10 boys is 64 kg and the mean weight of 5 girls is 59 kg. Find the mean weight for the whole group.	10. Three classes A, B and C have 15, 21 and 12 students respectively. The three classes A, B and C have a mean mark of 52, 71 and 48 respectively. Calculate the mean mark for the three classes combined together.
11. Three classes P, Q and R have 20, 15 and 25 students respectively. The three classes P, Q and R have a mean mark of 58, 74 and 52 respectively. Calculate the mean mark for the three classes combined together.	12. A group of 10 boys ran an average of 22.4 km during a marathon. If the whole competition consisted of 25 participants and their mean distance ran was 19.6 km, determine the mean distance ran by the girls.
13. Given that the mean of $5, 10, 15, x$ and 25 is 15, determine the value of x.	14. The set of values $5, 8, 8, x, 13, 15$ is arranged in ascending order. If the median is 9, state the value of x.
15. For the set of numbers $10, 12, 14, x, 14, y$, determine the value of x and y if the range is 10 and the median is 14.	16. For the set of numbers $x, 10, 10, y, z, 14$, determine the value of x, y and z if • the range is 9 • the median is 10 • The mean is 10.

17. For the set of numbers arranged in ascending order
$$a, b, 5, 7, 9, c, d, 12, 12, 17$$
- The range is 15
- The median is 10
- The mode is 12
- The mean is 9.

Work out the values of a, b , c and d.

$a = 2$, $b = 3$, $c = 11$, $d = 12$

18. For the set of numbers arranged in ascending order
$$5, 6, a, 8, b, 11, 12, 15, c, d$$
- The range is 15
- The median is 10.5
- The mode is 20
- The mean is 11.4

Work out the values of a, b , c and d.

$a = 7$, $b = 10$, $c = 20$, $d = 20$

CHAPTER 3

MEASURES OF DISPERSION OR SPREAD

3A CALCULATING STANDARD DEVIATION BY LONG HAND

The standard deviation calculation tells us how spread out the numbers is in a set of data. Mathematically, standard deviation is the average amount by which the scores in a data differ from the mean. A smaller value of the standard deviation implies that the data set has more consistency whereas the larger the standard deviation means that there is more spread among the data.

The symbol for standard deviation is σ (read as sigma) and can be calculated using the following formulae:

$$\sigma_x = \sqrt{\frac{\sum(x-\bar{x})^2}{n}} \text{ or } \sqrt{\frac{\sum x^2}{n} - (\bar{x})^2}$$

To calculate standard deviation by hand follow these simple steps:
- Find the mean
- Subtract the mean from each score
- Square each of these differences
- Find the mean of these differences
- Find square root of this mean

EXAMPLE 1

Find the standard deviation of the numbers 5, 10, 15, 20 and 25.

SOLUTION

Mean $= \bar{x} = \frac{\sum x}{n} = \frac{75}{5} = 15$

The easiest way to find the standard deviation is to use a table as shown.

Scores (x)	mean ($\bar{x}$)	$x - \bar{x}$	$(x - \bar{x})^2$
5	15	-10	100
10	15	-5	25
15	15	0	0
20	15	5	25
25	15	10	100
			$\sum(x-\bar{x})^2 = 250$

Using the formula $\sigma_x = \sqrt{\frac{\sum(x-\bar{x})^2}{n}} = \sqrt{\frac{250}{5}} = 7.07 \ (2 \ d.p)$

We can also use the second formula $\sigma_x = \sqrt{\frac{\sum x^2}{n} - (\bar{x})^2}$ to calculate standard deviation in a much quicker way.

EXAMPLE 2

Calculate the standard deviation of 2, 3, 5, 7, 8.

SOLUTION

This time we are going to use the formula $\sigma_x = \sqrt{\frac{\sum x^2}{n} - (\bar{x})^2}$ to compute standard deviation.

Mean $= \bar{x} = \frac{\sum x}{n} = \frac{25}{5} = 5$

Scores (x)	x^2
2	4
3	9
5	25
7	49
8	64
	$\sum x^2 = 151$

$\sigma_x = \sqrt{\frac{151}{5} - (5)^2} = 2.28$

EXERCISE 3A

Calculate the standard deviation of each of the following using the long hand method. Use the given tables to show your workings.

1. 4, 6, 8, 10, 12

Scores (x)	mean ($\bar{x}$)	$x - \bar{x}$	$(x - \bar{x})^2$
			$\sum(x - \bar{x})^2 =$

2. 10, 10, 10, 10, 10

Scores (x)	mean ($\bar{x}$)	$x - \bar{x}$	$(x - \bar{x})^2$
			$\sum(x - \bar{x})^2 =$

3. 10, 12, 15, 18, 20, 21

Scores (x)	x^2
	$\sum x^2 =$

4. 40, 42, 45, 48, 51, 62

Scores (x)	x^2
	$\sum x^2 =$

5. 6, 6, 6, 6

Scores (x)	x^2
	$\sum x^2 =$

6. 10, 20, 30, 60

Scores (x)	x^2
	$\sum x^2 =$

7. 2, 10, 17, 23, 28

Scores (x)	mean ($\bar{x}$)	$x - \bar{x}$	$(x - \bar{x})^2$
		$\sum (x - \bar{x})^2 =$	

Without making any calculations determine which graph in each question has the greater mean or greater standard deviation.

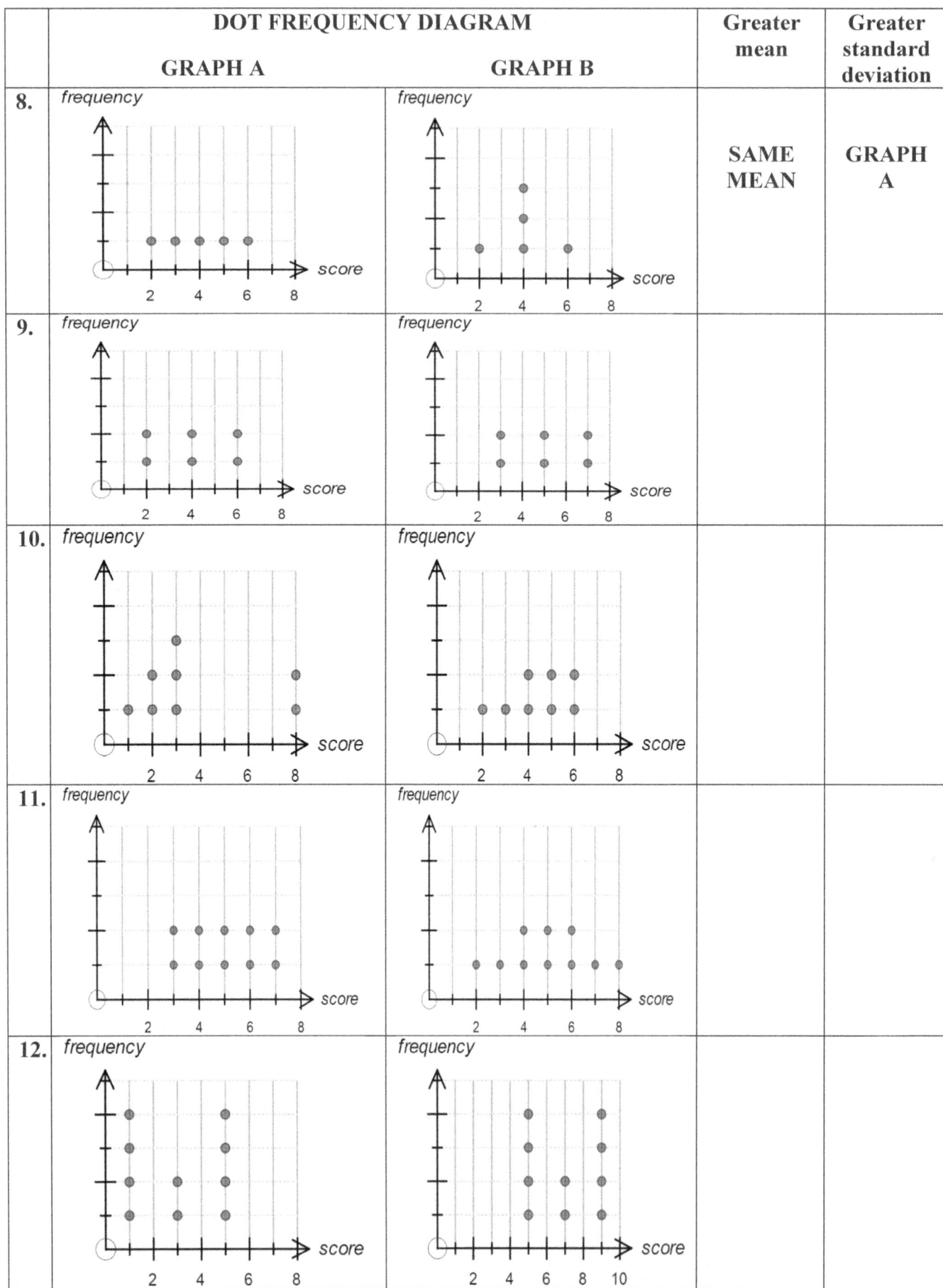

3B USING TECHNOLOGY TO CALCULATE STANDARD DEVIATION OF RAW NUMBERS

Now we are going to make use the statistical capabilities of our calculators to determine the standard deviation in a much quicker and less painstaking way.

EXAMPLE

Use the statistical capability of your calculator to find the standard deviation of the following set of raw data. Also state the value of the variance.

$$11 \quad 13 \quad 15 \quad 14 \quad 21 \quad 28 \quad 29 \quad 35 \quad 14 \quad 14 \quad 28$$

SOLUTION

Use the following steps on your CAS to find the averages of raw numbers.

step 1 : Menu
step 2 : Statistics
step 3 : Enter the scores in list 1
step 4 : Calc
step 5 : One-variable
step 6 : OK

$\bar{x} = 20.181818$
$\Sigma x = 222$
$\Sigma x^2 = 5178$
$\boldsymbol{\sigma_x = 7.9637609}$
$S_x = 8.3524629$
$n = 11$
$minX = 11$
$Q_1 = 14$
$Med = 15$

All the answers will be displayed as shown on the right.
Hence standard deviation (shown in bold) = 7.96

Variance is the square of the value of the standard deviation.

∴ Variance = $\sigma^2 = 7.96^2 = 63.4$

EXERCISE 3B

1. For each of the following, find the mean ($\bar{x}$) and the standard deviation (σ_x) and the variance $\sigma^2(x)$ using your calculators. Give your answers to 1 decimal place.

		mean	Standard deviation	Variance
1.	2,3,4,5,8,8,10,15			
2.	1,5,4,7,8,11,15,20			
3.	4,10,15,20,25,50			
4.	6,8,7,8,9,10,11,13			
5.	10,12,14,15,16,17,18			
6.	-4,-5,-4,-7,-8,-9,-7,-11			
7.	11,13,11,15,11,20,25,28			

For each of the following determine the mean and the standard deviation using your calculator.

8.

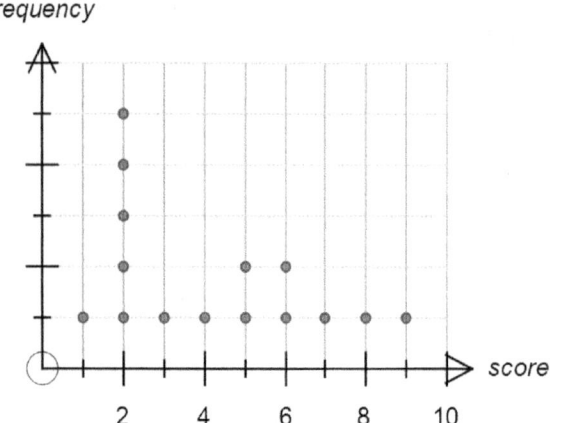

9.

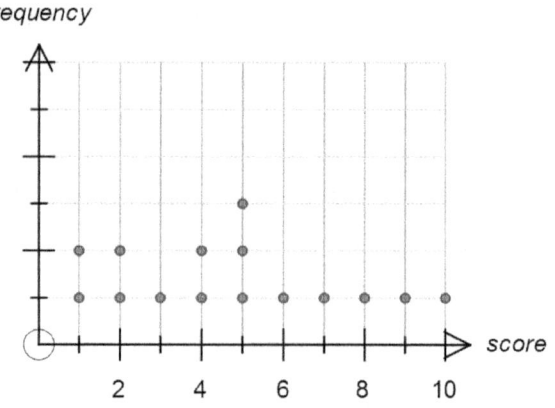

10. During the Olympics diving competition, three athletes recorded the following number of points in their 7 jumps during their final round.

Jumps	J1	J2	J3	J4	J5	J6	J7
Alex	8.5	8.5	9.0	8.5	8.5	8.0	7.5
Zhiao	7.0	7.5	8.0	8.5	7.5	7.5	8.5
Mitcham	10	9.5	10	9	10	9.5	9.5

(a) Rank the three athletes by calculating their mean scores over their 7 jumps.

(b) Which of the three divers had a more consistent result? Show workings to support your answer.

3C USING TECHNOLOGY TO CALCULATE STANDARD DEVIATION FOR FREQUENCY TABLES

In this section, we use two columns instead of one to determine the standard deviation. The steps that the reader needs to follow on their calculator together with the calculator display have been presented in a well set example below.

EXAMPLE

Use the statistical capability of your calculator to find the standard deviation of the frequency table given below.

Score on die	1	2	3	4	5	6
Frequency	7	11	13	18	5	2

SOLUTION

Use the following steps on your CAS to find the averages.

step 1 : Menu
step 2 : Statistics
step 3 : Enter the score on die in list 1
step 4 : Enter frequency in list 2
step 5 : Calc
step 5 : One-variable   XList : list1
                                       Freq : list2
step 6 : OK

Hence standard deviation = 1.29

	list1	list2	list3
1	1	7	
2	2	11	
3	3	13	
4	4	18	
5	5	5	
6	6	2	
7			
8			
9			

$\bar{x} = 3.1607143$
$\sum x = 177$
$\sum x^2 = 653$
$\boldsymbol{\sigma_x = 1.2925167}$
$S_x = 1.3042139$
$n = 56$
$minX = 1$
$Q_1 = 2$
$Med = 3$

EXERCISE 3C

Use the statistical capability of your calculator to calculate the mean and standard deviation of the frequency tables given below.

1.

Scores (x)	Frequency (f)
0	8
1	11
2	12
3	15
4	10

$\bar{x} = $ .... $\sigma_x = $ ....

2.

Scores (x)	Frequency (f)
2	5
4	11
6	10
8	9
10	6

$\bar{x} = $ .... $\sigma_x = $ ....

3.

Scores (x)	Frequency (f)
8	6
9	11
10	13
12	4
15	15

$\bar{x} = $ .... $\sigma_x = $ ....

4.

Scores (x)	Frequency (f)
5	2
10	4
15	3
20	7
25	18

$\bar{x} = $ .... $\sigma_x = $ ....

5.

Scores	1	3	4	6	8
frequency	1	2	5	4	12

$\bar{x} = $ .... $\sigma_x = $ ....

Use the statistical capability of your calculator to calculate the mean, standard deviation of the following dot frequency diagrams.

6.

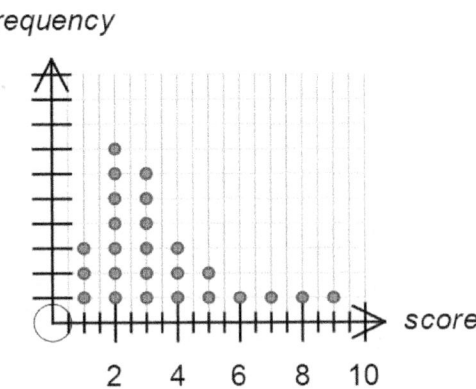

7.

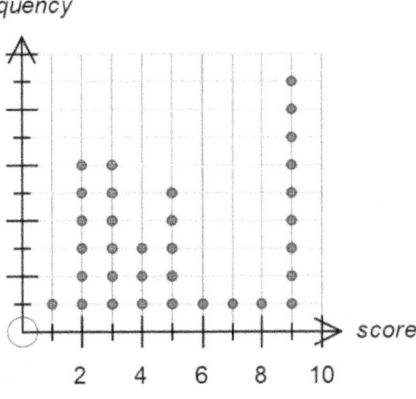

8. Tickets for a show at the Writing Cinemas cost $10, $12, $14, $20 or $32. The number of tickets sold during the premiere of a popular horror movie at each price is shown in the table.

Price ($)	10	12	14	20	32
Number of tickets sold	110	50	30	180	12

Determine the mean and the standard deviation of the ticket prices to the nearest integer.

9. A survey was carried out to find the number of pets in each of 120 families living in a village. The results are shown in the table below.

Number of pets owned	0	1	2	3	4	5	6
Number of families	10	25	15	10	40	13	7

Determine the mean and the standard deviation of the number of pets owned by the villagers.

10. The table below shows the number of run scored by the 15 cricketers during a friendly match.

Number of runs	0	10	20	25	50	75	100
Number of players	2	5	0	4	2	1	1

Determine the mean and the standard deviation of the number of runs scored by the cricketers.

11. Petra rolled an octahedral die 50 times and recorded her scores below.

$$\begin{array}{cccccccccc}
2 & 5 & 1 & 3 & 4 & 8 & 5 & 5 & 7 & 2 \\
1 & 3 & 3 & 2 & 3 & 5 & 3 & 3 & 6 & 2 \\
4 & 8 & 3 & 3 & 3 & 4 & 5 & 4 & 7 & 8 \\
1 & 2 & 3 & 4 & 5 & 6 & 7 & 8 & 3 & 4 \\
7 & 3 & 4 & 3 & 3 & 5 & 3 & 3 & 3 & 3
\end{array}$$

(a) Complete the table in the answer space.

Number on top of die	1	2	3	4	5	6	7	8
Frequency							4	4

(b) For this distribution, find the mean and standard deviation.

(c) Do you consider this die to be fair or biased? Justify your answer.

12. A special six-sided die is rolled 80 times. The results are tabulated below:

Number on top of die	1	2	3	4	5	6
Frequency	7	10	3	25	15	20

(a) Determine:

(i) the mean of these data.

(ii) the standard deviation of these data.

(b) What conclusion can you draw about the die? Is it fair or biased?

CHAPTER 3 : MEASURES OF DISPERSION OR SPREAD

3D USING TECHNOLOGY TO CALCULATE STANDARD DEVIATION FOR GROUPED FREQUENCY TABLES.

By now the reader must be very familiar as to how the statistical capabilities of the calculator help to compute the value of the standard deviation. For a grouped frequency table, we still use two columns, the first column being the midpoints of each class interval and the second column used for frequency as shown.

EXAMPLE

Use the statistical capability of your calculator to find the mean and standard deviation of the grouped frequency distribution.

Score	10-19	20-29	30-39	40-49	50-59	60-69
Frequency	5	11	23	4	1	7

SOLUTION

Use the following steps on your CAS to find the averages.

step 1 : Menu
step 2 : Statistics
step 3 : Enter midpoint of each class in list 1
step 4 : Enter frequency in list 2
step 5 : Calc
step 5 : One-variable   XList : list1
                         Freq : list2
step 6 : OK
Hence   mean $\bar{x} = 35.68$,
     standard deviation = 14.09

One Variable

	list1	list2	list3
1	14.5	5	
2	24.5	11	
3	34.5	23	
4	44.5	4	
5	54.5	1	
6	64.5	7	
7			
8			

One variable

$\bar{x} = 35.676471$
$\Sigma x = 1819.5$
$\Sigma x^2 = 75042.75$
$\sigma_x = 14.093116$
$S_x = 14.233349$
$n = 51$
$minX = 14.5$
$Q_1 = 24.5$
$Med = 34.5$

Note that a two variable statistics will give two sets of averages. Thus, by using two variable statistics we would obtain two means, two modes, two standard deviations etc.. one for each column.

EXERCISE 3D

Use the statistical capabilities of your calculator to the mean ($\bar{x}$) and the standard deviation (σ_x) of the following grouped frequency tables.

1.

Scores (x)	Midpoint of class	Frequency (f)
0-10		2
10-20		11
20-30		13
30-40		10
40-50		8

2.

Scores (x)	Midpoint of class	Frequency (f)
20-30		6
30-50		11
50-70		15
70-100		4
100-120		10

3.

Scores (x)	midpoint	Frequency (f)
10-19		7
20-29		12
30-39		25
40-49		6
50-59		10

4.

Scores (x)	midpoint	Frequency (f)
20-30		5
30-40		4
40-60		10
60-70		8
70-100		13

5. The heights of 50 trees in a nursery are shown in the table below.

Height (x cm)	$60 < x \leq 70$	$70 < x \leq 80$	$80 < x \leq 90$	$90 < x \leq 100$
Number of tress	8	22	13	7

Determine the mean and the standard deviation of the height of trees correct to 1 decimal place.

6. The time taken by a group of 100 participants to swim a 200m race is given in the table below.

Time in seconds	$35 < x \leq 45$	$45 < x \leq 55$	$55 < x \leq 65$	$65 < x \leq 75$
Number of participants	12	18	20	50

Determine the mean and the standard deviation of the time taken by those participants correct to one decimal place.

7. The table below shows the age distribution of the teaching staff in a certain school in the metropolitan area. Determine the mean and the standard deviation of the ages of the staff.

Age (x years)	$20 < x \leq 30$	$30 < x \leq 40$	$40 < x \leq 50$	$50 < x \leq 60$
Number of staff	8	12	15	5

8. The length of time spent by shoppers at a shop during Christmas time were recorded and tabulated below. Determine the mean and the standard deviation of the time taken by those shoppers.

Time (x hours)	$0.5 < x \leq 1.5$	$1.5 < x \leq 2.5$	$2.5 < x \leq 3.5$	$3.5 < x \leq 4.5$
Number of people	200	150	50	20

Determine an estimate of the mean ($\bar{x}$) and standard deviation (σ_x) for the following histogram.

9.

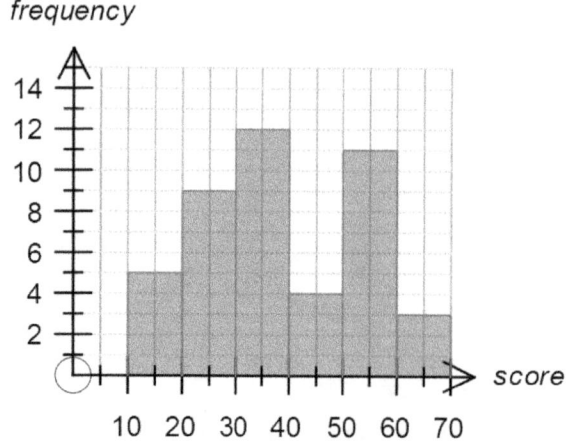

10. The table below shows the prices and number of properties sold in two different states.

		State A	State B
Prices ($000)	$250 \leq P < 300$	10	1
	$300 \leq P < 350$	8	18
	$350 \leq P < 400$	15	3
	$400 \leq P < 450$	5	10
	$450 \leq P < 500$	16	15
	$500 \leq P < 550$	2	5

(a) Calculate the mean and standard deviation of the property prices in both states.

(b) Write down the modal class for State A.

(c) Determine the median class for State B.

(d) In which of the two states were the prices less variable? Explain.

(e) State A sold a property for $980 000 but was not recorded in the table. What effect would this have to your answers in (a)?

11. Table 1 below shows a summary statistics for maximum daily temperatures in March 2013 and March 2014 for a New Zealand town. The maximum daily temperatures (°C) in March 2013 for the town are summarised in Table 2.

Table 1 : Maximum daily temperatures (°C), March 2013-2014		
	March 2013	March 2014
Mean		22.6
Standard deviation		5.1

Table 2 : Maximum daily temperature (°C), March 2013	
Temperature T (°C)	Frequency
$15 \leq T < 19$	4
$19 \leq T < 23$	9
$23 \leq T < 27$	1
$27 \leq T < 31$	5
$31 \leq T < 35$	12

(a) Use the data in Table 2 to

(i) Calculate the mean and standard deviation of the temperatures for March 2013 and enter the results in Table 1.

(ii) State the modal class.

(iii) In which class interval would the median lie?

(b) In which of the two years were the March temperatures in the town less variable? Explain your choice.

(c) Melissa is a new statistician recruit. She had a quick glimpse of the data and concluded that the year 2014 tended to be cooler than the year 2013. Is she correct? Justify your answer.

3E   STANDARD DEVIATION FROM MEAN

Very often we come across question such as "how many scores lie within 2 standard deviations from the mean" or "how many scores are 0.5 standard deviation above the mean" and so on. In this part of the chapter, an attempt has been made to answer all those queries.

EXAMPLE

An ordinary die was rolled 50 times and the results recorded in the table below.

Score on die (x)	1	2	3	4	5	6
frequency	3	10	7	15	11	4

(a) Determine the mean and the standard deviation for the above frequency table.
(b) How many scores are 0.5 standard deviation above the mean?
(c) How many scores are 1 standard deviation below the mean?
(d) How many scores are 2 standard deviations within the mean?

Solution

(a) $\bar{x} = 3.66$ $\sigma_x = 1.38$

(b) Multiply the standard deviation by 0.5 and add to the mean as shown

$$\bar{x} + 0.5\,\sigma_x = 3.66 + 0.5(1.38) = 4.35$$

Find from the table the number of scores greater than 4.35 (scores 5 or 6) = 11 + 4 = 15.

(c) Subtract one standard deviation from the mean as shown

$$\bar{x} - \sigma_x = 3.66 - 1.38 = 2.28$$

Now find from the table the number of scores less than 2.28 (scores 1 and 2) = 3 + 10 = 13.

(d) Add and subtract 2 standard deviations from the mean as shown

$$\bar{x} - 2\,\sigma_x = 3.66 - 2(1.38) = 0.9$$
$$\bar{x} + 2\,\sigma_x = 3.66 + 2(1.38) = 6.42$$

Now find from the table the number of scores lying between 0.9 and 6.42, which are obviously all 50.

SUMMARY

Hence to find out how many scores lie above, below or within the mean, the following table will definitely help.

Above the mean	ADD
Below the mean	SUBTRACT
Within the mean	ADD & SUBTRACT

CHAPTER 3 : MEASURES OF DISPERSION OR SPREAD

EXERCISE 3E

1.

Score on die	1	2	3	4	5	6
frequency	2	8	10	7	11	2

(a) Determine the mean and the standard deviation for the above frequency table.

(b) How many scores are 0.5 standard deviation above the mean?

2.

Score	1	2	3	4	5
frequency	3	11	9	4	3

(a) Determine the mean and the standard deviation for the above frequency table.

(b) How many scores is one standard deviation below the mean?

3. Mr Andrews had a maths test out of 8 marks and the class results have been tabulated below.

Test Score	0	1	2	3	4	5	6	7	8
frequency	2	3	7	1	2	5	6	1	2

(a) How many students took the test?

(b) Use your calculator to determine the mean and the standard deviation.

(c) How many scores are 2 standard deviations **within** the mean?

4. Nine Applications Mathematics students' performance in a recently held examination is given below.

Mercury	54	Eli	55	Valerie	7
Venus	52	Amy	39	John	13
Alex	13	Norman	16	Peppa	21

   (a) Calculate the mean and standard deviation of the students 'marks.

   (b) An outlier for this data is defined to be a score which is at least two standard deviations above or below the mean. Which, if any, of the above student can be classified as an outlier? Show all working.

5. In a study of family size at New Idea High School, a teacher gathers information from students in a particular year group.

No. of children per family	1	2	3	4	5	8
No. of families	7	45	35	8	2	1

   (a) Find the total number of children questioned in this survey.

   (b) Find the mean number of children per family, giving your answer to 2 decimal places.

   (c) Determine the standard deviation of the number of children per family.

   (d) An outlier for this data is defined to be a score which is at least 1.5 standard deviations above the mean. How many families can be classified as outliers?

   (e) Describe what happens to the mean and standard deviation if the student from the 8-child family leaves the group.

CHAPTER 4

BOX PLOTS AND DESCRIBING DISTRIBUTIONS

4A BOX AND WHISKER PLOTS

A box plot also referred to as a box and whisker plot is a graph that summarises a set of data along a number line. It uses values called quartiles, which divide the data into four equally sized groups.

To make a box plot from a set of data, we need what is called the five-number summary consisting of

- The smallest value
- The highest value
- The lower quartile
- The median
- The upper quartile

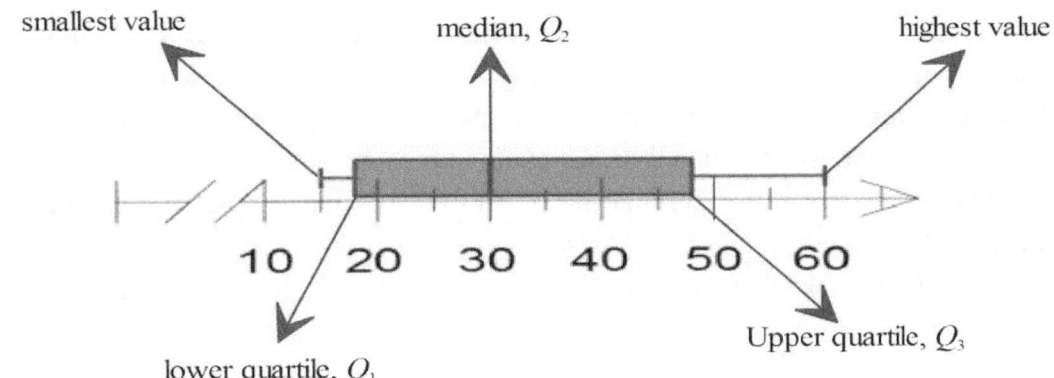

Each section in a box plot represents 25% of the whole data set as shown below.

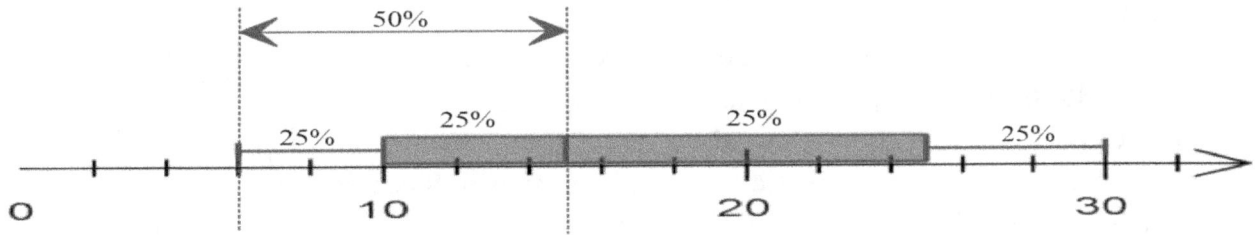

For example, if the above box plot represents the results of 40 students in a Mathematics Applications Assessment then we can draw the following conclusions:

- ❖ 25% (means 10 students) scored 10 marks or lower.
- ❖ 50% scored less than 15 marks.
- ❖ The top 25% students scored between 25 and 30 marks.
- ❖ 100% scored in the range 6 – 30 marks.
- ❖ Inter quartile range which is the width of the box and it is the central 50% of the scores can be calculated as
  IQR = $Q_3 - Q_1$ = 25 – 10 = 15.

EXAMPLE 1

Draw a box plot for the following set of scores

22, 23, 25, 27, 29, 35, 38, 40, 42, 45, 50

SOLUTION

First, be sure that your data is arranged from least to greatest.

Then we find the median or middle value that splits the data set into two equal groups. The 6th value of 35 divides this list into two equal groups of 5 numbers each as shown below.

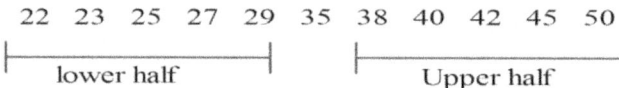

Next we find the median of the lower half of the data set known as the lower quartile (Q_1). Similarly, we find the median of the upper half of the data set. The latter is termed as upper quartile (Q_3).

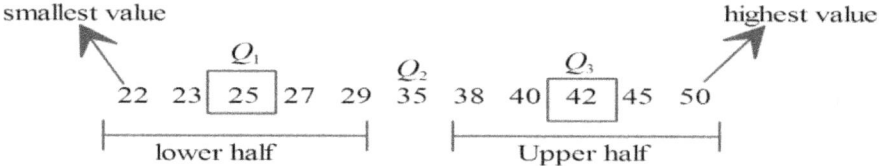

Next we find the smallest and the highest value of the data set. These 5 values as shown above and previously called as the five-number summary is everything we need to construct a box plot. To construct the actual box plot, we will need to draw an ordinary number line that extends far enough in both directions to include all the numbers in the data set.

The lower quartile, median and upper quartiles form the box. We join the lowest value of 22 to the box by means of a line to obtain our left whisker. Similarly, joining 50 to the box produces the right whisker as shown. Hence the diagram below is our box plot showing the five-number summary.

EXAMPLE 2

Draw a box and whisker plot for the set of 12 scores given below.

15, 11, 16, 21, 36, 20, 25, 38, 42, 48, 40, 29

SOLUTION

First, we re-arrange the numbers in ascending order as shown. Since there are 12 scores there will two groups of six numbers each as shown. The median lies between 25 and 29. So we find the average of these two numbers. Hence $Q_2 = \frac{25+29}{2} = 27$. Similarly $Q_1 = 18$ and $Q_3 = 39$.

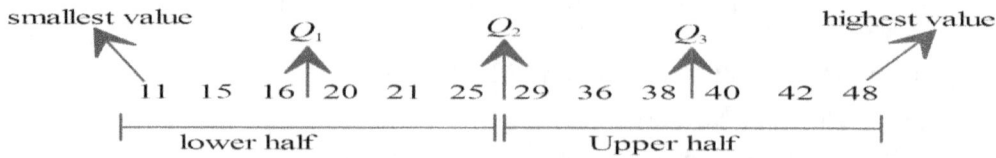

With 11 being the least value and 48 the largest, we can construct our box plot as shown below.

CHAPTER 4 : BOX PLOTS AND DESCRIBING DISTRIBUTIONS

EXERCISE 4A

Complete the table below stating the smallest value (S), the largest value (L), the lower quartile (Q_1), the median (Q_2) and the upper quartile (Q_3) and interquartile range (IQR) for each box plot.

		S	L	Q_1	Q_2	Q_3	IQR
1.	box plot with scale 0 to 50						
2.	box plot with scale 0 to 50						
3.	box plot with scale 0 to 50						
4.	box plot with scale 0 to 20						

5. Three classes had a common investigation (out of 20 marks) on box plots and the results shown below.

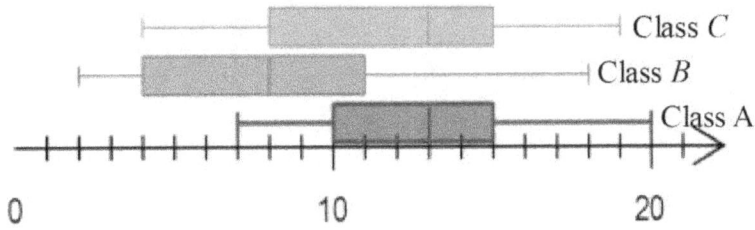

   (a) In one of the classes one student scored 100%. Which class was he/she in?

   (b) Which class had the lowest range?

6. The box plots show the performance, measured in minutes, of two clubs P and Q at a swimming event held during the summer school holidays.

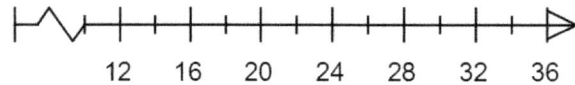

Tick true or false for each of the following:

Statement	True	False
Club Q had the fastest swimmer		✓
Club P has a larger median		
Both clubs have the same range of swimming times		
Interquartile range for Club P is 10		
50% of the swimmers in Club P took more than 35 minutes to finish their race.		

7. Draw box plots for each of the following data sets.

   (a) 12, 15, 15, 18, 20, 22, 26, 28, 28, 30, 34

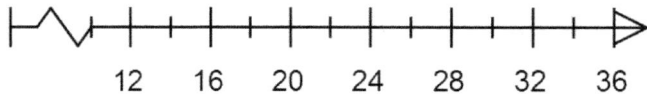

   (b) 11, 12, 15, 16, 18, 18, 18, 22, 26, 30

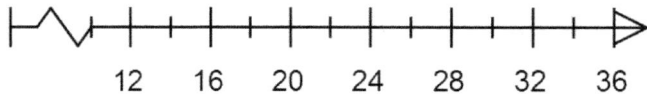

   (c) 22, 20, 23, 28, 35, 22, 26, 38, 45, 46, 40

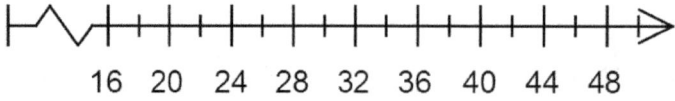

   (d) 20, 16, 22, 28, 46, 40, 30, 20, 18, 36, 35, 30

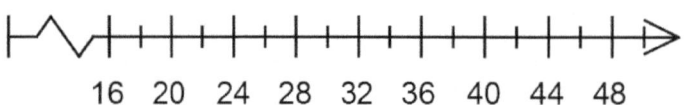

4B   IDENTIFYING OUTLIERS

An outlier can be defined as an extremely high or extremely low value in our data set. Outliers can be identified by using one of the following formulae:

Any number greater than $Q_3 + 1.5 \times IQR$ or $Q_3 + 1.5(Q_3 - Q_1)$ will be an outlier.
Or if a number is less than $Q_1 - 1.5 \times IQR$ or $Q_1 - 1.5(Q_3 - Q_1)$ it will be considered an outlier too.

EXAMPLE
Peter recorded the number of phone calls he received during the first 12 days at work.

10, 11, 12, 11, 14, 11, 15, 14, 13, 22, 14, 16
Determine if his data contains any outliers.

SOLUTION
Re-arranging the numbers in ascending order, we have

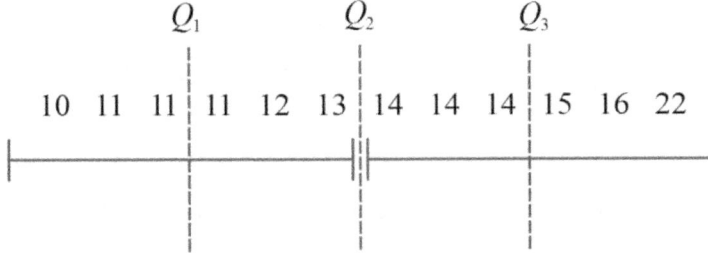

$\therefore Q_1 = 11, Q_2 = 13.5 \text{ and } Q_3 = 14.5.$

To find out if any of these 12 scores are extremely high in value, we compute
$Q_3 + 1.5 \times IQR = Q_3 + 1.5(Q_3 - Q_1)$
$\qquad\qquad\qquad\quad = 14.5 + 1.5 \times (14.5 - 11) = 19.75$
Any number greater than 19.75 will be an outlier.
From the data set only 22 is greater than 19.75, therefore 22 is an outlier.

Similarly, to find out if any of the 12 scores are extremely low in value, we compute
$Q_1 - 1.5 \times IQR = Q_1 - 1.5(Q_3 - Q_1)$
$\qquad\qquad\qquad\quad = 11 - 1.5 \times (14.5 - 11) = 5.75$
So any number less than 5.75 will be called an outlier.
From the data set 10 being the smallest data, there is nothing less than 5.75. Hence there is no lower outlier.

A usual box plot for the above data will be as follows. The highest value will be 16 instead of 22. The outlier 22 will be marked with a dot on the diagram.

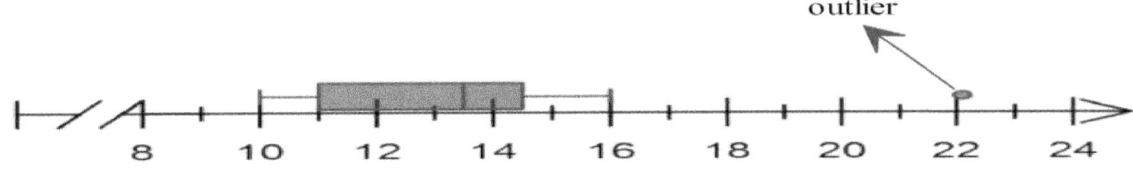

EXERCISE 4B

For the following sets of data, determine if there are any outliers.

1. 111, 21, 29, 19, 121, 25, 29, 27, 17	2. 41, 31, 49, 199, 32, 35, 39, 37, 43
3. 222, 121, 212, 242, 225, 209, 277, 237	4. 19, 1, 2, 3, 5, 4, 7, 3, 2, 1, 20
5. 44, 41, 49, 19, 42, 45, 49, 47, 46	6. 56, 76, 12, 82, 53, 70, 53, 53

CHAPTER 4 : BOX PLOTS AND DESCRIBING DISTRIBUTIONS

4C   SKEWNESS

Skewness can be defined as a measure of asymmetry in a statistical distribution, in which the curve appears distorted or skewed either to the left or to the right. In this part of the chapter, an attempt has been made to analyse three different types of scenarios: symmetric distributions, positively skewed distributions and negatively skewed distributions.

To establish skewness, we are going to use three different methods
- ❖ Measures of central tendency
- ❖ The quartiles
- ❖ Graphical representation

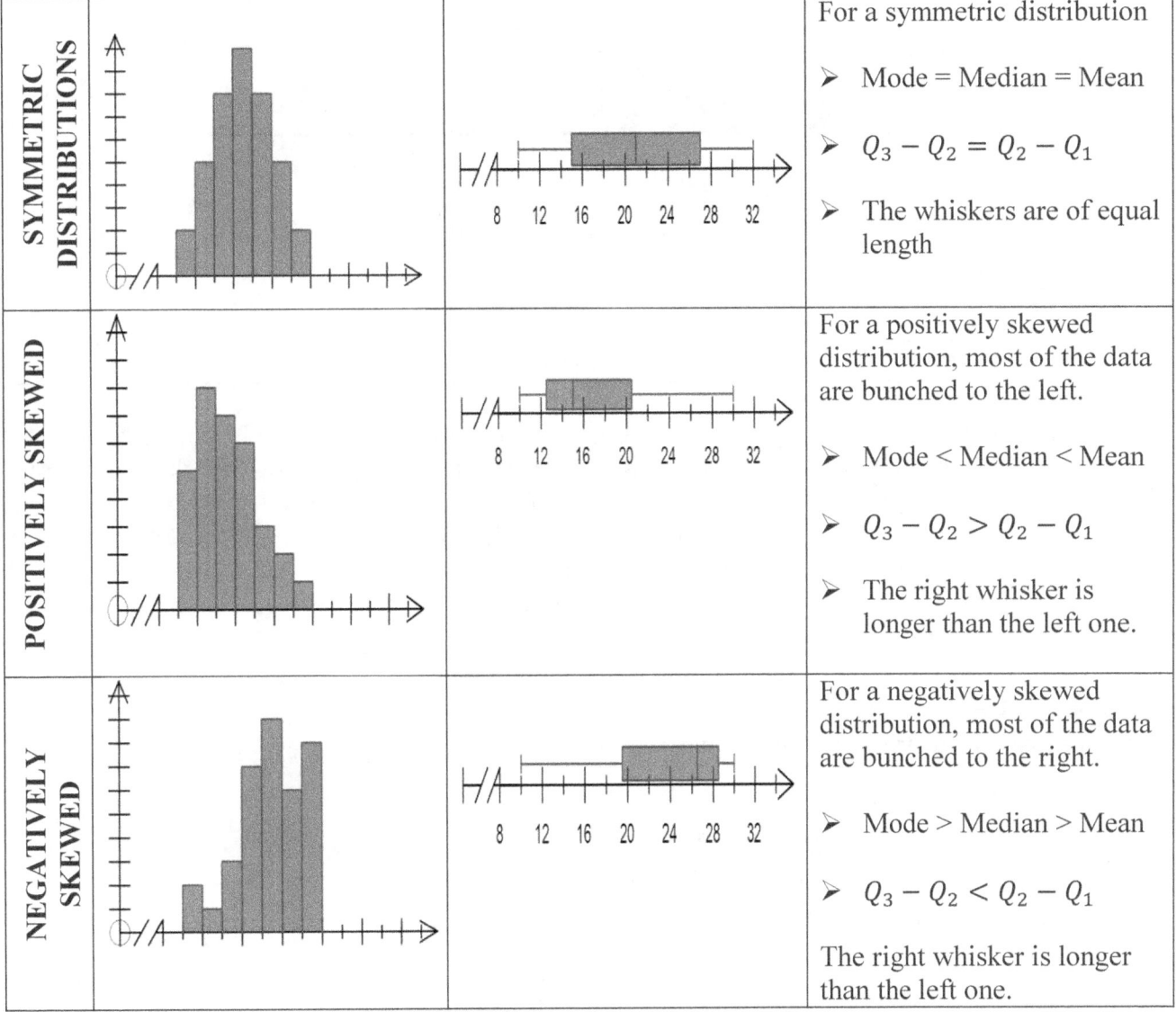

SYMMETRIC DISTRIBUTIONS

For a symmetric distribution

- ➢ Mode = Median = Mean
- ➢ $Q_3 - Q_2 = Q_2 - Q_1$
- ➢ The whiskers are of equal length

POSITIVELY SKEWED

For a positively skewed distribution, most of the data are bunched to the left.

- ➢ Mode < Median < Mean
- ➢ $Q_3 - Q_2 > Q_2 - Q_1$
- ➢ The right whisker is longer than the left one.

NEGATIVELY SKEWED

For a negatively skewed distribution, most of the data are bunched to the right.

- ➢ Mode > Median > Mean
- ➢ $Q_3 - Q_2 < Q_2 - Q_1$

The right whisker is longer than the left one.

EXERCISE 4C

Without doing any statistical computations, tick whether the following sets of data will be symmetrical, positively skewed or negatively skewed.

		symmetric	Positively skewed	Negatively skewed
1.	2, 5, 5, 6, 7, 8, 8, 10, 16, 20			
2.	4, 4, 6, 6, 8, 8, 10, 10, 12, 12			
3.	2, 5, 8, 15, 18, 21, 23, 25			
4.	10, 12, 12, 15, 15, 18, 18, 20			

4D COMPARING BOX PLOTS

To compare box plots comment on the following:

- Median
- Spread : include range or interquartile range
- Skewness : symmetrical, positively skewed or negatively skewed

EXAMPLE
Compare the annual rainfall of two states A and B. The data for the past ten years has been plotted as box plots.

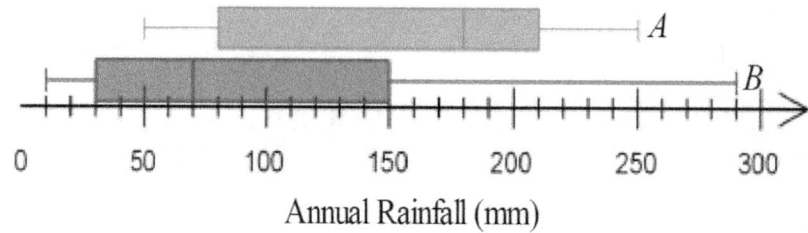

Annual Rainfall (mm)

SOLUTION

Median	spread	skewness
• Median for state A is greater than median for state B	• State B has a greater range (280mm) compared to State A (200mm) • IQR for State A (130 mm) is greater than IQR for State B (70 mm)	• State B's box plot is positively skewed whereas state A's box plot is negatively skewed.

EXERCISE 4D

1. Compare the distributions of the two box plots.

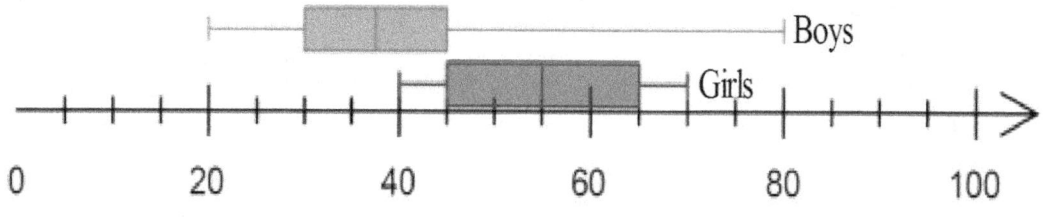

Median	spread	skewness

2. The box plots below shows the running time, in minutes, of two groups A and B in a marathon. Compare the distributions of the two box plots and recommend which group is better.

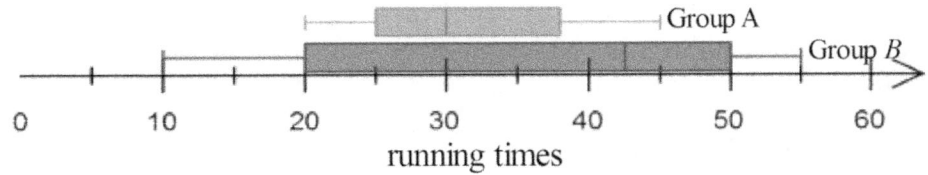

Median	spread	skewness

3. The box plot below shows the service waiting time in two stores in the city. Compare the distributions of the two box plots and recommend which store is better.

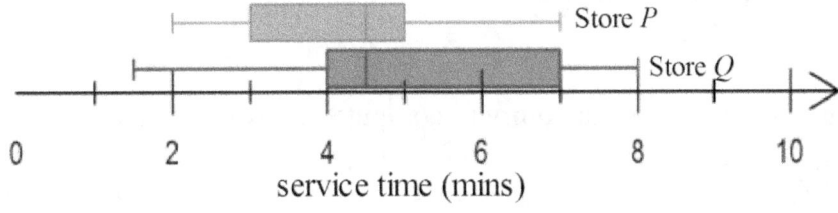

Median	spread	skewness

4E DESCRIBING OTHER DISTRIBUTIONS

In this part of the chapter, the reader will be given histograms, dot plots, bar graphs or stem and leaf plots and their objective would be to describe those using appropriate statistical terms as explained through an example below.

To describe a given distribution of scores, comment on each of the following to have a complete overview of the distributions.

- **Modality**: Uni or multi modal
- **Location**: calculate the value of the mean and median using your calculator.
- **Spread**: state the range and calculate the standard deviation, or interquartile range.
- **Shape**: symmetric versus positively or negatively skewed. Mention about any other feature of the graph such as gaps, clusters, outliers etc..

EXAMPLE

The histogram below shows the amount of pocket money received by a group of students in a particular class. Give a brief description of the distribution of the scores.

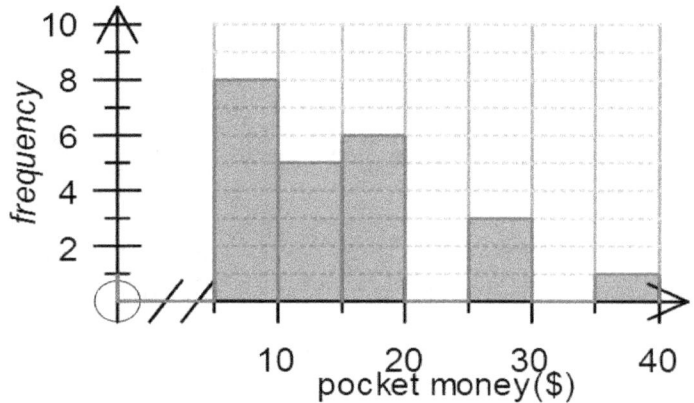

	list1	list2	list3
1	7.5	8	
2	12.5	5	
3	17.5	6	
4	22.5	0	
5	27.5	3	
6	32.5	0	
7	37.5	1	
8			
9			

SOLUTION

Use your calculator and the statistical capability to compute most of the measures given below.

Modality
✓ Clearly the distribution is uni modal

Location
✓ The mean pocket money is $ 15.10
✓ the median class is $10 - 15

Spread
✓ The scores have a range of 5–40
✓ The standard deviation is 8.06

Shape
✓ There is gaps at 20-25 and 30-35
✓ Clusters between 5-20
✓ Positively skewed

$\bar{x} = 15.108696$
$\Sigma x = 347.5$
$\Sigma x^2 = 6743.75$
$\sigma_x = 8.0581535$
$S_x = 8.2392582$
$n = 23$
$minX = 7.5$
$Q_1 = 7.5$
$Med = 12.5$

CHAPTER 4 : BOX PLOTS AND DESCRIBING DISTRIBUTIONS

EXERCISE 4E

1. The histogram below shows the distances, in kilometres, walked monthly by a group of athletes in a particular sporting club in Bunbury. Describe the distribution of the distance walked.

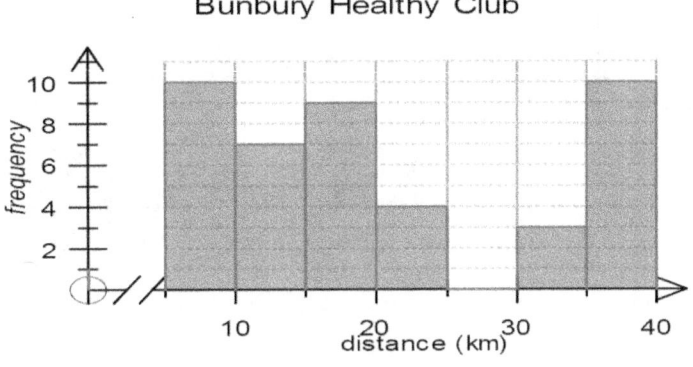

SOLUTION

Modality	Location
Spread	**Shape**

2. For the dot frequency diagram on the right, describe the distribution of the numerical data.

SOLUTION

Modality	Location
Spread	**Shape**

MATHEMATICS APPLICATIONS UNIT 2

3. The histogram below shows the speed, in km/h, of vehicles in a busy highway (speed limit 70 km/h) at 9 a.m. Give a brief description of the distribution of the speed of the vehicles.

Final Destination Highway

(Histogram: frequency vs speed km/h. Bars approximately: 40–50: 20; 60–70: 30; 70–80: 50; 80–90: 40; 90–100: 5)

SOLUTION

Modality	Location
Spread	**Shape**

4. For the given stem plot, describe the distribution of the numerical data.

SOLUTION

Maths Test Score

```
4 | 0 2 5 8 9
3 | 2 2 5 8 8 9
2 | 1 5
1 | 1
0 | 9
```

Modality	Location
Spread	**Shape**

CHAPTER 5

STATISTICAL INVESTIGATION PROCESS

As a statistical project or investigation, one of the tasks may be listed as follows:

- ❖ Are left handers tennis players better than right handers?
- ❖ Are boys better than girls in Mathematics?
- ❖ Children's favourite pet.
- ❖ Fitness level of soccer players of varying ages and gender.
- ❖ Genre of movies preferred by youngsters.

To be able to come up with a great statistical report, the reader must try and put into practice majority of the statistical terms and concepts learnt in the previous four chapters of this book. An attempt has been made to enumerate the different steps that will help during the investigation process.

STEP 1 – IDEAS TO EXPLORE

Carefully decide upon an idea that would be of your interest or your group investigating. Choose a topic that you are fond and familiar with. This will help and lend itself to data collection and statistical analysis. For example

- People's preferred novels: action, drama, horror etc..
- Types of vehicles at a busy intersection at different times.
- Popularity of Mathematics in Years 10 - 12.
- Characteristics of advertisement shown on television.

STEP 2 – SAMPLE SELECTION

Choose a sampling technique appropriate to your situation. Stratified sampling, random sampling, draw names out of a hat … and always choose a suitable sample size. The sample must be representative of the whole population. For example, if we are find out about the popularity of Mathematics in Years 10-12, stratified sampling might be the best where each year level would have some say in the sample.

STEP 3 - STATISTICAL INSTRUMENTS

Design the tools and instruments that are going to be used to gather the required data. For example,
- ➢ Questionnaires
- ➢ interviews
- ➢ Surveys
- ➢ Experiments and so on.

STEP 4 – DATA COLLECTION AND ORGANISATION

Collect and organise your data the simplest possible way so that it can be understood easily. For example make use of
- ➢ Tables

- Lists
- charts
- diagrams etc.

STEP 5 – STATISTICAL REPRESENTATION

Use statistical measures and graphs where appropriate. Now it's time for you to use all the statistics you have learnt so far. Try and include:

- Bar charts, pie graphs, dot frequency diagrams.
- Make use of technology to calculate all the measures of central tendency (mean, mode and median) wherever they are applicable.
- Frequency distributions and grouped frequency distributions, histograms, stem and leaf plots, box and whisker plots.
- Scatter graphs, tree diagrams, two-way tables.

STEP 6 – STATISTICAL REPORT

With the aid of the statistical representation mentioned above
- make reasonable conclusions and observations about the data you have gathered
- Evaluate the statistical instruments you used to gather the data.

Ask yourself and seek help from the members of your group

- Are your data reliable?
- Are the data collected valid or irrelevant?
- What could be done to improve the exercise?

STEP 7 – THE FINAL PRODUCT

Last you need to collate your project into a logical, neat and concise statistical report. To give your work some more weight, it is advisable that you include a title page and a good quality final copy of all your data, statistics and conclusions.

EXERCISE 5A

Using the 7 steps described above, choose one of the following topics to investigate.

1. Are rich people happier?

2. Are boys better than girls in Physics?

3. Children's favourite toy.

4. Fitness level of footy players of varying ages.

5. Genre of movies preferred by mature people.

6. Do children coming from split families tend to have lower academic progress?

7. Should IPAD be banned from classes?

CHAPTER 6

SOLVING EQUATIONS

6A SOLVING EQUATIONS USING TECHNOLOGY (CAS)

In this section, we are going to make use of the solve facility in our calculator to solve equations. In the subsequent stage of the chapter, it is recommended to try each question by a step by step approach and try to isolate the unknown.

To solve equations in CAS make use of the following steps:

MAIN → ACTION → ADVANCED → SOLVE

Most calculators have similar solve facilities available which the reader can try and adapt accordingly.

EXAMPLES

Use the solve facility on your calculators to solve the following equations.

(a) $5x + 12 = 2(x - 3) + 9$

$$Solve\ (5x + 12 = 2(x - 3) + 9, x)$$
$$\{x = -3\}$$

(b) $2(3x + 7) - 2(x - 1) = 5(x + 3)$

$$Solve\ (2(3x + 7) - 2(x - 1) = 5(x + 3), x)$$
$$\{x = 1\}$$

(c) Given that $v^2 = u^2 + 2as$, find the value of a when $v = 20$, $u = 8$ and $s = 12$.

MENU → NUMSOLVE → INSERT EQUATION → INPUT GIVEN VALUES → SOLVE

Remember to click the bubble at a as we are solving for a.

$$a = 14$$

Equation:
$v^2 = u^2 + 2 \times a \times s$

○ $v = 20$
○ $u = 8$
● $a =$
○ $s = 12$

Lower = -9E+999

Upper = 9E+999

MATHEMATICS APPLICATIONS UNIT 2

EXERCISE 6A

Use the solve facility on your calculator to solve the following equations.

1. $5x + 3 = -17$	2. $2x + 5 = x + 11$
3. $5(x - 4) = 3(x + 6)$	4. $3x + 4 = x + 16$
5. $\dfrac{2x+3}{5} = \dfrac{x-1}{3}$	6. $\dfrac{5}{x} = 10$
7. $\dfrac{x}{2} + \dfrac{x+1}{5} = \dfrac{2}{3}$	8. $\dfrac{x}{3} + \dfrac{x}{2} = 10$

Use Num Solve to find the missing pronumerals in each case.

9. Given $v = u + at$, find a given that $v = 55, u = 25$ and $t = 3$.	10. If $E = \dfrac{1}{2}mv^2$, find the value of m when $E = 125$ and $v = 5$.
11. In the formula $y = mx + c$, find m given that $y = 18, x = 10$ and $c = 13$.	12. In the formula $E = mc^2$, find m given that $E = 2000$ and $c = 5$.
13. Given that $v^2 = u^2 + 2as$, find the value of s when $v = 30, u = 12$ and $a = 10$.	14. Given that $A = \dfrac{a+b}{2} \times h$, find the value of h when $A = 88, a = 11$ and $b = 5$.
15. $V = \pi r^2 h$, find the value of h given that $V = 200$ and $r = 5$.	16. $V = 2\pi r^2 + 2\pi rh$, find the value of h given that $V = 500$ and $r = 4$.

6B SIMPLE LINEAR EQUATIONS

A linear equation is a polynomial of degree 1. In order to solve for the unknown variable, we have to isolate the variable as shown in the examples below.

CLASS ACTIVITY
Solve the following linear equations.

1. $3x + 1 = 16$ **SOLUTION** Subtract 1 on both sides $\qquad 3x + 1 - 1 = 16 - 1$ $\qquad 3x = 15$ $\qquad x = 5$	2. $10 - 2x = 30$ **SOLUTION** Subtract 10 on both sides $\qquad 10 - 2x - 10 = 30 - 10$ $\qquad -2x = 20$ $\qquad x = -10$
3. $4x - 3 = 17$	4. $2x - 7 = 11$
5. $3(x + 4) = 9$	6. $5(2x - 1) = 15$
7. $\dfrac{x}{5} = 6$	8. $\dfrac{y}{2} = -8$
9. $\dfrac{z}{4} = -6$	10. $\dfrac{20}{x} = 5$
11. $\dfrac{x - 6}{3} = 2$	12. $\dfrac{x + 4}{2} = 3$

13. $$2(x+3) - 8 = 10$$	**14.** $$3(1-5x) = 15$$
15. $$2x + 3x = 50$$	**16.** $$3y + 4y - 2 = 19$$
17. $$2(3x+1) - 3(1-5x) = -43$$	**18.** $$4(x+2) - 7x = -16$$
19. $$\frac{2x-1}{5} - 3 = 7$$	**20.** $$\frac{10-2x}{5} = 4$$
21. $$\frac{7x+10}{5} - 2 = 7$$	**22.** $$\frac{12-3x}{6} = 5$$

6C   EQUATIONS HAVING UNKNOWN ON BOTH SIDES

EXERCISE 6C

Solve the following linear equations.

1. $3x + 7 = x + 13$ **SOLUTION** Subtract x on both sides $\quad 3x - x + 7 = x - x + 13$ $\quad 2x + 7 = 13$ Subtract 7 on both sides $\quad 2x + 7 - 7 = 13 - 7$ $\quad 2x = 6$ $\quad \therefore x = 3$	2. $5x - 3 = 2x + 12$ **SOLUTION** Subtract 2x on both sides $\quad 5x - 2x - 3 = 2x - 2x + 12$ $\quad 3x - 3 = 12$ Add 3 on both sides $\quad 3x - 3 + 3 = 12 + 3$ $\quad 3x = 15$ $\quad \therefore x = 5$
3. $2x + 11 = 3x + 5$	4. $6x - 1 = 4x + 11$
5. $2(3x - 4) = 5x + 6$	6. $3(2x + 5) = 4x - 10$
7. $7(2x + 5) = 3(x + 4)$	8. $5(x + 3) = 4(x + 1)$

6D SOLVING EQUATIONS USING CROSS MULTIPLICATION

EXERCISE 6D

Solve the following linear equations.

1.
$$\frac{3x+1}{2} = \frac{2x-6}{3}$$

SOLUTION
Cross multiplying, we have
$$3(3x+1) = 2(2x-6)$$
Expand both sides
$$9x+3 = 4x-12$$
Subtract $4x$ on both sides
$$9x-4x+3 = 4x-4x-12$$
$$5x+3 = -12$$
Subtract 3 on both sides
$$5x+3-3 = -12-3$$
$$5x = -15$$
$$\therefore x = -3$$

2.
$$\frac{2x+1}{7} = \frac{x-1}{4}$$

3.
$$\frac{10-3x}{5} = \frac{2x+1}{2}$$

4.
$$\frac{x+4}{3} = \frac{5-2x}{4}$$

5.
$$\frac{4}{2x-1} = \frac{5}{x+3}$$

6.
$$\frac{8}{2x+3} = \frac{2}{x+1}$$

CHAPTER 6 : SOLVING EQUATIONS

6E   USING LCM TO SOLVE EQUATIONS

Many equations cannot be solved by the method of cross multiplication. The easiest way would be to find the LCM of the denominators and multiply each term by the LCM.

EXERCISE 6E

1. Solve
$$\frac{x}{2} + \frac{x+1}{5} = \frac{2}{3}$$

SOLUTION
The LCM of 2, 3 and 5 is 30.
Multiplying each term by 30, we have
$$\frac{x}{2} \times 30 + \frac{x+1}{5} \times 30 = \frac{2}{3} \times 30$$

$$15x + 6(x+1) = 20$$
$$15x + 6x + 6 = 20$$

$$21x = 14$$

$$\therefore x = \frac{14}{21} = \frac{2}{3}$$

2. Solve
$$\frac{4x-3}{6} = \frac{x-1}{3} + 2$$

SOLUTION
LCM of 3 and 6 is 6.
So we multiply each term by 6.

$$\frac{4x-3}{6} \times 6 = \frac{x-1}{3} \times 6 + 2 \times 6$$

$$4x - 3 = 2(x-1) + 12$$

$$4x - 3 = 2x - 2 + 12$$
$$4x - 3 = 2x + 10$$
$$2x = 13$$
$$x = 6.5$$

3. Solve $\dfrac{x}{2} + \dfrac{2x+1}{3} = \dfrac{1}{4}$

4. Solve $\dfrac{3x+2}{5} - 1 = \dfrac{4x-1}{4}$

5. Solve $\dfrac{3x-4}{2} + \dfrac{1}{3} = \dfrac{x}{4}$

6. Solve $\dfrac{x}{2} + \dfrac{x}{3} = 5$

6F APPLICATIONS

1. John is a full time electrician and the cost (C $) for hiring him is made up of a callout fee of $80 and a fixed rate of $42 per hour. The equation for hiring John for h hours is given by $C = 0.95(80 + 42h)$.

   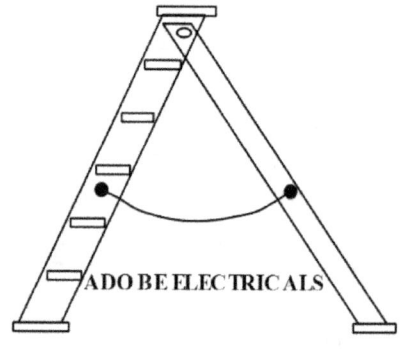

   (a) John offers a discount to all his clients. State how much discount he allows to his clients.

   (b) Calculate the cost of hiring John for 6 hours.

   (c) Mrs Will paid $335.35 for some repairs undergone. For how many hours did she hire the electrician?

2. Julia is employed by JK Real Estates. She earns $300 per week plus 1.25% commission on selling a property.
   Her monthly revenue ($R) for selling a property worth $P can be modelled by the equation
   $$R = 3000 + 0.0125P$$

   (a) In a particular week she sold a property for $280 000. Determine her revenue for that week.

   (b) During another week, she earned $6050 when she managed to sell one property only. Calculate the price the property was sold for.

3. When an agent of a particular software company sells IT products to the general public, his monthly salary is given by
   $S = 2500 + 3N$, where N is the amount he sells in thousands of dollars.
   Find the agent's salary if he sells
   (a) $125 000 worth of goods during the month of July,

   (b) $90 000 worth of IT products in August.

   (c) During the month of December his salary was $2980. Find out the value of the products he sold.

4. Jimmy is a broker and is self-employed. His income depends on the value of the loan he is able to get approved for his clients. His income can be summarised by the formula
   $I = 400 + 1.5L$, where L is the value of the loan in hundreds of dollars.
   (a) Calculate Jimmy's income if he settles a loan of $250 000.

   (b) For one of his transaction, Jimmy was paid $5200. Calculate the value of the loan he got approved for his client.

5. The centripetal force F acting on a particle travelling at a speed of v m/s, having mass m kg and radius r metres is given by $F = \dfrac{mv^2}{r}$.

   (a) Determine the centripetal force of a car of 800 kg if it travels a curve of radius 400m at 12 m/s.

   (b) If the centripetal force of an object of mass 60 kg travelling in a circular path of radius 30m is 200N, determine the particle's velocity.

6. Appleby Senior High usually hires a school bus to pick and drop its students. The cost ($C) associated with hiring the bus for d days and travelling k kilometres is given by
$C = 50 + 25d + 10k$

   (a) Explain the significance of 50 in the above equation.

   (b) Calculate the cost of hiring the bus for 10 days and travelling a total distance of 165 km.

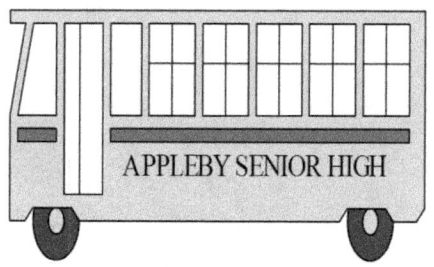

   (c) In a particular month, the school hired the bus for 21 days for a total cost of $3005. Determine the number of kilometres the bus travelled.

   (d) The school Principal was concerned that the cost of hiring the school bus was way past the available budget. She appointed all the top mathematics teachers and assigned them a task to work out a minimal spanning tree whereby all the students using the school bus could be picked up and dropped by travelling the least possible distance. If the school's budget is $2500 for 21 days, determine the maximum distance that the bus could travel without exceeding the budget.

CHAPTER 7

USING EQUATIONS TO SOLVE PROBLEMS

7A   ALGEBRAIC EQUATIONS AND EXPRESSIONS

1. Complete the following table by converting the mathematical statements into **algebraic expressions**. Use x to represent the number.

Mathematical statement	Algebraic Expression
(a) Add six to the number	$x + 6$
(b) Multiply the number by three then subtract five	
(c) Double the number then add three	
(d) Twice a number divided by five	
(e) Two less than five times a number	
(f) Six more than five times a number	
(g) Three times a number minus eleven	
(h) Two less than three times a number	
(i) Five times the sum of x and 3	
(j) Three less than twice the sum of x and five	

2. Complete the following table by converting the mathematical statements into **algebraic equations**. Use x to represent the number.

Mathematical statement	Algebraic Equation
(a) Think of a number, add two and the result is 10	$x + 2 = 10$
(b) Think of a number, subtract five and the result is 11	
(c) When a number is decreased by six and the result multiplied by four, the final answer is 32	
(d) When a number is increased by two and the result multiplied by three, the final answer is 15	
(e) Think of a number, multiply it by three, subtract four from the result and the final answer is 8	
(f) When a certain number is subtracted from 25 and the result multiplied by six, the final result is 60.	
(g) Think of a number, add one then divide the result by two and the final result is 4	
(h) When two consecutive numbers are added together the answer is 25.	
(i) The sum of three consecutive even numbers is 36.	

3. For these questions, rewrite the statements as equations involving x and hence solve them.

(a) When a number is decreased by four and the result multiplied by three, the final answer is fifteen.

$$3(x - 4) = 15$$
$$3x - 12 = 15$$
$$3x = 27$$
$$\therefore x = 9$$

> **OR USING TECHNOLOGY**
>
> $Solve\ (3(x - 4) = 15, x)$
>
> $\{x = 9\}$

(b) When a number is increased by nine and the result multiplied by five, the final answer is 50.

(c) When two consecutive numbers are added together the answer is 101.

(d) Think of a number, multiply it by two, subtract seven from the result and the final answer is 11.

(e) The sum of three consecutive even numbers is 30.

(f) When a certain number is subtracted from ten and the result multiplied by three, the final result is -30.

(g) The sum of two consecutive odd numbers is 44.

(h) Think of a number, add eleven then divide the result by two and the final result is 10.

(i) Think of a number, add one then divide the result by four and my final answer is 7.

7B MATH PYRAMIDS

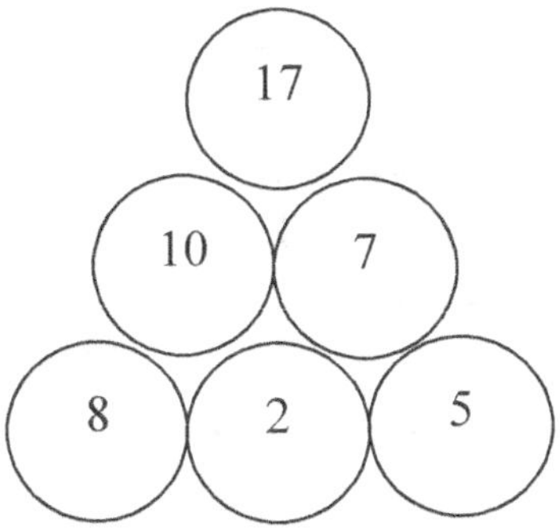

How does the pyramid work?

Add the 2 numbers next to each other in the bottom row and write the answer in the circle above those 2 numbers as shown on the left.

EXERCISE 7B

Complete the following pyramids.

1.

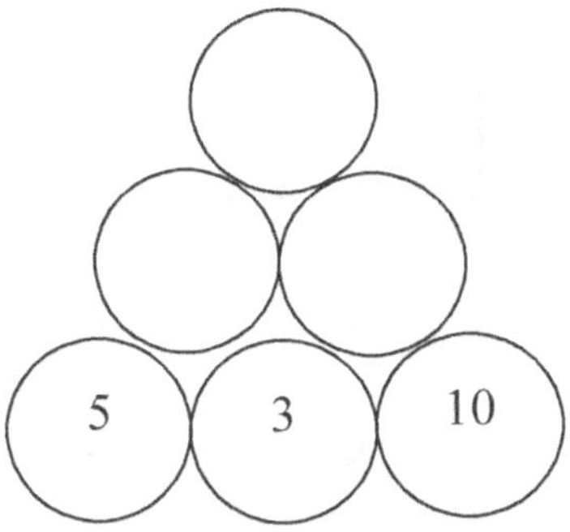

2.

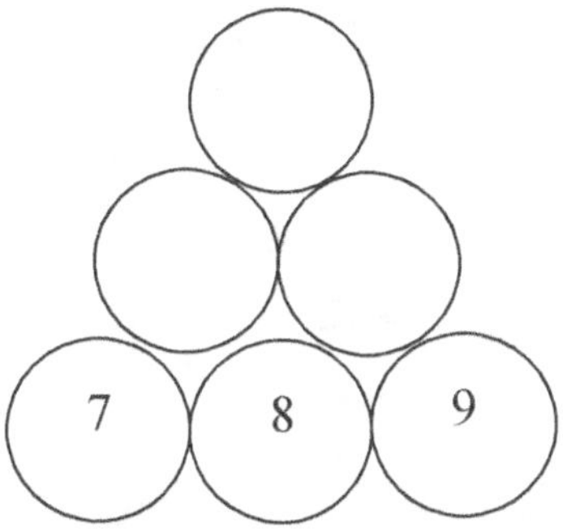

3.

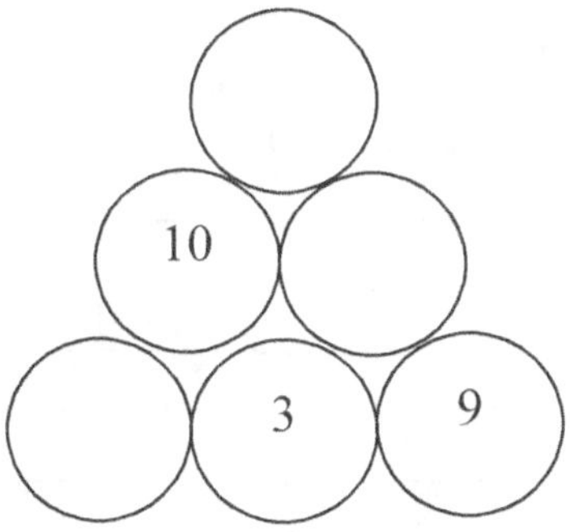

4.

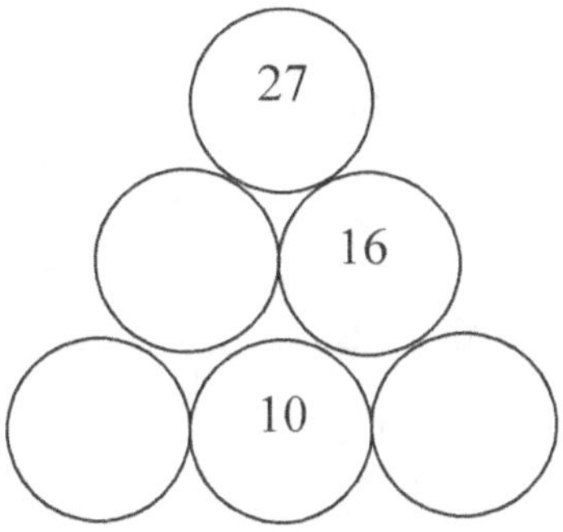

5.

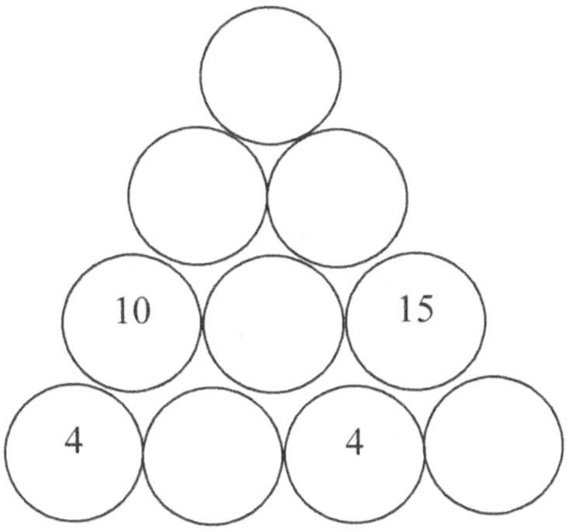

6.

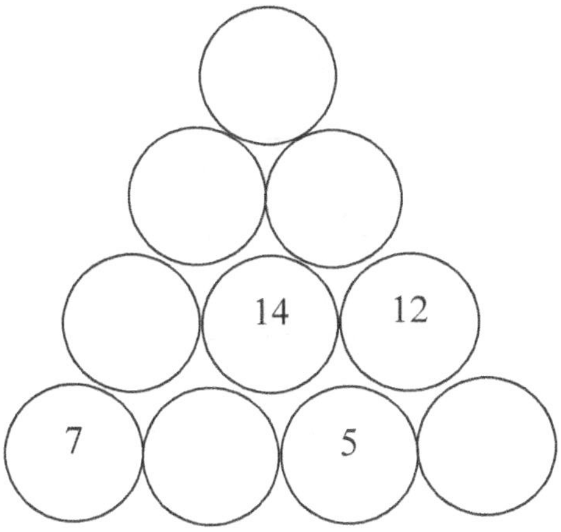

In the following pyramids work out the value of x. Question 7 has been done as an example.

7.

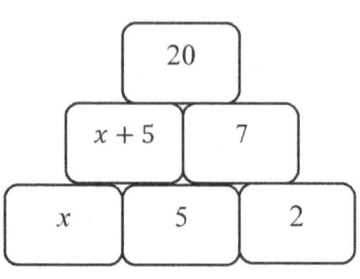

$x + 5 + 7 = 20$
$x + 12 = 20$
$\therefore x = 8$

8.

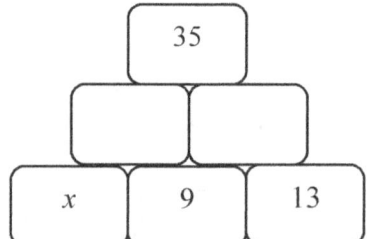

9.

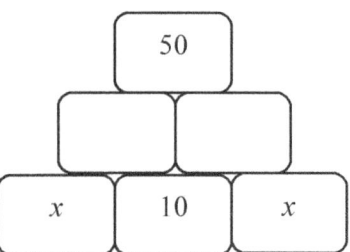

10.

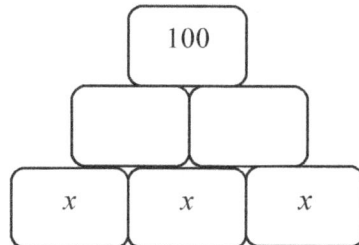

11.

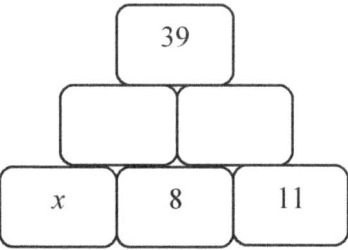

12.

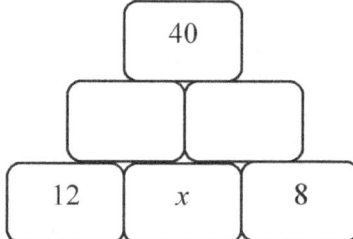

13.

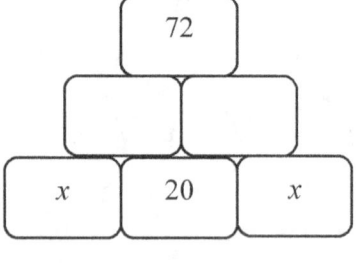

14.

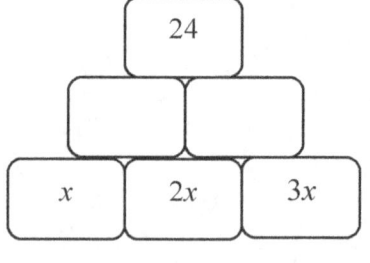

15. Study the two number patterns below carefully and then complete the sentences.

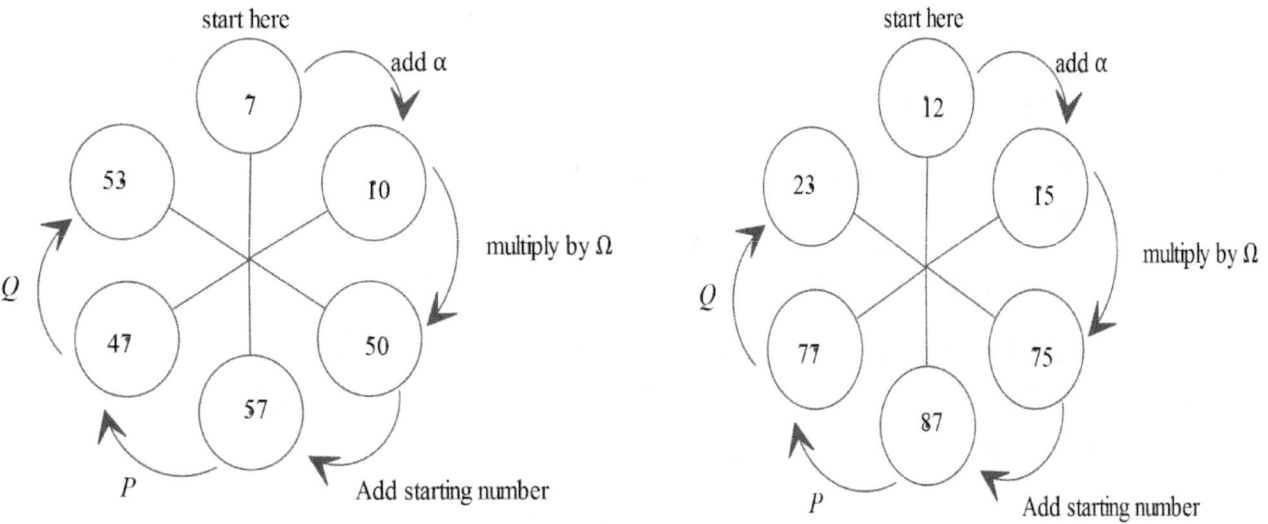

(a) The value of α =

(b) The value of Ω =

(c) What instruction must P represent?

(d) What does Q stand for?

Now complete the following number patterns diagrams.

(e) 10

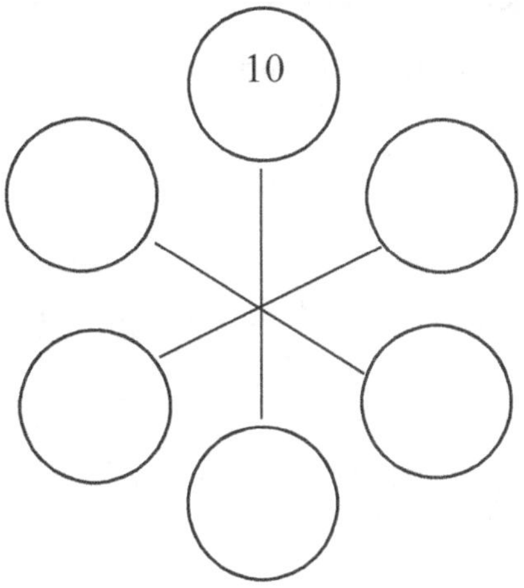

(f) 5

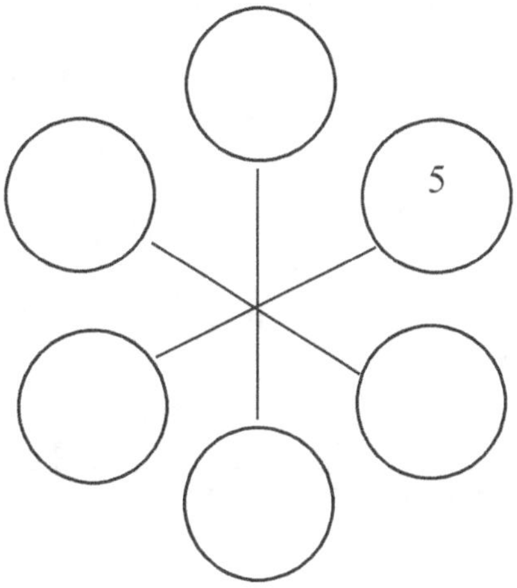

(g) 85

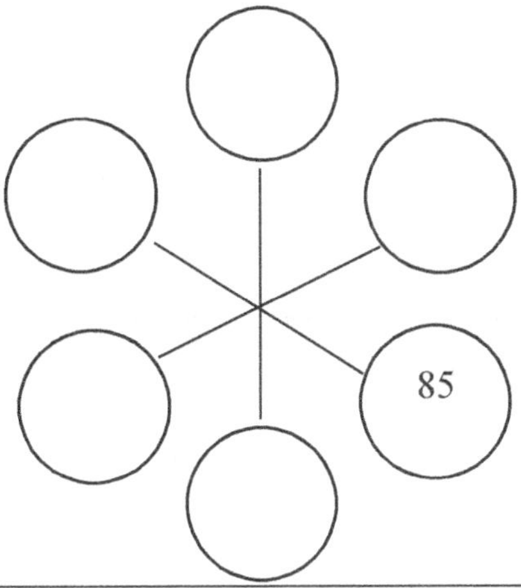

(h) 81

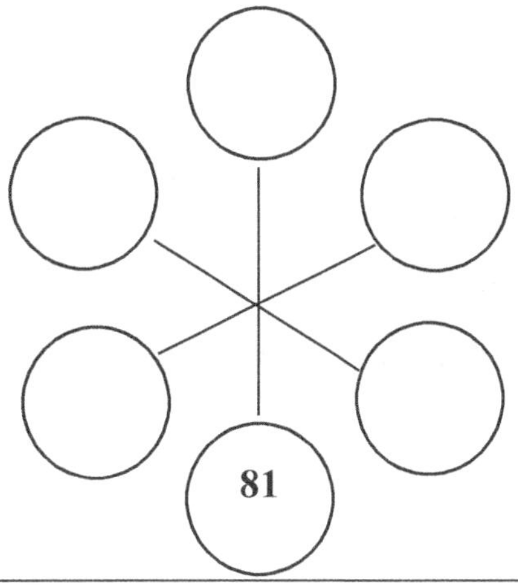

7C APPLICATIONS

EXAMPLE 1
Peter is 6 years younger than his sister Jane. In ten years' time, the sum of their ages will be 40 years. Find their present ages.

SOLUTION
First, we express the information in a table as shown.

	Peter	Jane
now	x	x + 6
In 10 years' time	x + 10	x + 6 + 10 = x + 16

Since the sum of their ages will be 40 years after 10 years,

$$x + 10 + x + 16 = 40$$
$$2x + 26 = 40$$
$$2x = 14$$
$$x = 7$$

Hence Peter is 7 years old and his Jane is 13 years of age.

EXAMPLE 2

Tickets for a show were priced at $4, $7 and $12. The number of $4 tickets sold was three times the number of $7 tickets. The number of $12 tickets sold was 20 more than the number of $7 tickets.

The number of $7 tickets sold was x.

(a) Find an expression, in terms of x, for the total sum of money received from the sale of the tickets.

Price	$4	$7	$12
Number of tickets	3x	x	x + 20

Total sum of money received $= 4 \times 3x + 7 \times x + 12(x + 20)$
$= 12x + 7x + 12x + 240$
$= 31x + 240$

(b) Given that $1790 was received from the sale of the tickets, form an equation in x. Solve this equation and hence find the total number of tickets that were sold.

Total number of tickets sold
$= 3(50) + 50 + (50 + 20) = 270$ tickets.

$$solve\ (31x + 240 = 1790, x)$$
$$\{x = 50\}$$

EXERCISE 7C

1. David is 3 years younger than his sister Rita. Given that the sum of their ages is 31 years, determine their respective ages.

2. Rahul is x years old. John, his father, is twice as old as him. In two years' time the sum of their ages will be 79 years. Form an equation in terms of x and hence determine their present ages.

3. The length of a rectangle is 5 cm more than its width. If the perimeter of rectangle is 38 cm, find the dimensions of the rectangle.

4. A 240 m long wire is used to fence a rectangular plot whose length is twice its width. Find the length and width of the plot.

5. The diagram shows an equilateral triangle of side x cm and a square whose edge is 2cm greater than that of the triangle. Given that the perimeter of square is 20 cm more than the perimeter of the triangle, determine the value of x.

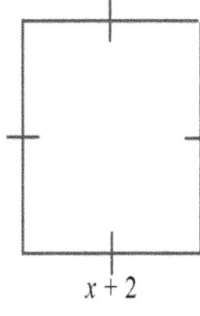
$x + 2$

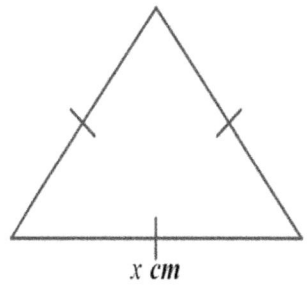
x cm

6. A cell phone cap plan costs $30 per month for unlimited calls plus $0.12 per text message.
   (a) Write a linear model that represents the monthly cost of this cell phone plan if the user sends x text messages.

   (b) A particular user paid $102 during the month of December. How many texts did he send?

7. A computer salesperson earns a basic salary of $30 000 per year plus a commission of $190 for every computer he sells.
   (a) Write an equation that shows the total amount of income the salesperson earns, if he sells x computers in a year.

   (b) How many computers would the salesperson need to sell to earn a total income of $87000?

8. At an outdoor cinema, children's tickets cost $4 each and adult tickets cost $7 each. The total amount of money earned from ticket sales equals $395. The number of children was 5 more than twice the number of adults. Work out how many adults attended the movie show.

9. Al, Ben and Connor are three friends and run a small business in the city. At the end of the financial year, they made a profit of $42 000. Sharing the profits, Ben earns twice as much as Al and Connor earns $2000 more than Al. Determine how much each partner received.

10. Tickets for a school play were priced at $5, $8 and $10. The number of $5 tickets sold was twice the number of $8 tickets. The number of $10 tickets sold was 80 more than the number of $8 tickets. The number of $8 tickets sold was x.
    (a) Find an expression, in terms of x, for the total sum of money received from the sale of the tickets.

    (b) Given that $4160 was received from the sale of the tickets, form an equation in x.
    Solve this equation and state the number of $8 tickets that were sold.

7D SOLVING EQUATIONS USING RATIOS

Ratios can be used to solve simple algebraic equations. It can also be used to solve similar shapes problems. The examples that follow attempt to show how ratios can be used efficiently to work out unknowns in equations.

EXAMPLES

Find the value of the pronumeral in each case.

(a) $x:2 = 5:1$

SOLUTION

Express the ratios as fractions

$$\frac{x}{2} = \frac{5}{1}$$

Cross multiply, we have

$$x = 10$$

(b) $3:y = 18:9$

SOLUTION

Express the ratios as fractions

$$\frac{3}{y} = \frac{18}{9}$$

Cross multiply, we have

$$18y = 27$$
$$\therefore y = 1.5$$

(c) Use ratios to find the height of Tom.

SOLUTION

In $\triangle ABC$ and $\triangle ADE$,

$\angle CAB = \angle DAE$ (Common angle)

$\angle CBA = \angle DEA = 90°$

Hence $\angle ACB = \angle ADE$ (sum of angles in a triangles = 180°)

$\therefore \triangle ABC \sim \triangle ADE$

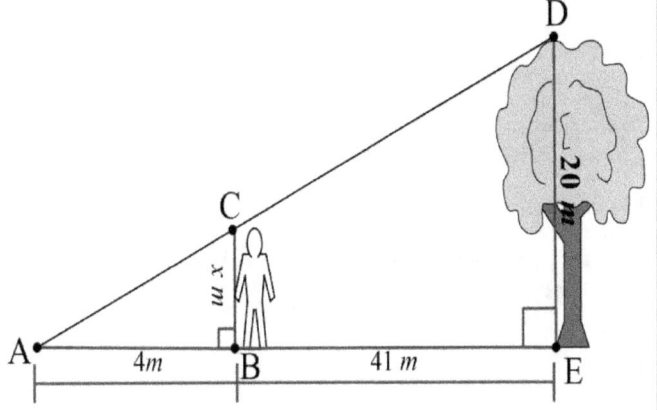

Using ratios, we have

$x:20 = 4:45$

Express the ratios as fractions

$$\frac{x}{20} = \frac{4}{45}$$

Cross multiply, we have

$45x = 80 \quad \therefore x = 1.78m$

Hence Tom is 1.78m tall.

EXERCISE 7D

Find the value of the pronumeral in each case.

1. $x:6 = 1:3$	2. $2:y = 36:3$	3. $x:18 = 1:9$
4. $2.5:x = 12:60$	5. $5:y = 7:8$	6. $x:3 = 6:1.5$
7. The distance run by Peter and Pan is in the ratio 4:5. The distance run by Peter is 14km. Find the distance run by Pan.	8. The ratio of boys to girls in a Year 11 Mathematics class is 3:4. If the number of boys is 9, find the number of girls.	
9. Given that $w:6 = 2:5$, find the value of w.	10. Given that $5x = 3y$, find the ratio $x:y$.	

11. Determine the height of the tree using ratios given that △ABC ~ △ADE.

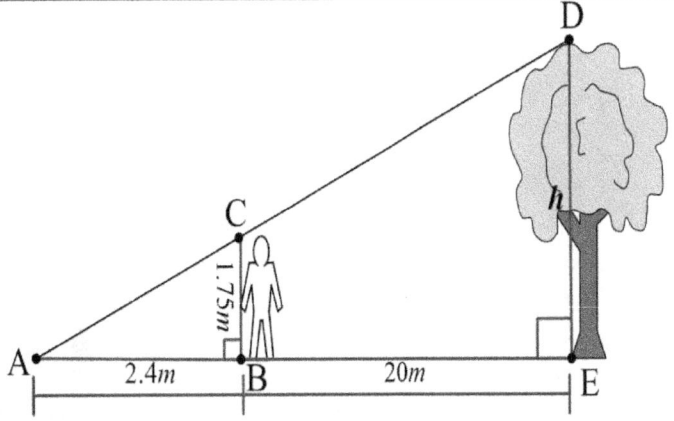

12. Use ratios to determine the height of the coconut tree given that the two given triangles are similar.

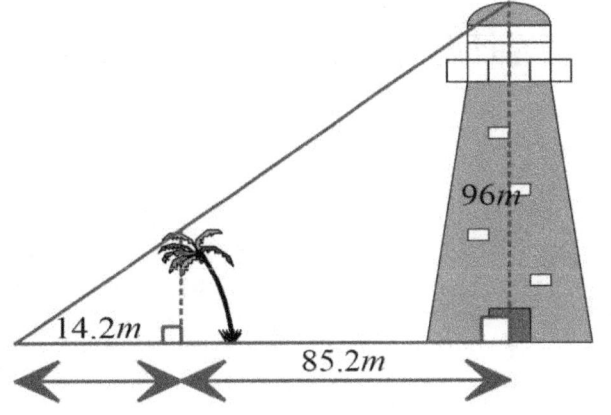

13. The diagram shows a 2.5m high sailing boat and a huge ship. Using the measurements determine the height of the ship.

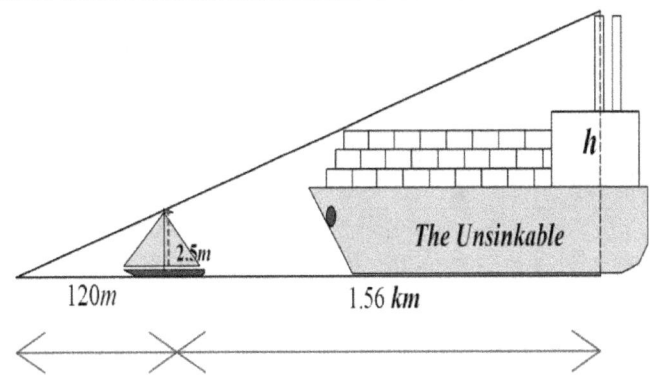

14. Use ratios and similar triangles to determine the height of the truck.

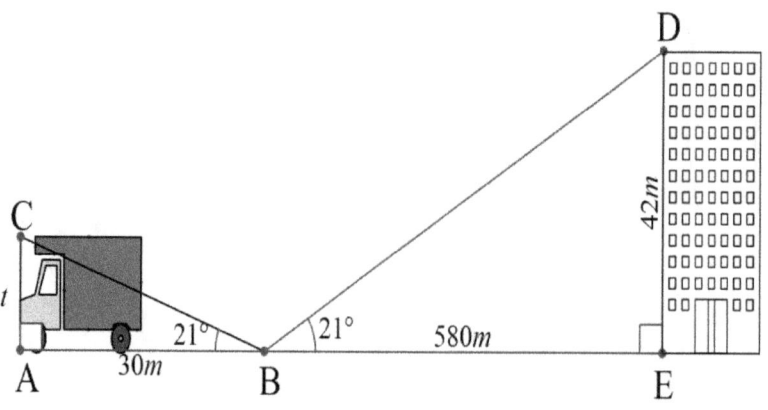

15. Find the depth of the submarine under water level.

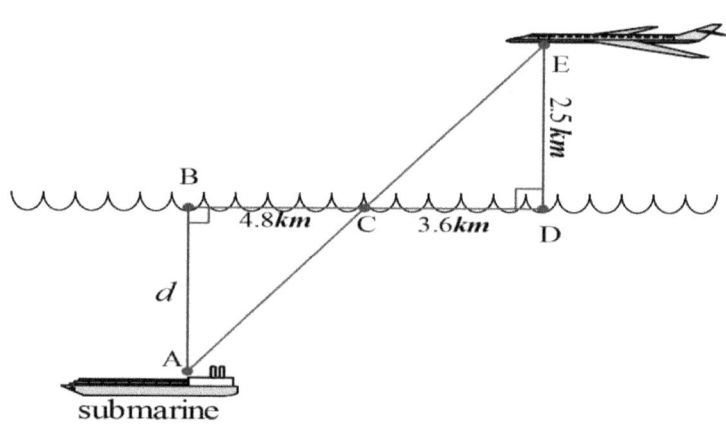

CHAPTER 8

LINEAR RELATIONSHIPS

8A   GRADIENT

Consider the four diagrams below, all showing ladders held in a different direction and angle.

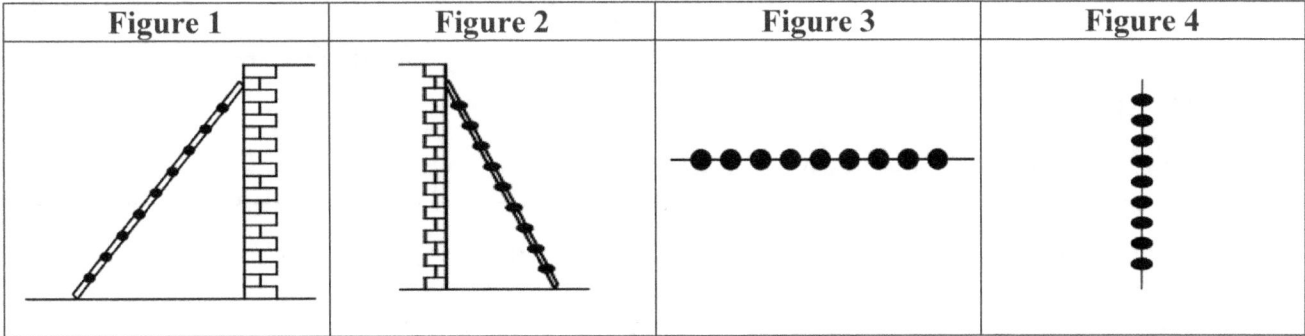

Comparing Figures 1 and 2, obviously Figure 2 seems steeper as it has a greater slope. The slope of the ladders represents the gradient of the ladder. Note that,

- Figure 1 has a positive gradient, sloping upward from left to right.
- Figure 2 has a negative gradient sloping downwards from left to right.
- Figure 3 has no slope as it is horizontal. The gradient is zero.
- Figure 4 being vertical, we say the gradient is undefined.

EXERCISE 8A

For each of the lines below, determine whether it has a positive, negative, undefined or zero gradient.

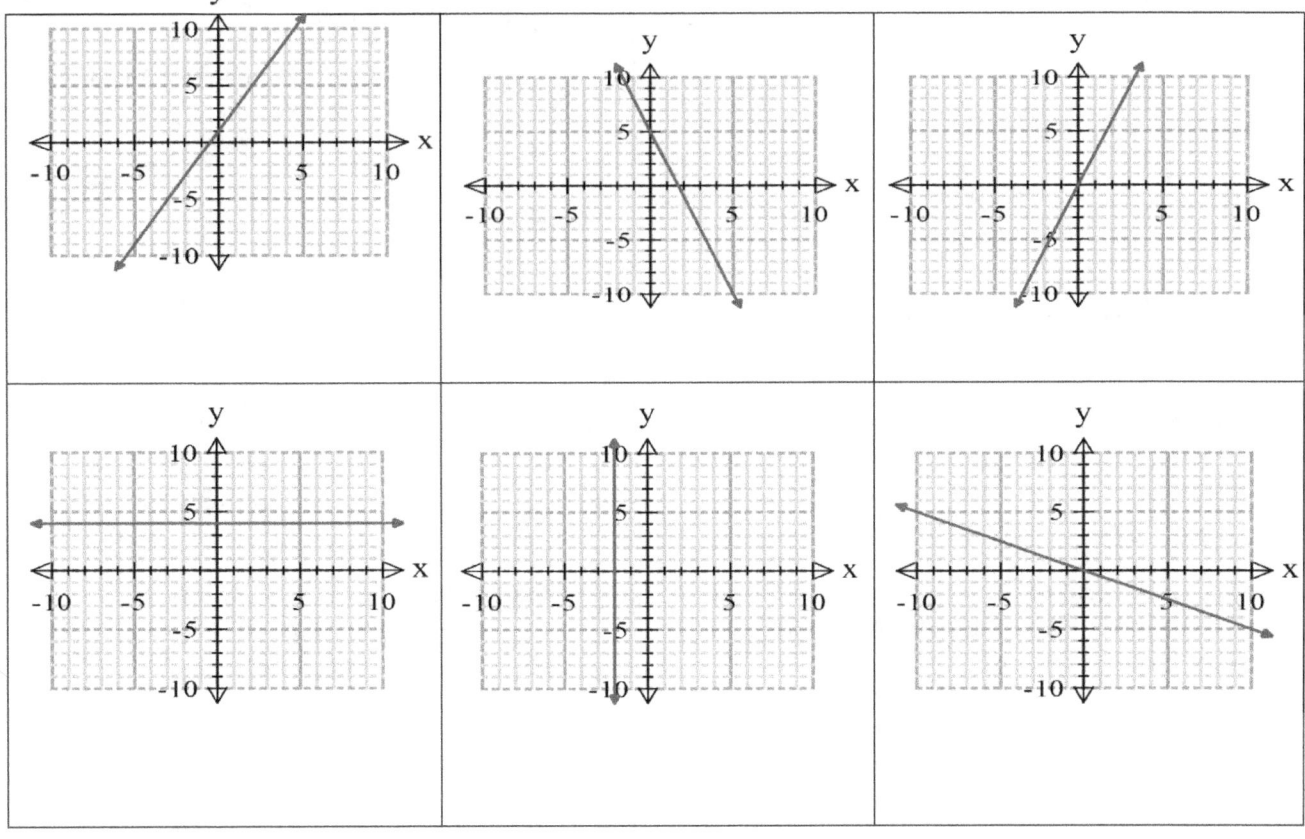

8B DETERMINING GRADIENT OF A LINE FROM A GRAPH

As we have seen earlier, the slope of the ladder represents the gradient. But how do we calculate the gradient?

Gradient is defined as the ratio of the vertical increase (rise) to the horizontal increase (run).

Gradient of AB = $\frac{RISE}{RUN}$ = $\frac{length\ of\ BC}{length\ of\ AC}$

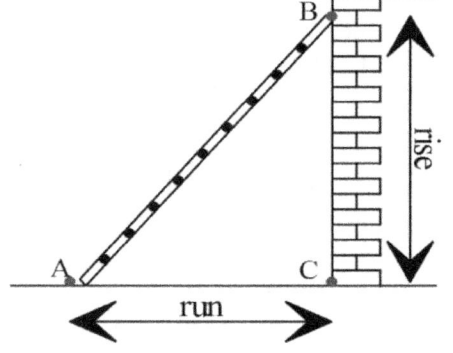

EXAMPLE

Determine the gradient of the following straight lines.

1.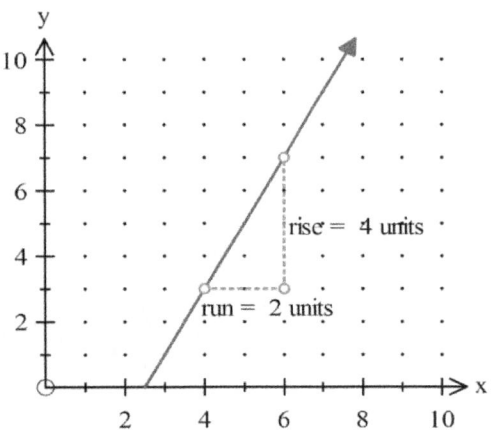

SOLUTION

To determine the gradient, we can draw any size triangles as far the points lie exactly on the given line.

Gradient = $\frac{rise}{run}$ = $\frac{4}{2}$ = 2.

Now, a gradient of 2 implies every one unit we move to the right we have to move two units up to reach the line again.

2.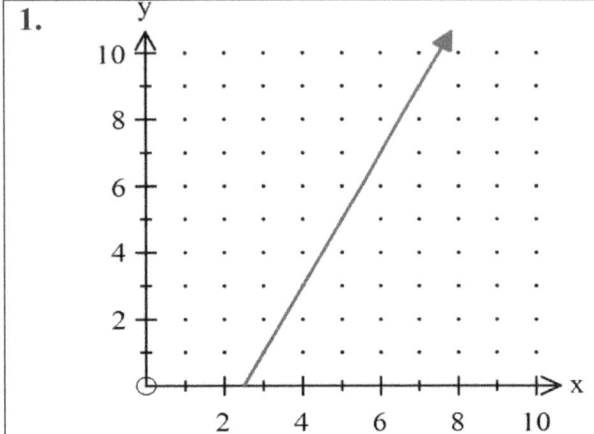

SOLUTION

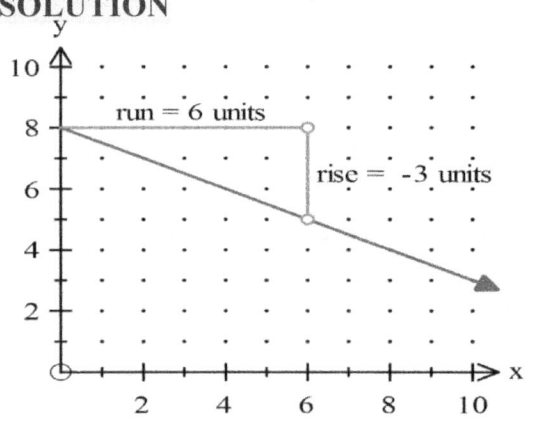

As we can see for each unit we move right, the graph is not rising as such, it is going down. This is the reason the rise is negative and thereby giving a negative gradient as a result.

Gradient = $\frac{rise}{run}$ = $\frac{-3}{6}$ = $-\frac{1}{2}$.

Now, a gradient of -0.5 implies every one unit we move to the right we have to move down 0.5 unit to touch the graph again.

EXERCISE 8B

Determine the gradient (m) of the following straight lines.

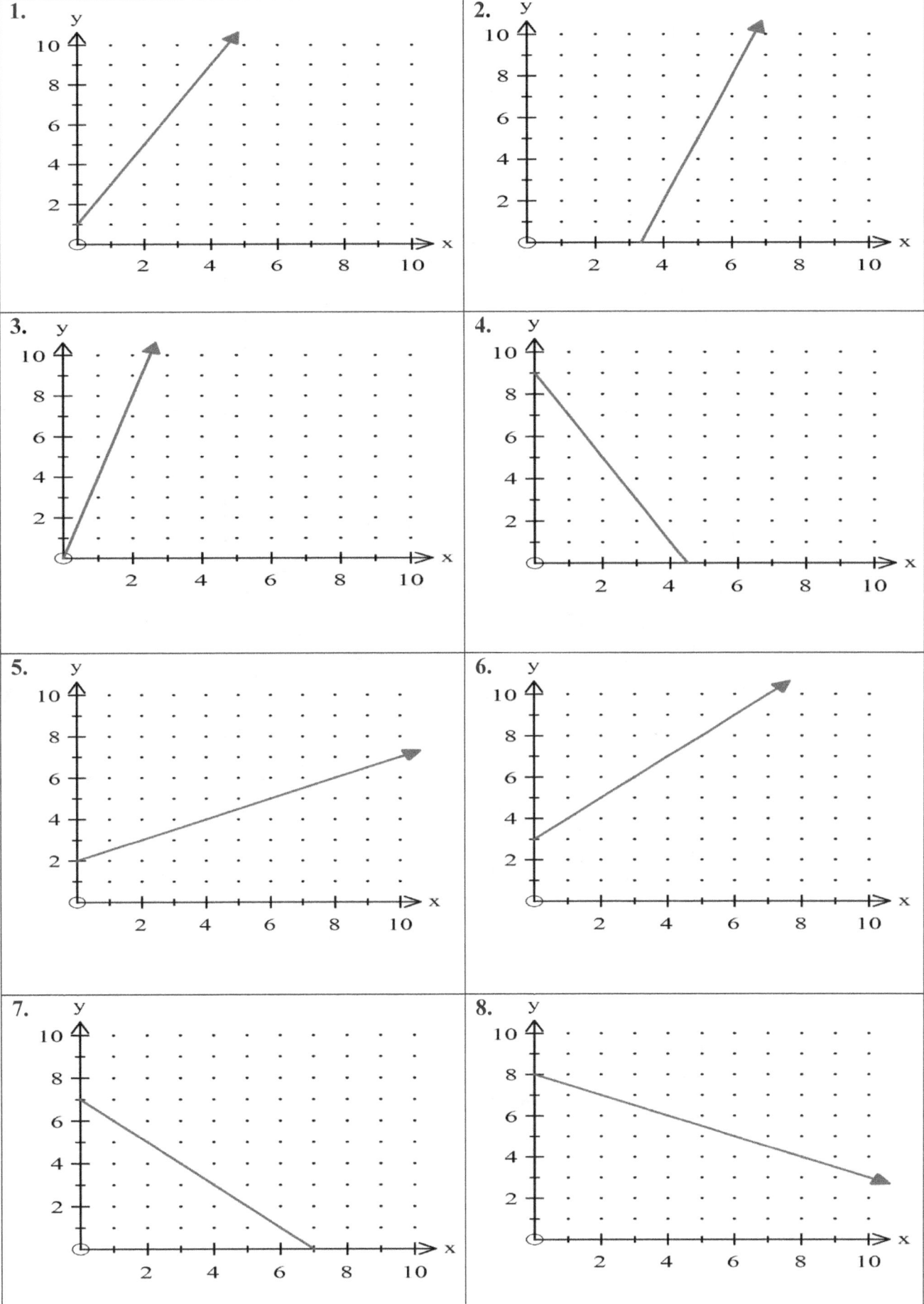

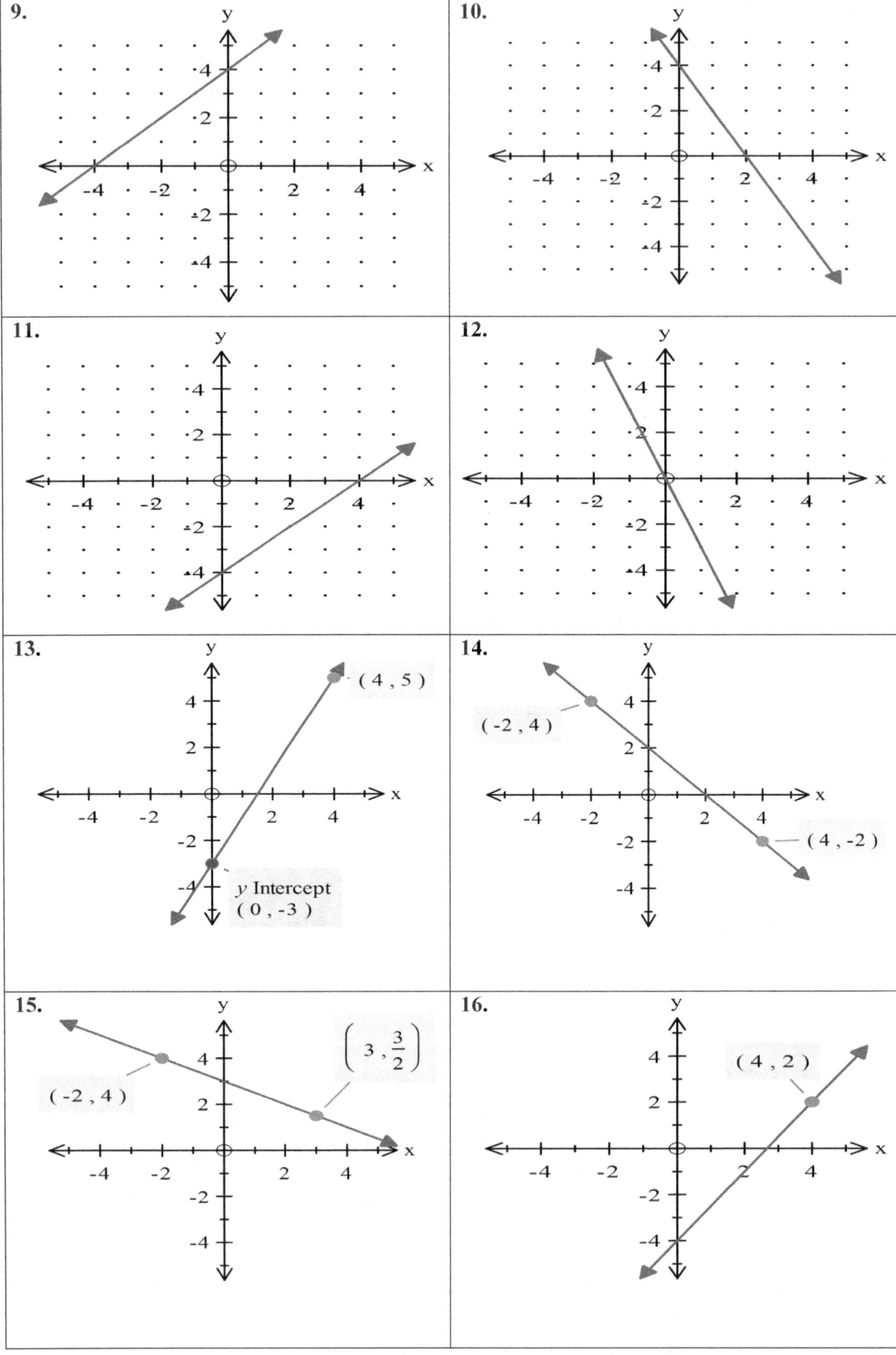

8C DETERMINING GRADIENT GIVEN TWO POINTS

Now we are going to derive the general rule when we have to find the gradient of a line segment joining two points.

To find the gradient (**m**) of a line joining two points A(x_1, y_1) and B(x_2, y_2), use the formula

$$m = \frac{y_2 - y_1}{x_2 - x_1}$$

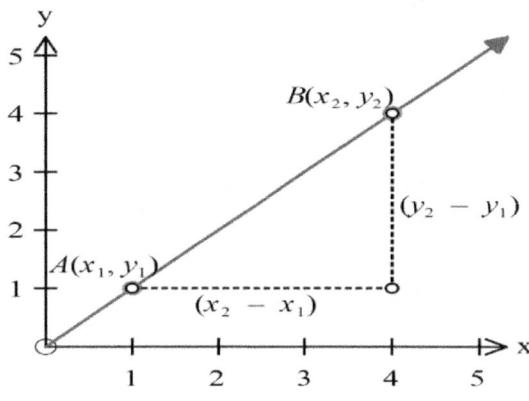

EXERCISE 8C

Find the gradient (m) of the line joining the points.

Points	Gradient (m)	Points	Gradient (m)
1. (2, 3) and (4, 7)	$m = \frac{7-3}{4-2} = \frac{4}{2} = 2$	2. (0, 2) and (3, 5)	
3. (1, -4) and (3, 6)	$m = \frac{6-(-4)}{3-1} = \frac{10}{2} = 5$	4. (2, -3) and (1, 0)	
5. (1, 2) and (5, 6)		6. (1, 2) and (8, 10)	
7. (1, 3) and (3, 9)		8. (5, 4) and (3, -2)	
9. (-1, 6) and (3, 10)		10. (0, 2) and (3, 5)	
11. (-2, 10) and (3, 10)		12. (2, 4) and (2, 8)	

8D FINDING GRADIENT AND Y-INTERCEPT FROM AN EQUATION

EXAMPLES

Express each of the following in the form $y = mx + c$, where necessary. Hence state the gradient (**m**) and the y-intercept (**c**).

1. $y = 3x + 5$ Compare this equation with $y = mx + c$. Clearly $m = 3$ and $c = 5$ So, $y = 3x + 5$ has a gradient of 3 and a y-intercept of 5.	2. $y = 10 - 2x$ This equation can be re-written as $$y = -2x + 10$$ Compare this equation with $y = mx + c$. Clearly $m = -2$ and $c = 10$ So, $y = 10 - 2x$ has a gradient of -2 and a y-intercept of 10.
3. $2y = 8x + 1$ Remember to make y the subject of formula before stating the value of m and c. Since the coefficient of y is 2, we divide each term by 2 and get $$\frac{2y}{2} = \frac{8x}{2} + \frac{1}{2} \quad \therefore y = 4x + 0.5$$ Compare this equation with $y = mx + c$. Clearly $m = 4$ and $c = 0.5$ So, $2y = 8x + 1$ has a gradient of 4 and a y-intercept of 0.5.	4. $3y + 6x = 2$ Making y the subject of formula, we have $$3y = -6x + 2$$ $$\frac{3y}{3} = \frac{-6x}{3} + \frac{2}{3} \quad \therefore y = -2x + \frac{2}{3}$$ Compare this equation with $y = mx + c$. Clearly $m = -2$ and $c = \frac{2}{3}$ So, $3y + 6x = 2$ has a gradient of -2 and a y-intercept of $\frac{2}{3}$.

USING TECHNOLOGY

We can also make use of technology to make y the subject of the formula as shown in the examples below.

Use the following steps on your CAS:

- ❖ Main
- ❖ Action
- ❖ Advanced
- ❖ Solve

5. For the equation, $4y - 12x - 15 = 0$ state the gradient and the y-intercept. **SOLUTION** $Solve\ (4y - 12x - 15 = 0, y)$ $\{y = 3x + 3.75\}$ $m = 3$ and $c = 3.75$	6. For the equation, $3x + 2.5y + 9 = 0$ state the gradient and the y-intercept. **SOLUTION** $Solve\ (3x + 2.5y + 9 = 0, y)$ $\{y = -1.2x - 3.6\}$ $m = -1.2$ and $c = -3.6$

EXERCISE 8D

For each of the following equations, state the gradient (**m**) and the y-intercept (**c**).

Equation (RULE)	Gradient (m)	y-intercept (c)	Equation (RULE)	Gradient (m)	y-intercept (c)
1. $y = 2x + 5$			2. $y = x + 6$		
3. $y = -3x + 7$			4. $y = 8 - 5x$		
5. $y = 5 - x$			6. $y = -x + 9$		
7. $y = 0.5x + 1$			8. $y = 4 - 0.75x$		
9. $y = -2$			10. $y = 3x$		
11. $y = 5$			12. $y = -7x$		
13. $2y = 4x + 10$			14. $2y = x + 8$		
15. $3y = 9x$			16. $4y = 2x$		
17. $2y = 6x - 1$			18. $3y = 4x + 3$		
19. $y - 2x = 10$			20. $y + 5x = 4$		
21. $y - 0.5x - 2 = 0$			22. $y + 2x - 7 = 0$		
23. $5y - 2x = 15$			24. $5y - 10x = 4$		

8E GIVEN A RULE HOW TO TABULATE

By using simple substitution learnt from Chapter 1 in Unit 1, we should be able to find a set of values of y for corresponding values of x.

EXERCISE 8E

Complete the following tables without using a calculator.

x	-2	-1	0	1	2	5
$y = 2x$	-4		0			10

x	-3	-2	0	1	4	10
$y = 3x$	-9		0		12	

x	0	1	2	3	4	5
$y = 7 - 2x$	7			1		

Complete the following tables using technology.

The following steps must be used while using the tabulating capabilities of your calculator.

- Menu
- Graphs & Table
- Insert the equation e.g $y_1 = 8 - 3x$
- Check the box
- Tap on the tabulate icon (second from top left hand corner)
- Tap on the icon to set the correct domain (values of x).

x	-2	-1	1	2	5	10
$y = 8 - 3x$	14	11	5	2	-7	-22

x	2	3	5	8	10	20
$y = 5x - 2$	8			38		

x	-2	-1	0	1	3	5
$y = 3x - 5$	-11					

x	2	3	5	8	10	20
$y = 4x + 1$		13		33		

8F LINEAR OR NOT?

We have seen that a linear function can be written in the form $y = mx + c$.

Examples of linear functions are:

$$y = 4x, \quad y = 5 - 3x, \quad y = \frac{x}{2}, \quad \frac{x}{4} + \frac{y}{5} = 2$$

For pairs of values of x and y as shown below, we can say it is linear if the first differences in the y-values are same.

x	2	3	4	5	6
y	7	10	13	16	19

First difference   +3   +3   +3   +3

As we can see the first differences in the y-values is constant and increasing by 3.

What is the rule then?

To figure out the rule, we need to work out the differences in the x-values as well. The differences in the y-values divided by the differences in the x-values gives the gradient.

Here the gradient is $m = \frac{3}{1} = 3$

To work out the value of c, we can use any pair of values from the table and substitute in the equation $y = 3x + c$.

Say we choose the point (3, 10) shown bold in the table. Substitute $x = 3$ and $y = 10$. We can then figure out the value of the y-intercept c.

$$10 = 3(3) + c \therefore c = 1$$

Hence the rule is $y = 3x + 1$.

USING TECHNOLOGY

To find the rule using your calculator, apply the following steps

- Menu
- Statistics
- Enter the x-values in List 1 and the y-values in List 2.
- Calc
- Linear Reg
- OK
- $y = ax + b$ appears, replace a and b in the equation which is the rule.

EXERCISE 8F

1. State which of the following are linear equations.

Function	Linear or Not	Function	Linear or Not
$y = 2x - 1$	YES	$y = 7 + 4x$	
$y = \dfrac{3}{4}x$		$y^2 = 2x$	
$y = x^2 + 4$		$y = \dfrac{5}{x}$	
$\dfrac{x}{2} = y - 4$		$y = 0.25x$	
$xy = 10$		$y = 25 - 0.2x$	

2. Investigate which of the following points will lie on a straight line if plotted. If yes, find the rule without a calculator.

(a)

x	1	2	3	4	5	6
y	3	5	7	9	11	13

(b)

x	-1	0	1	2	3	4
y	7	10	16	19	31	46

(c)

x	3	5	7	9	11	13
y	10	18	26	34	42	50

(d)

x	1	2	3	4	5	6
y	10	5	0	-5	-10	-15

(e)

x	10	12	14	16	18	20
y	10	20	40	80	160	320

(f)

x	5	6	7	8	9
y	-10	-13	-16	-19	-22

3. Find the rule of the following linear functions using technology.

(a)

x	4	5	6	7	8	9
y	-2	-5	-8	-11	-14	-17

(b)

x	2	4	6	8	10	12
y	5	10	15	20	25	30

(c)

x	-1	2	5	8	11
y	10	16	22	28	34

8G EQUATION OF LINE : GIVEN GRADIENT AND Y-INTERCEPT

The equation of a line is given by $y = mx + c$, where m is the gradient and c is the y–intercept. To find the equation of a line we need

1. the gradient
2. the y-intercept

EXAMPLES

1. Find the equation of the line having gradient 5 and crossing the y-axis at 3. **SOLUTION** $$y = 5x + 3$$	2. Find the equation of the line having slope -6 and crossing the y-axis at (0,4). **SOLUTION** $$y = -6x + 4$$

EXERCISE 8G

1. Find the equation of the line having gradient 7 and crossing the y-axis at 2.	2. Find the equation of the line having gradient 3 and crossing the y-axis at -2.
3. Find the equation of the line having slope -1 and y-intercept 6.	4. Find the equation of the line having gradient 9 and crossing the y-axis at 7.
5. Find the equation of the line having slope -2 and y-intercept 10.	6. Find the equation of the line having slope -5 and y-intercept 1.
7. Find the equation of the line having gradient 1 and crossing the y-axis at (0,5).	8. Find the equation of the line having gradient 4 and crossing the y-axis at (0,-3).

8H   EQUATION OF A LINE: GRADIENT GIVEN AND A POINT

EXAMPLES

1. Find the equation of the line having gradient 2 and passing through A(3,5).
   SOLUTION
   Let $y = mx + c$ be the equation of the line
   Here $m = 2$ $x = 3$ and $y = 5$
   Substitute these values in the equation $y = mx + c$ to determine the value of c
   $$5 = 2(3) + c$$
   $$5 = 6 + c$$
   $$c = -1$$
   $$\therefore y = 2x - 1$$

2. Find the equation of the line having gradient -0.5 and passing through A(-10,7).
   SOLUTION
   Let $y = mx + c$ be the equation of the line
   Here $m = -0.5$, $x = -10$ and $y = 7$
   Substitute these values in the equation $y = mx + c$ to determine the value of c
   $$7 = -0.5(10) + c$$
   $$7 = -5 + c$$
   $$c = 12$$
   $$\therefore y = -0.5x + 12$$

EXERCISE 8H

1. Find the equation of the line having gradient 3 and passing through (5,7).

2. Find the equation of the line having gradient 4 and passing through (1,2).

3. Find the equation of the line having slope 6 and passing through (-2,5).

4. Find the equation of the line having gradient -4 and passing through (1,5).

5. Find the equation of the line having gradient -5 and passing through (-2,3).

8I EQUATION OF A LINE PASSING THROUGH TWO POINTS

To find the equation of a line passing through two points we need to find

1. the gradient by using $\dfrac{y_2 - y_1}{x_2 - x_1}$
2. the y-intercept, by choosing any one of the two given points.

EXAMPLE

Find the equation of the line passing through A(3,1) and B(5,9)

Solution

Since the gradient is not given, we use the formula m = $\dfrac{y_2 - y_1}{x_2 - x_1}$ to find m.

$m = \dfrac{9-1}{5-3} = \dfrac{8}{2} = 4$

From the two points A and B, we can choose any point as the final answer would be same. Let us choose A (3,1) to find the y-intercept c.

Let $y = mx + c$ be the equation of the line

$$1 = 4(3) + c$$
$$1 = 12 + c$$
$$c = -11$$
$$y = 4x - 11$$

EXERCISE 8I

1. Find the equation of the line passing through A(2,5) and B(6,4).	2. Find the equation of the line passing through A(-2,3) and B(1,9).
3. Find the equation of the line passing through P(0,2) and B(2,8).	4. Find the equation of the line passing through A(-3,1) and B(1,5).

CHAPTER 8 : LINEAR RELATIONSHIPS

5. Find the equation of the line passing through A(3,11) and B(5,15).

6. Find the equation of the line passing through A(-2,16) and B(-1,13).

7. Find the equation of the line passing through P(4,10) and B(6,11).

8. Find the equation of the line passing through A(0,-3) and B(-1,-8).

9. Find the equation of the line passing through P(-5,-5) and B(2,16).

10. Find the equation of the line passing through A(4,6) and B(10,3).

8J DOES THE POINT LIE ON THE LINE?

To show a point lies on a given line,

- ❖ replace the *x* and y values in the equation of the line
- ❖ we have to obtain the constant value

EXAMPLES

1. Show that the point A(2,5) lies on the line with equation $4x + y = 13$. **SOLUTION** Replace $x = 2$ and $y = 5$ in the equation. We have to obtain the right hand side number. $4(2) + (5) = 13$ (**shown**)	2. Does the point B(3,-2) lie on the line with equation $2x + y = 10$? **SOLUTION** Replace $x = 3$ and $y = -2$ in the equation. $2(3) + (-2) = 4 \neq 10$ Therefore B does not lie on the line.
3. The point A(p,3) lies on the line $4x - 2y = 2$, find the value of p. **SOLUTION** Replace *x* by p and *y* by 3 in the given equation and solve for p. $$4(p) - 2(3) = 2$$ $$4p - 6 = 2$$ $$4p = 8$$ $$\therefore p = 2$$	

EXERCISE 8J

1. Show that the point (4,3) lies on the line with equation $3x + 2y = 18$.	2. Show that the point (5,-1) lies on the line with equation $x + 2y = 3$.
3. Does the point (2,5) lie on the line with equation $3x - y = 1$?	4. The point A(t,4) lies on the line $4x - 2y = 12$, find the value of t.
5. The point A(4,m) lies on the line $5x + 2y = 6$, find the value of m.	6. The point A(q,4) lies on the line $3x - 5y = -2$, find the value of q.

8K   HOW TO SKETCH A LINE WITHOUT A CALCULATOR?

To sketch a line, we can use one of the methods shown below. Note that each method works best depending the way the equation is presented to us. In some cases, we can make use of use of a table of values and in other cases we might be better off using the intercept-method.

EXAMPLES

1. Sketch the line $y = 2x + 3$
   SOLUTION: METHOD 1
   We make a table of values choosing a few values of x. Always choose numbers that you would be able to work out mentally.
   Note: two pairs of values are enough.

x	0	1	2
$y = 2x + 3$	3	5	7

   Now plot these pairs of values on the Cartesian plane to obtain the sketch of the line as shown.

   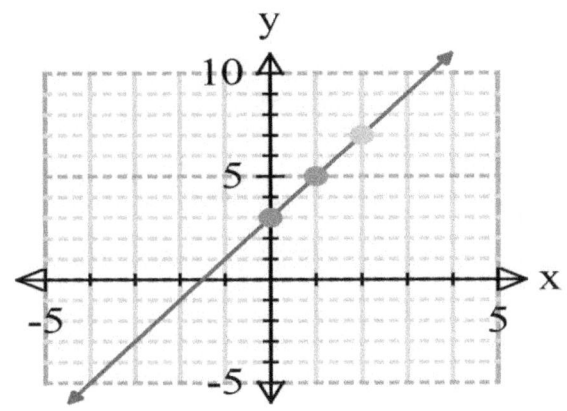

2. Sketch the line $y = 3x - 5$
   SOLUTION: METHOD 2
   Clearly the y-intercept is -5.
   First, we plot the point (0,-5) as we know that the line passes through this point.
   Now since the gradient is 3, for each unit we move right move 3 units up.
   Plot another couple points just to make sure no error is left behind.
   Now draw a smooth line passing through the y-intercept and the other two points.

   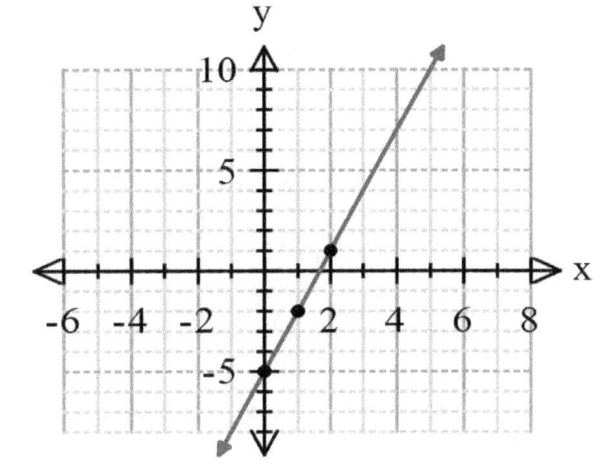

3. Sketch the line $2x + 3y = 12$
   SOLUTION
   This type of equation would be easier to sketch using the intercept method.
   When $x = 0$ (hide $2x$ with your finger)
   You'll see $3y = 12$
   $\therefore y = 4$
   Similarly, when $y = 0$ (hide $3y$ with your finger)
   Only $2x = 12$ will be visible
   $\therefore x = 6$
   Hence the line crosses the y-axis at 4 and the x-axis at 6,
   producing the sketch as shown.

   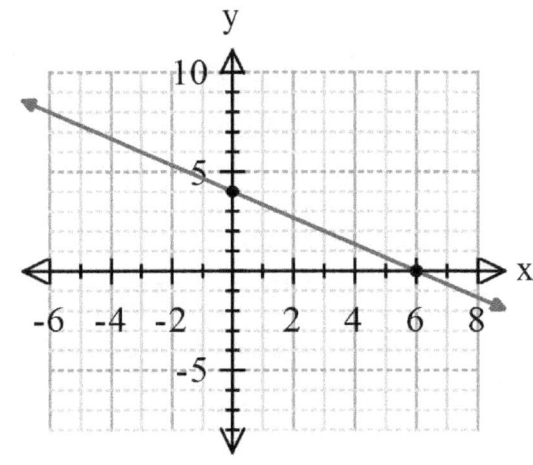

EXERCISE 8K

1. Sketch the line $y = x + 3$.

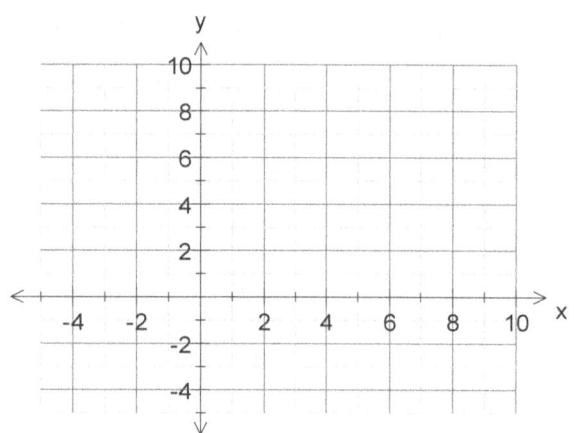

2. Sketch the line $y = 3x - 1$.

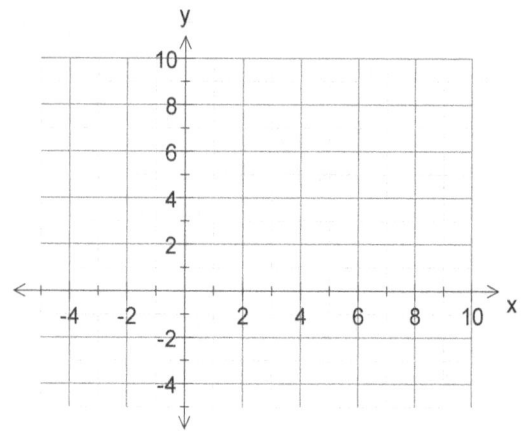

3. Sketch the line $y = 4 - x$.

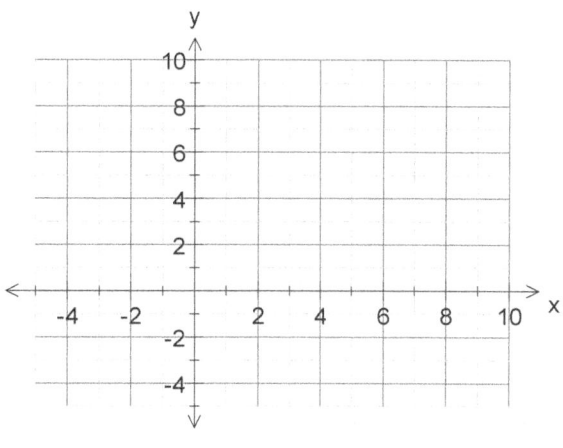

4. Sketch the line $y = x - 2$.

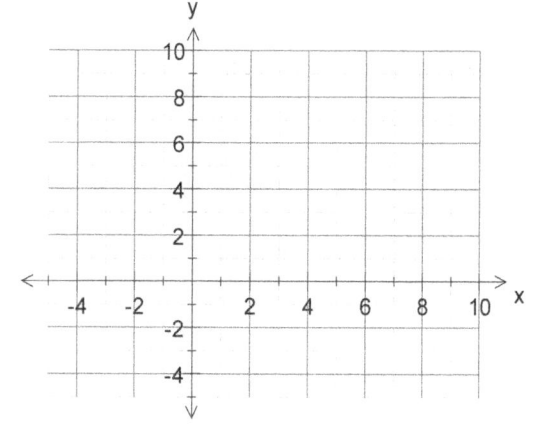

5. Sketch the line $y = 5 - 2x$.

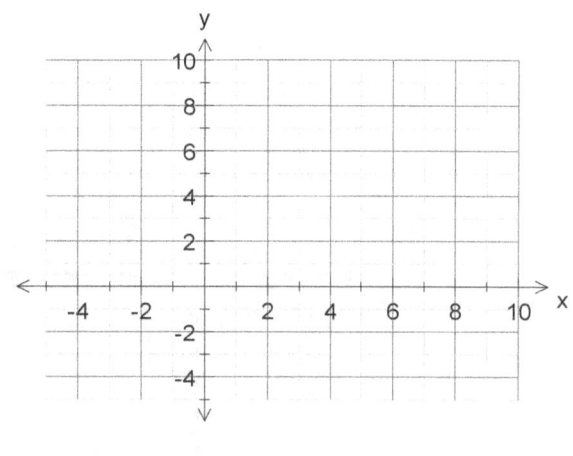

6. Sketch the line $y = 3x + 2$.

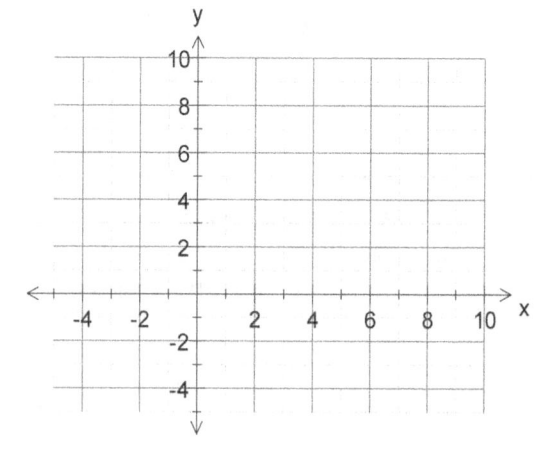

7. Sketch the line $y = 8 - 2x$.

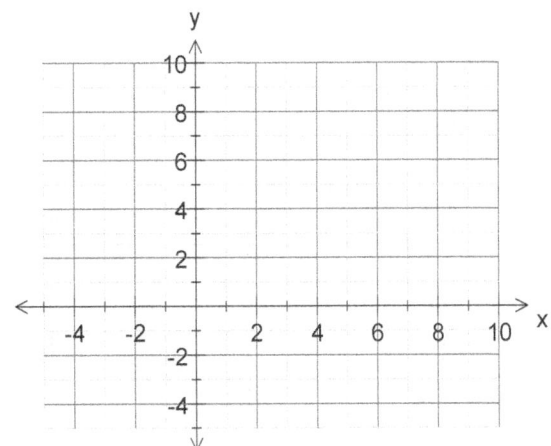

8. Sketch the line $y = 5x + 2$.

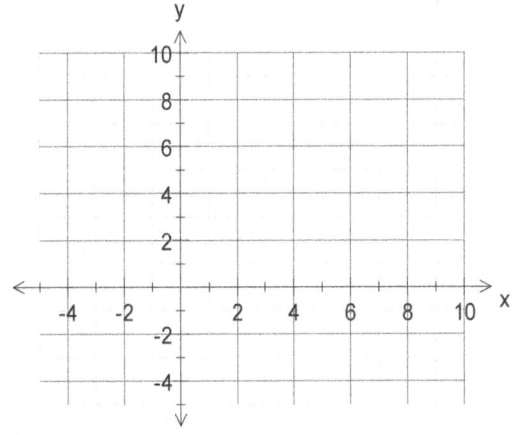

9. Sketch the line $2x + y = 8$.

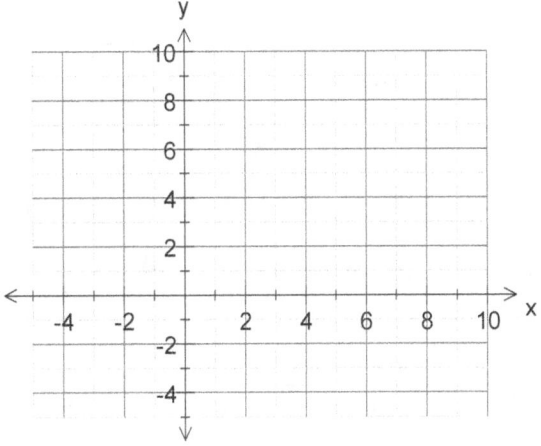

10. Sketch the line $x - 3y = 9$.

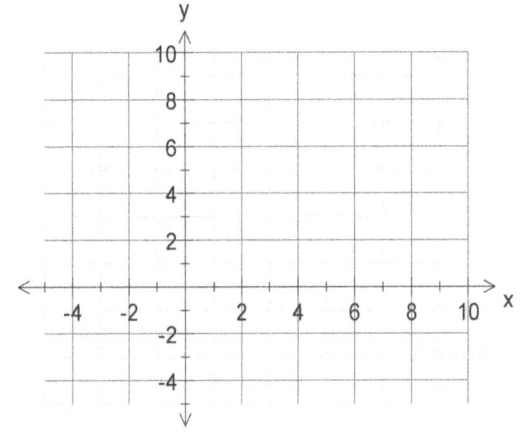

11. Sketch the line $2x + 3y = 12$.

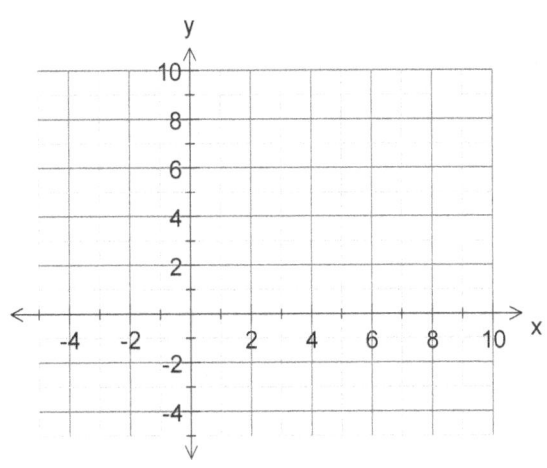

12. Sketch the line $2y - x = 4$.

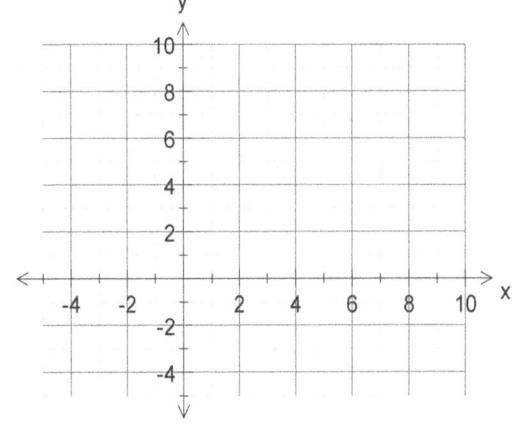

8L HORIZONTAL AND VERTICAL LINES

Consider the sketch on the right.
It is a vertical line passing through the points
(5,-4), (5,1) and (5,7).

Note that all the three points have x-coordinate 5.

Hence the equation of the line is $x = 5$.

Vertical lines are of the type $x = a$, where a is any number.

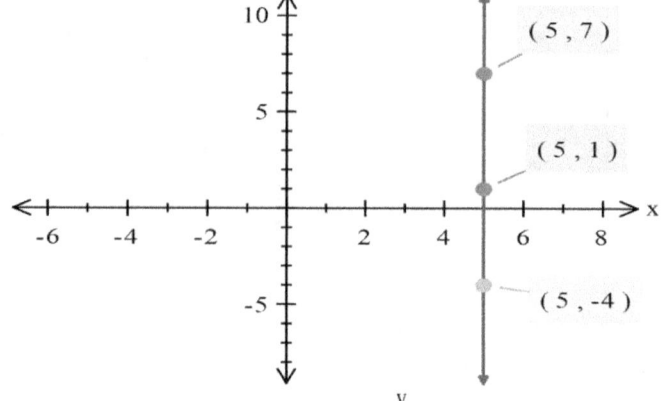

Similarly, in the diagram on the right it is a horizontal line passing through the points (-4,3), (1,3) and (6,3).

Note again that all the three points have y-coordinate as 3.
Hence the equation of the line is $y = 3$.

Horizontal lines are of the type $y = b$, where b is any number.

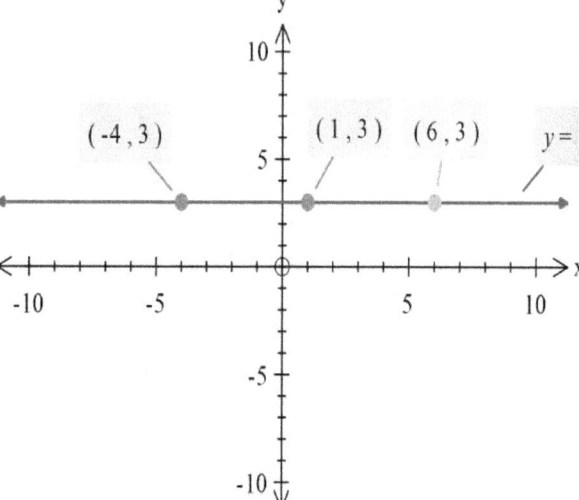

EXAMPLES

1. On the same set of axes sketch the line
$x = 4, x = -7, and\ y = 8$.

SOLUTION

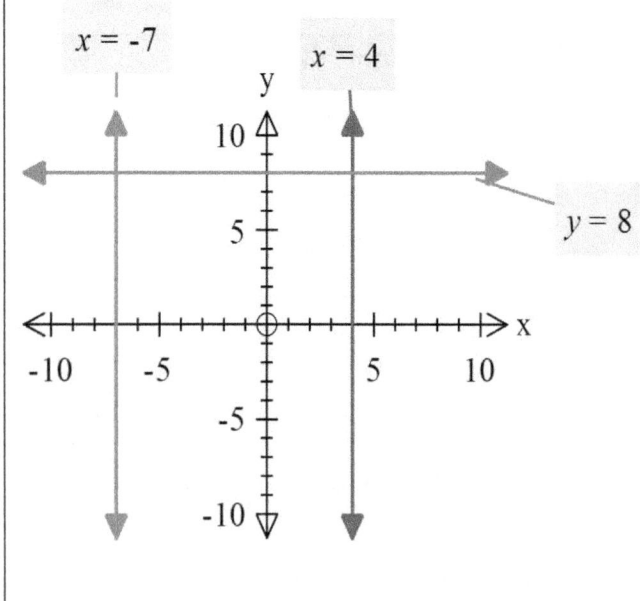

2. State the equation of each of the lines labelled A, B and C.

SOLUTION

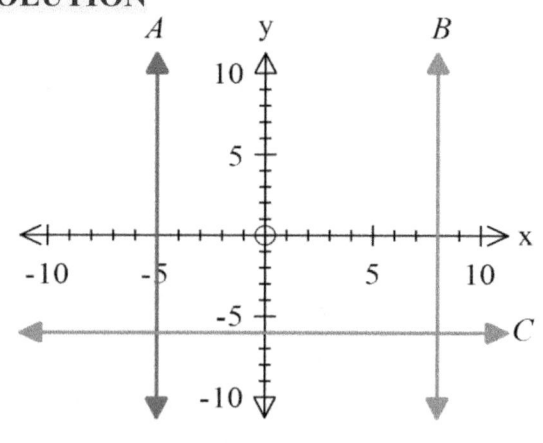

$A : x = -5$
$B : x = 8$
$C : y = -6$

EXERCISE 8L

1. On the same set of axes sketch the line $y = 4, y = -1,$ and $x = 2$.

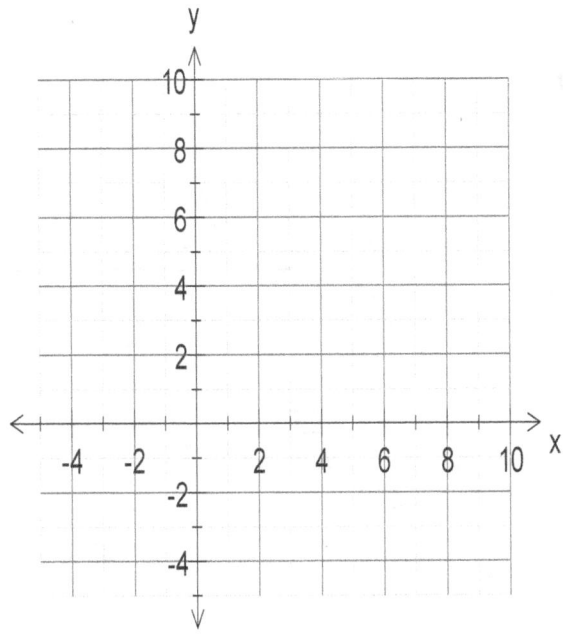

2. State the equation of each of the lines labelled A, B and C.

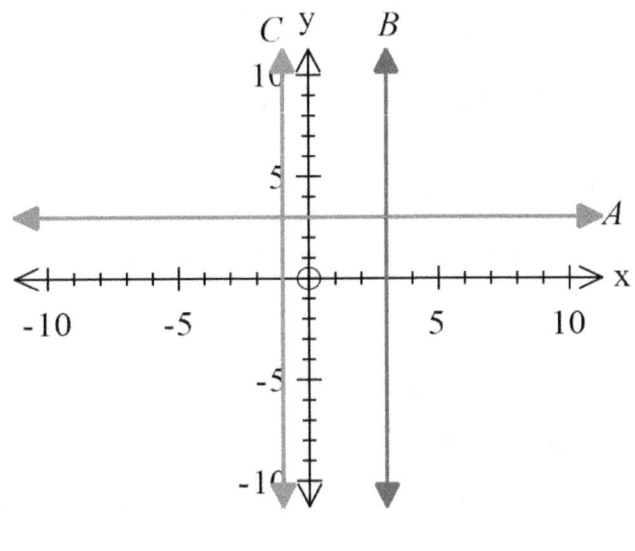

3. State the equation of each of the lines labelled A, B and C.

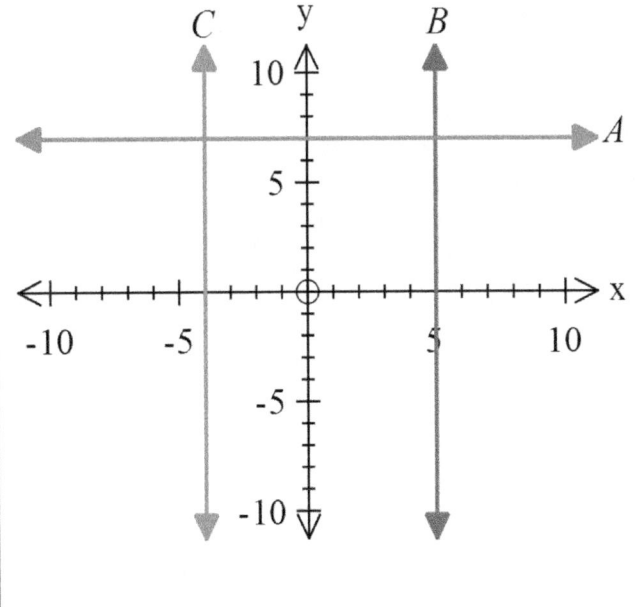

4. On the same set of axes sketch the line $x = 9, x = -2,$ and $y = 6$.

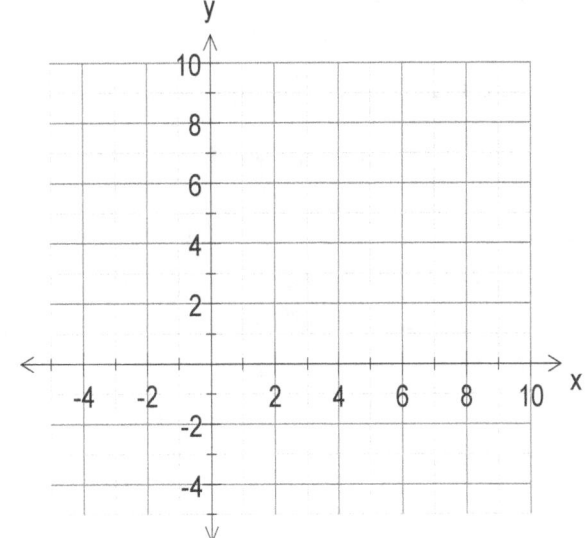

8M SKETCHING LINES USING TECHNOLOGY

To sketch a line or even a few lines, use the following steps on your calculator

- ❖ Menu
- ❖ Graph & Table
- ❖ Insert the equations
- ❖ check the box
- ❖ click on the graph icon (1st on top left)

EXAMPLES

Sketch the following using technology.

(a) $y = 3x + 5$
(b) $y = 4 - 3x$
(c) $y = 0.5x + 2$

SOLUTION

Your calculator must display a similar screen version to have the correct graphs.

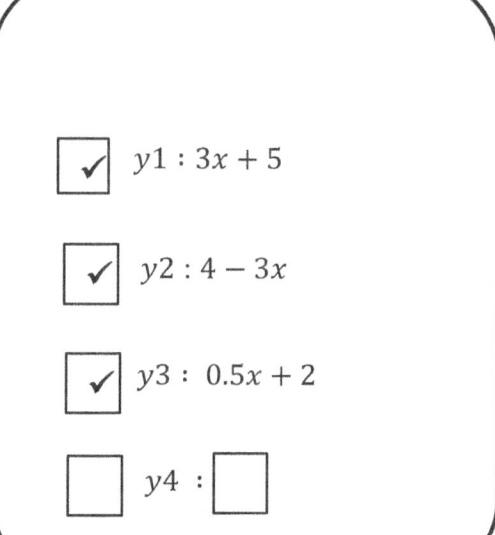

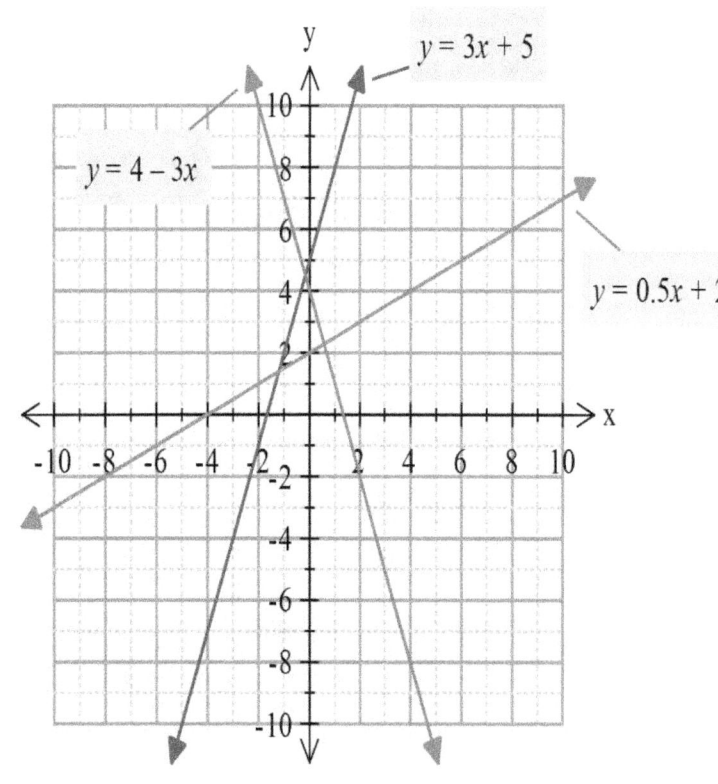

CLASS ACTIVITY

Sketch the following lines using technology. You don't have to graph them. Just practice!

1. $y = 2x + 9$
2. $y = 10 - 2x$
3. $y = 3x + 9$

4. $y = 0.25x + 4$
5. $y = 6 - 3x$
6. $y = x + 6$

7. $x + y = 8$
8. $2x + y = 8$
9. $x + 2y = 10$

8N HOW TO FIND THE EQUATION OF A LINE GIVEN THE SKETCH?

EXAMPLES

1. Find the equation of the line.

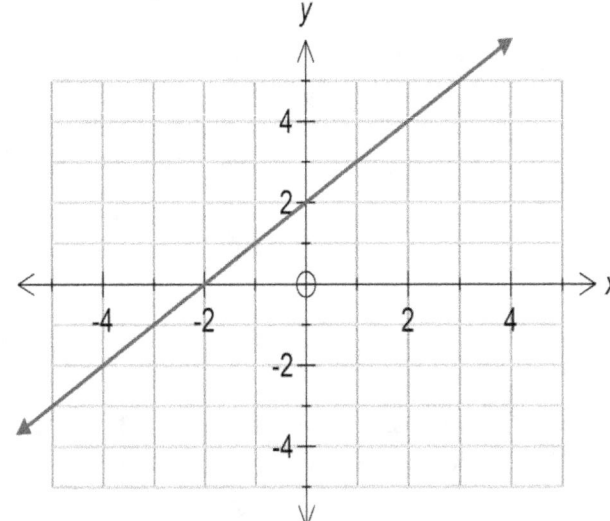

SOLUTION

This line has a y-intercept of 2.

The gradient $= \frac{rise}{run} = \frac{1}{1} = 1$

So the equation is $y = x + 2$.

2. Find the equation of the line.

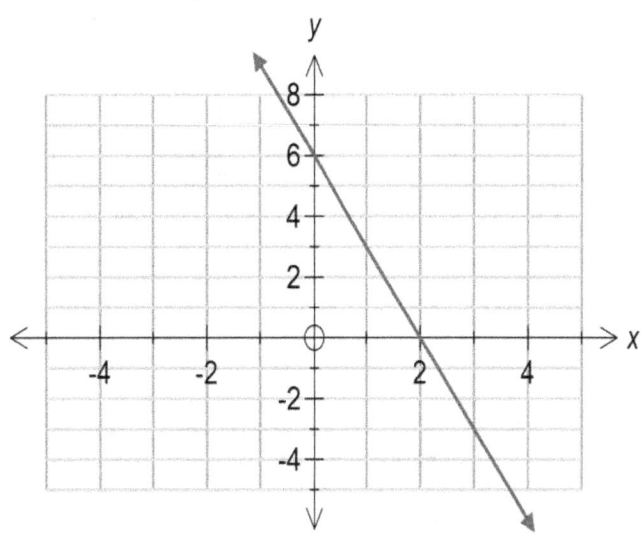

SOLUTION

This line has a y-intercept of 6.

The gradient $= \frac{rise}{run} = \frac{-6}{2} = -3$

So the equation is $y = -3x + 6$.

EXERCISE 8N

Find the equations of the following lines.

1.

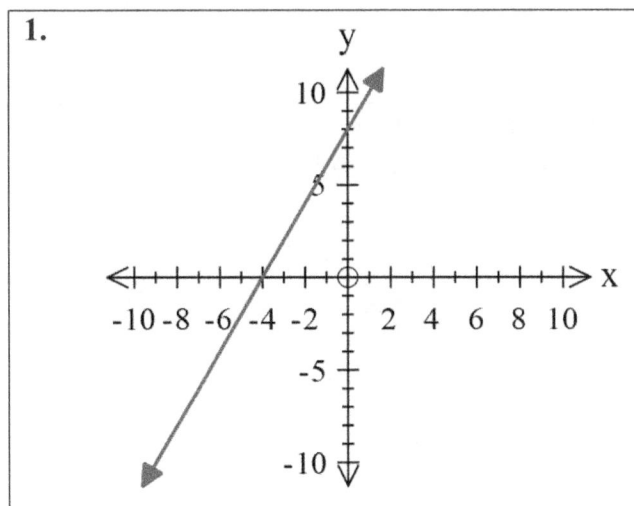

2.

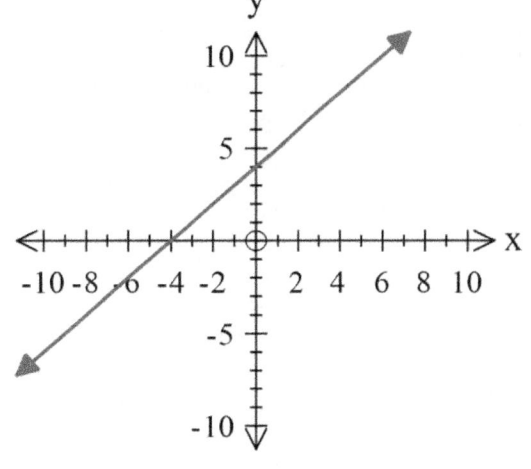

3.
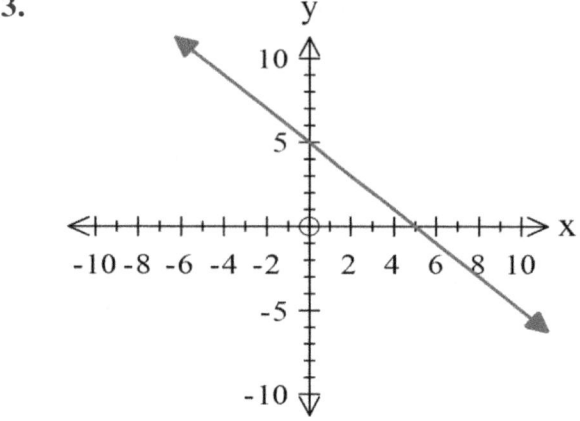

4.
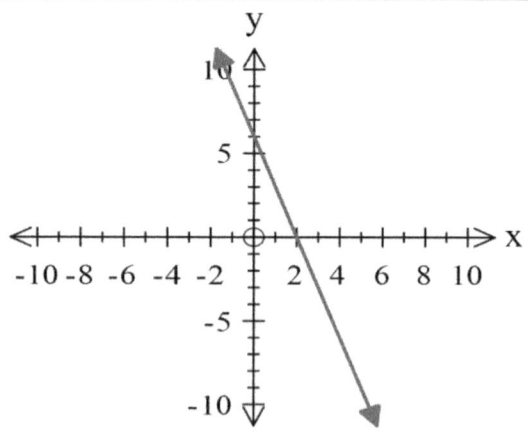

5. Write down the equation of the lines marked A and B.

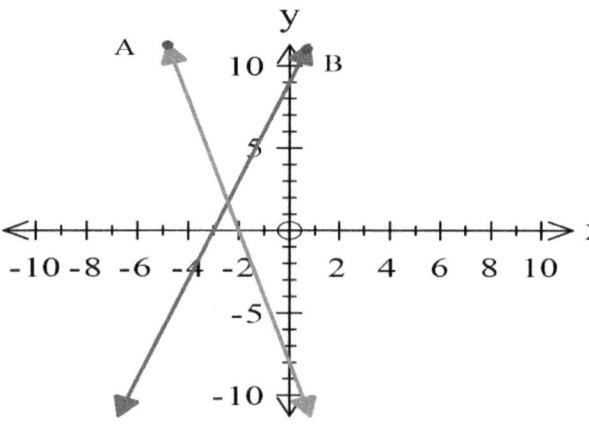

6. Write down the equation of the lines marked C and D.

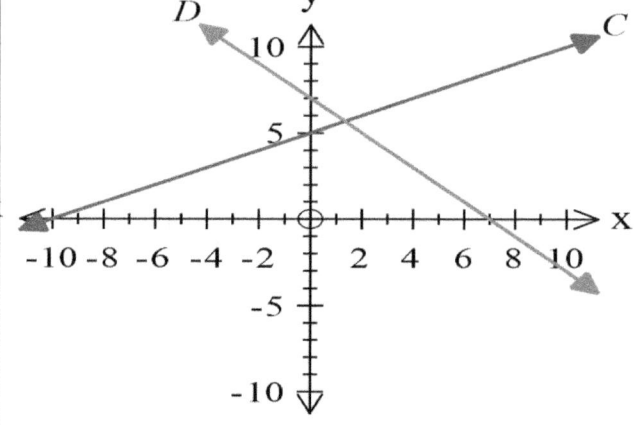

7. Write down the equation of the lines marked C, D and E.

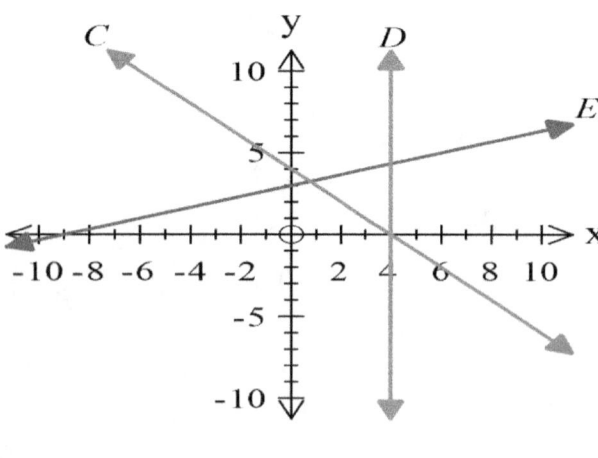

8. Write down the equation of the lines marked P, Q and R.

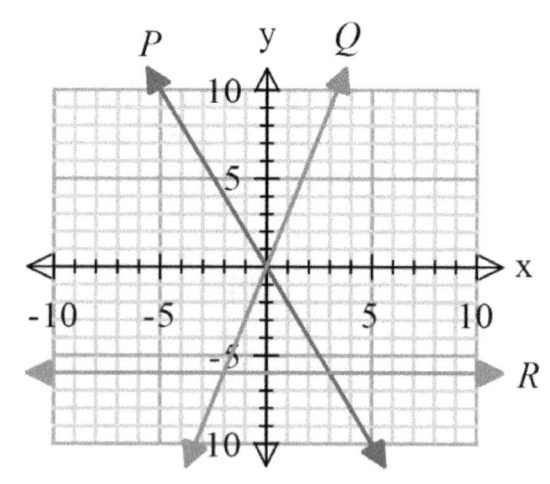

8O APPLICATIONS

In the last section of the chapter, an attempt has been made to familiarize the reader with all the concepts learnt so far and put them into practice by solving real world problems.

EXAMPLE

Tony is a professional electrician and charges an hourly rate of $60 per hour but no callout fee.

(a) Complete the table of values below to show the cost of having Tony complete jobs of varying lengths.

Time (t hours)	0	1	2.5	4	8
Cost ($C)	0	60	150	240	480

(b) On the axes below, plot the cost of Tony completing a job of length t hours. The cost of Grant, another electrician, has already been plotted.

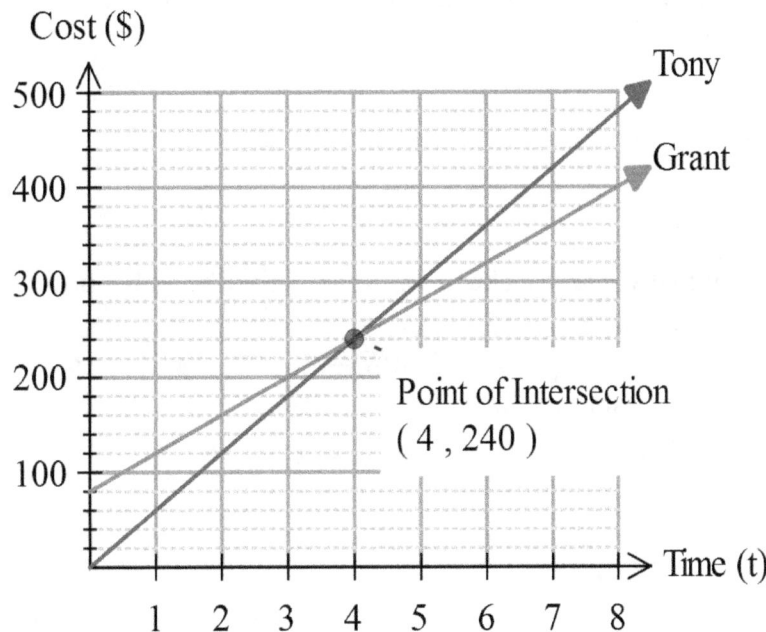

(c) Write a rule to calculate the cost of employing Tony for any length of time (t).

$$C = 60t$$

(d) Write a rule to calculate the cost of employing Grant for any length of time (t).

$$C = 80 + 40t$$

(f) You are keen to pay as little money as possible. For what interval of time would you employ Tony instead of Grant?

> When t = 4, the cost of employing both electrician is the same ($240).
> For any time less than 4 hours Tony is cheaper

EXERCISE 80

1. A mobile phone company charges a flat rate of $30 per month and an additional charge of 9c for each minute of service.

   (a) Write a linear equation for the monthly charge ($C) based upon the number of minutes (t) of service each month.

   (b) What will be the charge for 90 minutes of service?

   (c) Johanna can only afford a $45 of phone bill each month. How long can she afford to talk on the phone each month?

2. A small garment company makes cheap brand T shirts and the profit function P, of the company follows a linear model as shown below.
   $$P = 2.50N - 2000$$
   where N is the number of T shirts sold.

   (a) What does the figure 2000 represent?

   (b) What does 2.50 mean in the above context?

   (c) Calculate the profit of the company if they sell 1500 T shirts.

   (d) How many T shirts must the company sell to break even (no profit no loss)?

3. A cab company charges $3.80 boarding rate and an additional $1.10 for every kilometre.

   (a) Write a linear equation for the charge ($C) based upon the number of kilometres (d) of travel.

   (b) What will be the charge for 40 km of journey?

   (c) A random passenger paid $50 for catching a cab to his work place. What distance did the cab cover?

   (d) Mishka is a tourist and want to do some sight-seeing around the city. She has a $100 note. How many kilometres would she be able to travel in a cab?

4. A toy company makes little car toys. The Cost function $C, of the company follows a linear model as shown below.
$$C = 6.80N + 1250$$
   where N is the number of car toys produced.

   (a) What does the figure 1250 represent?

   (b) What does 6.80 mean in the above context?

   (c) Calculate the cost of the company if they produce 3000 cars.

   (d) The company sells each car for $10. How many cars must the company sell to break even?

MATHEMATICS APPLICATIONS UNIT 2

5. Michael is a plumber who charges $40 call out fee plus $35 per hour of labour.

(a) Complete the table below to show the cost of having Michael complete jobs of varying lengths.

Time (t hours)	0	2	3.5	6	10
Cost ($C)					

(b) On the axes below, plot the cost of Michael completing a job in t hours. The cost of Simpson, another plumber, has already been plotted.

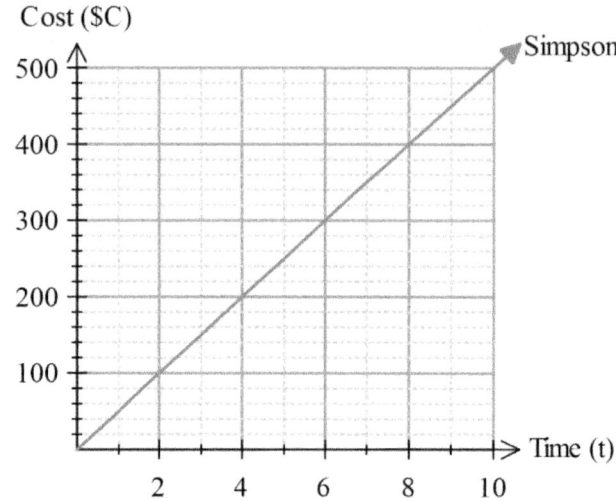

(c) Calculate the hourly cost of employing Simpson.

(d) Use your graph to determine the number of hours of labour for which the both plumbers would cost the same.

6. A white water rafting company rents kayaks at $12 per person for their guides fee and $8 dollars an hour to rent a kayak. Write an equation representing the cost per person, $C, of renting a kayak for x hours and work out the cost of renting a kayak for 5 hours?

7. A sequence of shapes is made of squares to form the shapes as shown below.

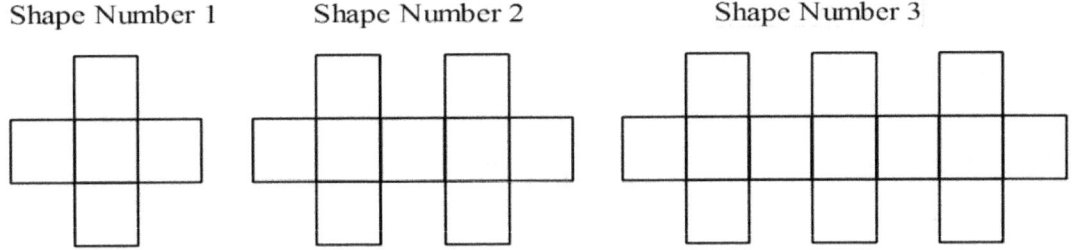

The table of results is shown below.

Shape Number (n)	1	2	3	4	5
Number of squares (s)	5	9	13	17	21

(a) Plot the above data on the axes below.

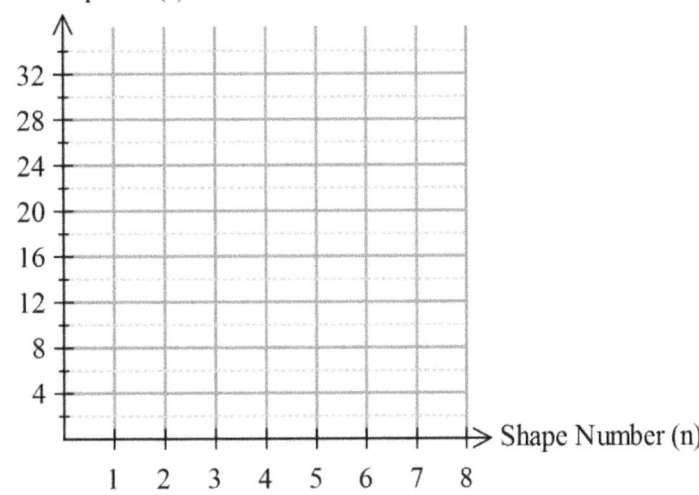

(b) Write a rule linking s and n, where s and n are as defined in the table.

(c) Determine the number of squares required for Shape Number 10.

(d) Justify that the point (15,61) lies on the line that would pass through the points plotted in (a).

(e) Mary has 80 squares. What is the biggest Shape Number she would be able to make?

8. Asian Motors will rent-a-car for $35 plus $0.20 per kilometre travelled.

(a) Write a function expressing rental cost ($C) as a function of the kilometres travelled (d).

(b) Find the cost of renting a car to travel 800 km.

(e) John's bill was $51.40. How far did he drive?

9. Dream Video Store charges a $25 membership fee and $3.50 for each DVD rented.

(a) Write a function expressing rental cost ($C) as a function of the number of DVDs (n) rented.

(b) Find the cost of renting 62 DVD's.

(c) On the axes below, plot the cost of renting DVD's.

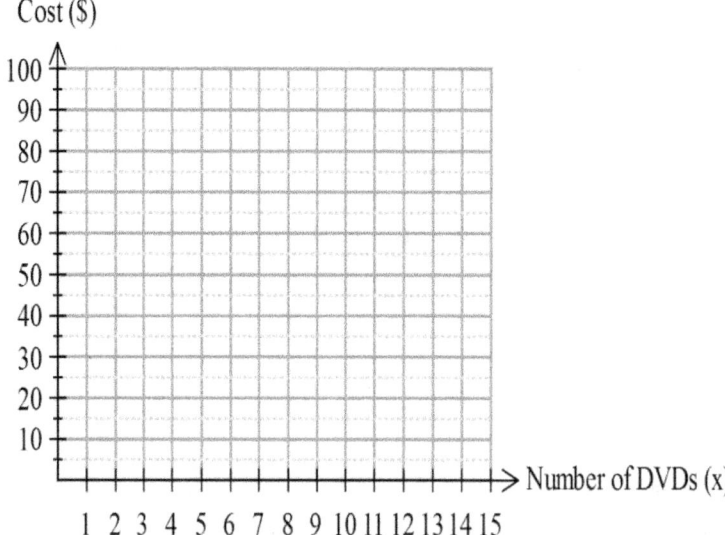

(d) If a new member paid the store $193 in the last 6 months, how many DVDs were rented?

CHAPTER 9

PIECEWISE LINEAR FUNCTIONS

9A   GRAPHING PIECEWISE FUNCTIONS

In chapter 8 we have learnt about lines and how they can be sketched by long hand and by using technology. In many real-life problems, however, functions are represented by a combination of equations, each corresponding to a part of the domain. Such functions are called piecewise functions.

For example, the piecewise function given by

$$f(x) = \begin{cases} x + 2 & \text{for } x < 3 \\ 2x - 1 & \text{for } x \geq 3 \end{cases}$$

is defined by two equations. One equation gives the values of $f(x)$ when x is less than 3 and the other equation gives the values of $f(x)$ when x is greater than or equal to 3.

If we sketch the above piecewise function, we have two lines as shown.
To the left of x = 3, we have the graph of $y = x + 2$ and to the right and including x = 3 the graph is given by $y = 2x - 1$.

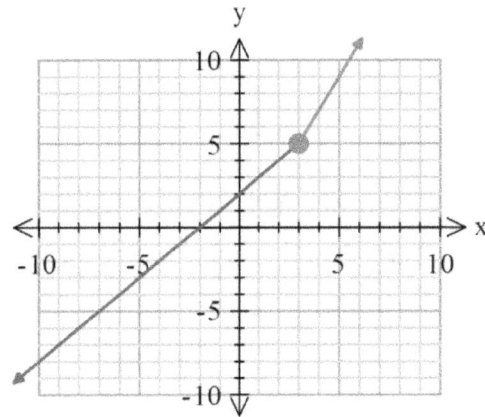

EXAMPLE

Graph the following piecewise function.

$$f(x) = \begin{cases} x - 4 & \text{for } x \leq 2 \\ 2 & \text{for } 2 < x < 5 \\ -x & \text{for } x \geq 5 \end{cases}$$

SOLUTION
This piecewise function will consist of 3 different lines.
The open ended circles in the horizontal lines indicate that 2 and 5 are not included in the domain.

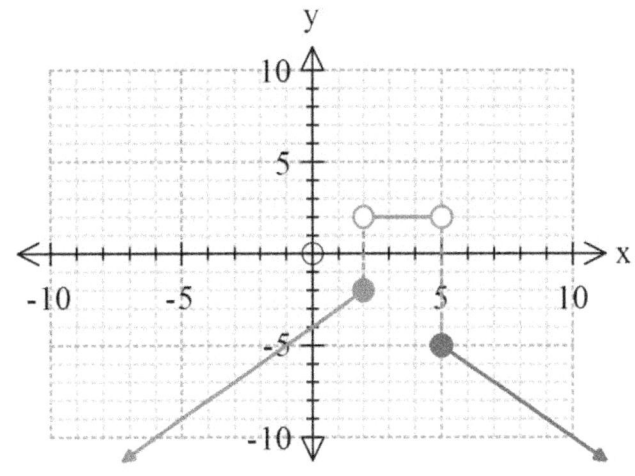

EXERCISE 9A

Graph the following piecewise functions.

1. $f(x) = \begin{cases} x+3 & \text{for } x < 1 \\ 3x+1 & \text{for } x \geq 1 \end{cases}$

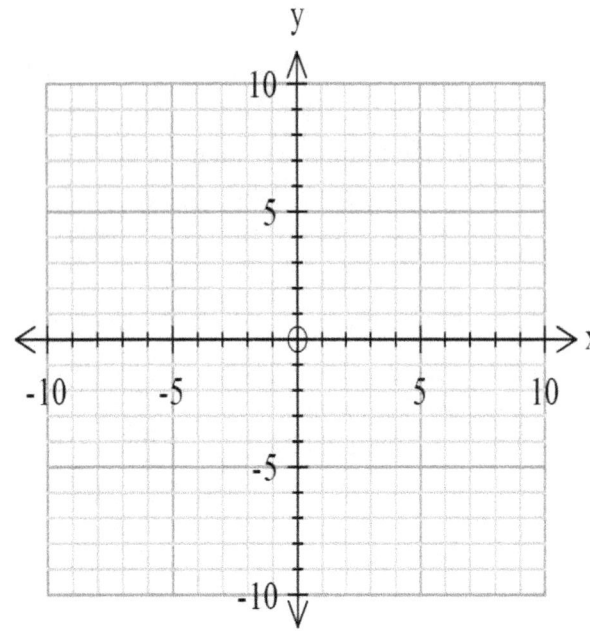

2. $f(x) = \begin{cases} 3x-1 & \text{for } x < -2 \\ x-5 & \text{for } x \geq -2 \end{cases}$

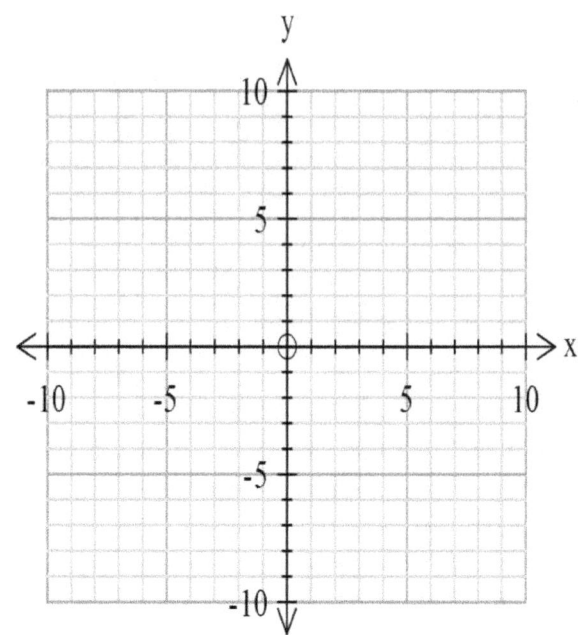

3. $f(x) = \begin{cases} -2x+3 & \text{for } x \leq -2 \\ 7 & \text{for } -2 < x \leq 3 \\ -x+10 & \text{for } x > 3 \end{cases}$

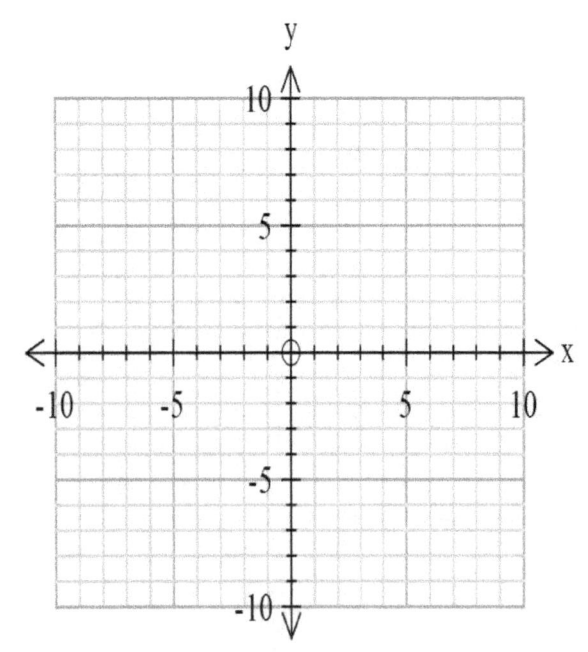

4. $f(x) = \begin{cases} 2x+3 & \text{for } x \leq -4 \\ -5 & \text{for } -4 < x < 0 \\ x & \text{for } x \geq 0 \end{cases}$

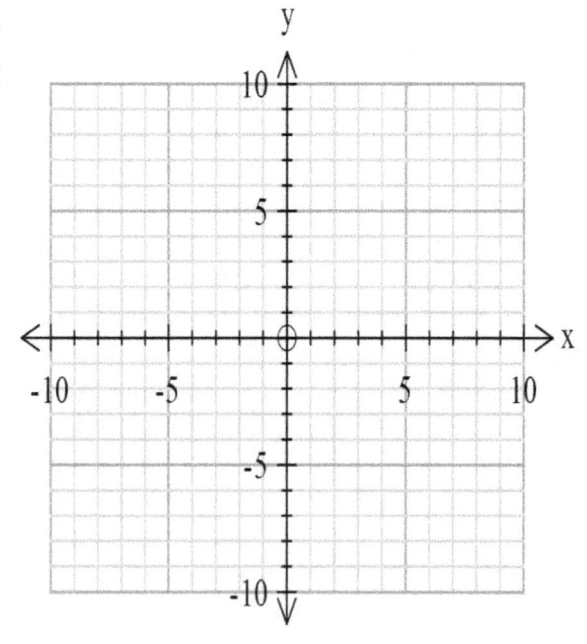

CHAPTER 9 : PIECEWISE FUNCTIONS   129

9B   DETERMINING EQUATIONS GIVEN GRAPHS OF PIECEWISE FUNCTIONS

The diagram shows a piecewise function. Fill in the missing equations for each given domain (values of x).

$$f(x) = \begin{cases} & \text{for } x \leq 0 \\ & \text{for } 0 < x < 3 \\ & \text{for } x \geq 3 \end{cases}$$

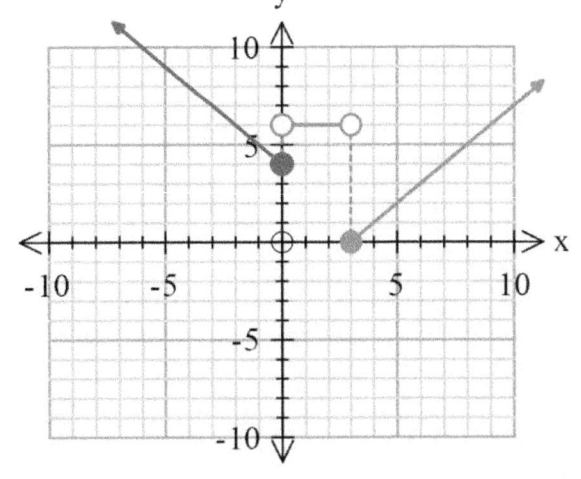

SOLUTION

For $x \leq 0$, the line has a gradient of -1 and a y-intercept of 4.
It's equation is $y = -x + 4$.

For $0 < x < 3$, it is a horizontal line with equation $y = 6$.

For $x \geq 3$, the line has a slope of 1 and if produced will cross the y-axis at -3. Hence its equation is $y = x - 3$.

Now we can define our whole piecewise function as under:

$$f(x) = \begin{cases} -x + 4 & \text{for } x \leq 0 \\ 6 & \text{for } 0 < x < 3 \\ x - 3 & \text{for } x \geq 3 \end{cases}$$

EXERCISE 9B

Complete the following piecewise function given their graphs.

1. $f(x) = \begin{cases} & \text{for } x \geq 0 \\ & \text{for } x < 0 \end{cases}$

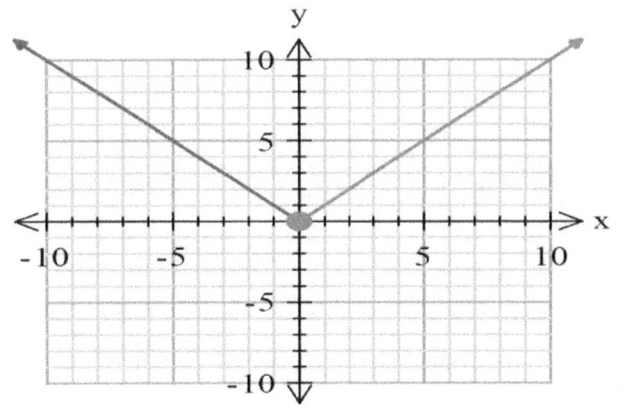

2. $f(x) = \begin{cases} & \text{for } x \geq 0 \\ & \text{for } x < 0 \end{cases}$

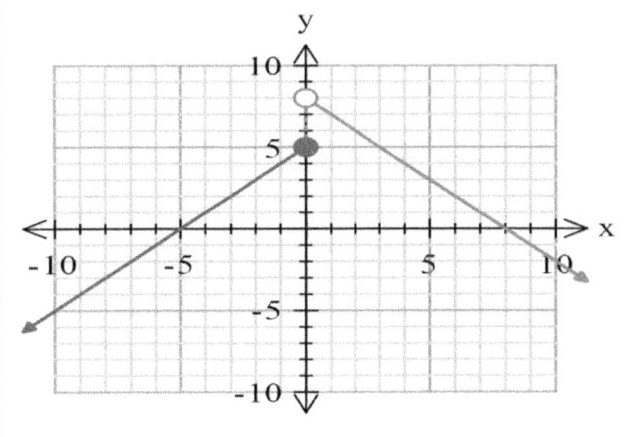

3. $f(x) = \begin{cases} & \text{for } x < -5 \\ & \text{for } -3 \le x \le 0 \\ & \text{for } x > 0 \end{cases}$

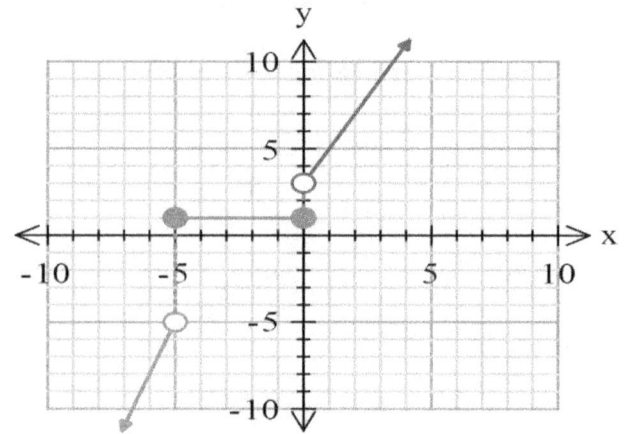

4. $f(x) = \begin{cases} & \text{for } x < -5 \\ & \text{for } -5 \le x \le 0 \\ & \text{for } x > 0 \end{cases}$

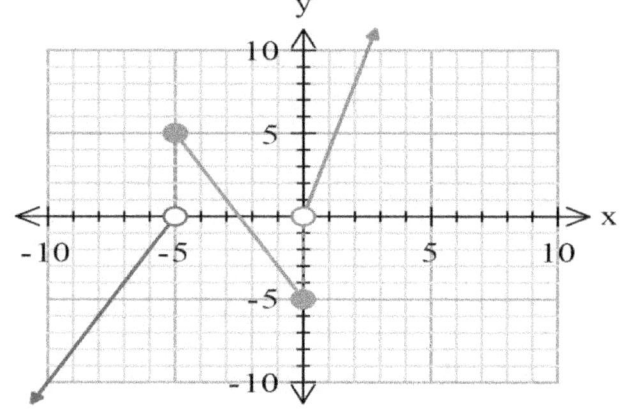

5. $f(x) = \begin{cases} & \text{for } x \le 0 \\ & \text{for } 0 < x < 3 \\ & \text{for } x \ge 3 \end{cases}$

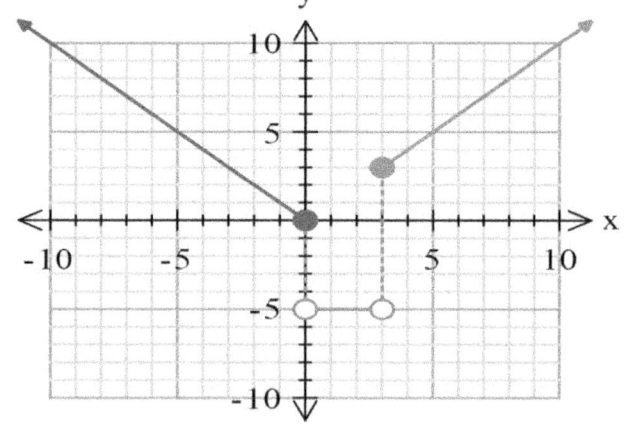

6. $f(x) = \begin{cases} & \text{for } x \le -3 \\ & \text{for } -3 < x < 5 \\ & \text{for } x \ge 5 \end{cases}$

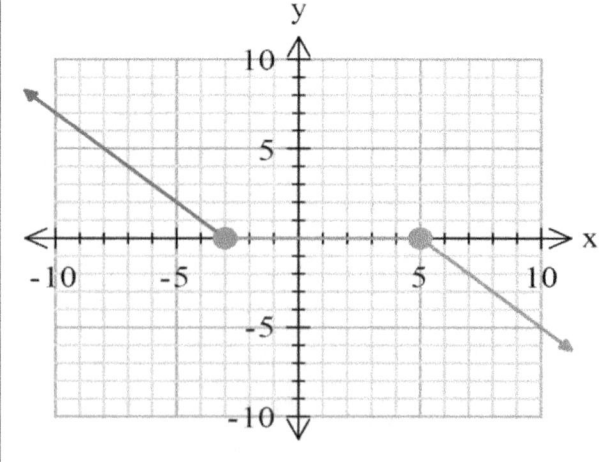

9C USING TECHNOLOGY TO GRAPH PIECEWISE FUNCTIONS

In this part of the chapter, we are going to make use of the capabilities of our calculators to graph linear piecewise functions. The steps have been outlined below.

- Menu
- Graph & table
- Keyboard
- Select the symbol $\left\{ \begin{matrix} \blacksquare \\ \square \end{matrix} \right.$, $\left\{ \begin{matrix} \square \\ \square \end{matrix} \right.$
- In the first column, insert all the equations you wish to graph. Remember do not type y =, just the expressions involving *x*. If we have more equations just tap on the same symbol mentioned above and it will create more rows.
- In the second column, insert the domain (the range of values of x).
- Tick the box
- Tap on the graph icon (first on top left hand corner)

EXERCISE 9C

Use your calculator to graph the following piecewise functions.

1. $f(x) = \begin{cases} x + 4 & \text{for } x < -4 \\ 2x + 8 & \text{for } x \geq -4 \end{cases}$

2. $f(x) = \begin{cases} -x + 5 & \text{for } x < -2 \\ 3x + 1 & \text{for } x \geq -2 \end{cases}$

3. $f(x) = \begin{cases} -x + 2 & \text{for } x \leq 0 \\ 4 & \text{for } 0 < x < 3 \\ x - 5 & \text{for } x \geq 3 \end{cases}$

4. $f(x) = \begin{cases} -2x + 4 & \text{for } x \leq 2 \\ 4 & \text{for } 2 < x < 5 \\ x - 3 & \text{for } x \geq 5 \end{cases}$

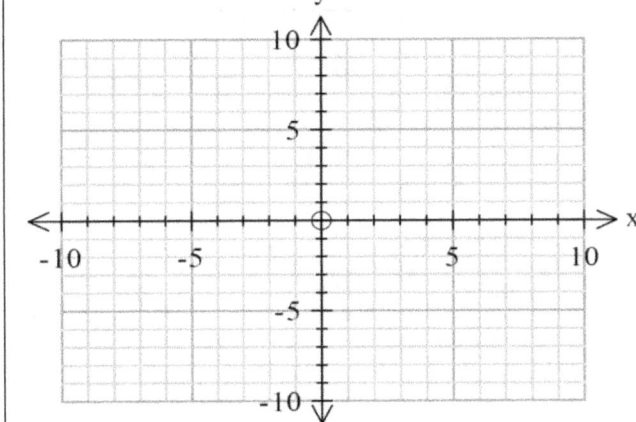
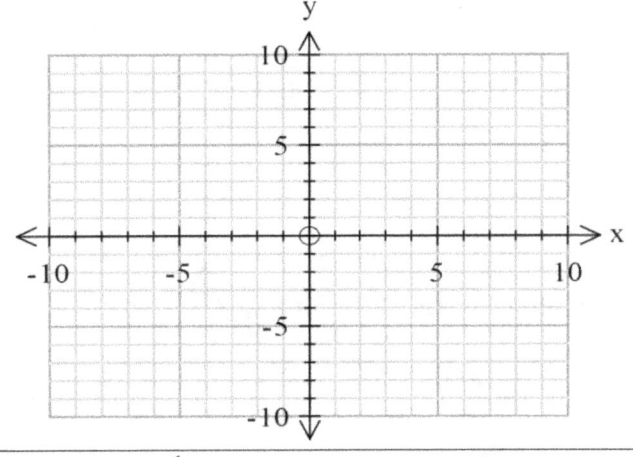
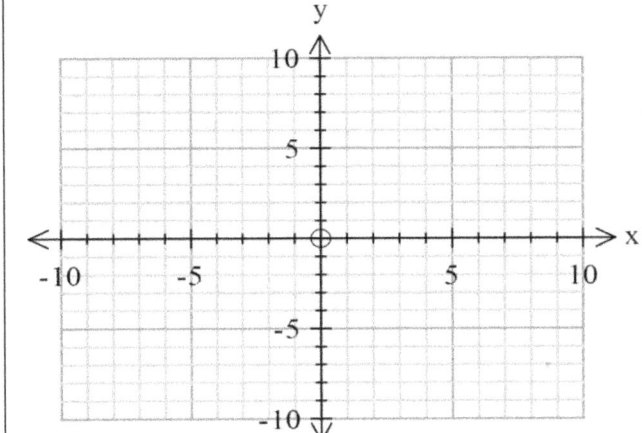
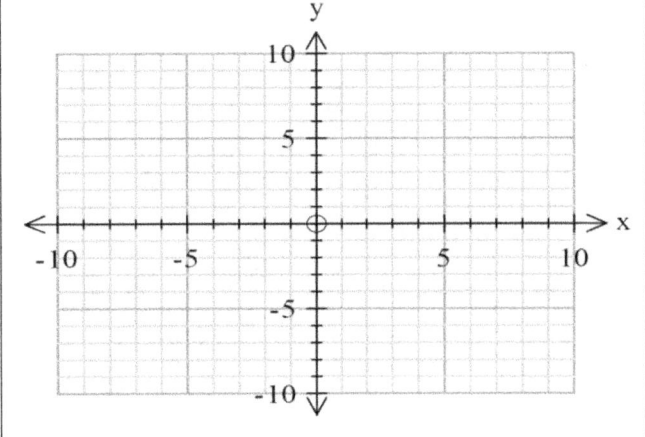

9D   APPLICATIONS

There are some real-life practical examples for studying piecewise linear functions. For example, it can be used to determine the tax payable by an individual. In other cases, we can use piecewise linear functions to calculate the electricity bills, commission paid to a salesman and so on.

EXAMPLE 1

The diagram on the right shows the distance time graph of a paper boy named Charlie. He uses his motorbike to deliver newspapers in the surrounding suburbs and part of the country as well.

(a) When did Charlie start his job?

2 a.m.

(b) Charlie stopped and had breakfast on his journey. At what time did he stop and for how long?

5 a.m. , for 1 hour

(c) Calculate Charlie's speed on the way home.

$$speed = \frac{distance}{time} = \frac{40}{3} = 13.\dot{3}\ km/h$$

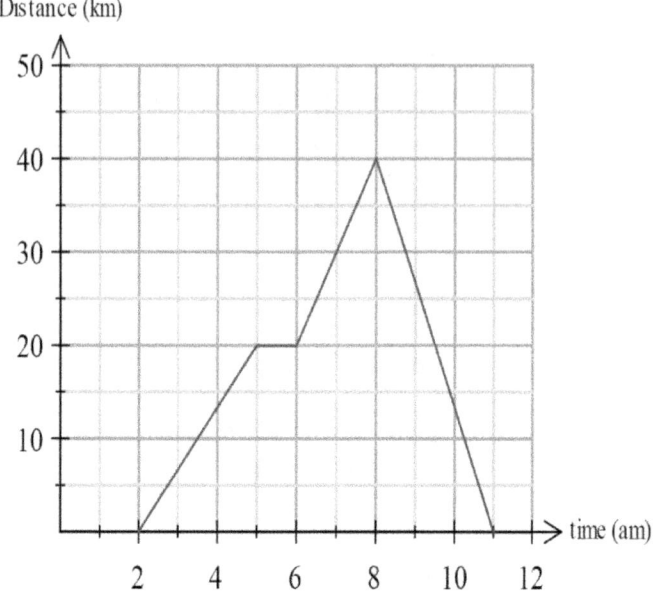

EXAMPLE 2

Home Care Electric Company has this rate for a family home:

- Monthly service charge $35.50
- First 100 kWh : $1.25 per kWh
- Over 100 kWh : $1.50 per kWh

Write down a piecewise model for the monthly charge C(x) as a function where x is the number of kWh.

SOLUTION

$$C(x) = \begin{cases} 35.50 + 1.25x & for\ 0 \leq x \leq 100 \\ 35.50 + 1.25 \times 100 + 1.50(x - 100) & x > 100 \end{cases}$$

which simplifies to

$$C(x) = \begin{cases} 35.50 + 1.25x & for\ 0 \leq x \leq 100 \\ 160.50 + 1.50(x - 100) & x > 100 \end{cases}$$

EXERCISE 9D

1. The diagram shows the round trip journey of a delivery truck travelling from Town A to Town B.

   (a) At what time did the truck depart?

   (b) Calculate the speed of the truck on the first leg of its journey.

   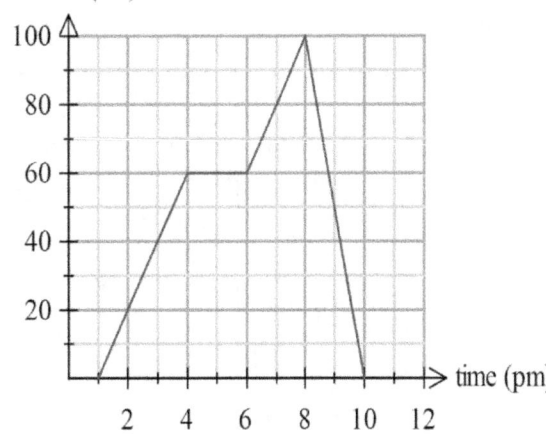

   (c) Kevin, the truck driver stopped for lunch on the way to Town B. For how long did he stop for?

   (d) On the way back Kevin drove at a constant speed. Trucks are only allowed a maximum speed of 40 km/h. There is a speed camera detector on the way from Town B to Town A. Do you think that Kevin's truck must have received a flash while passing the speed camera? Explain.

2. The graph on the right shows the round trip journey of a car and a van between Towns P and Q.

   (a) At what time did the van start its journey?

   (b) Explain the motion of the car between 10 a.m to 11 a.m.

   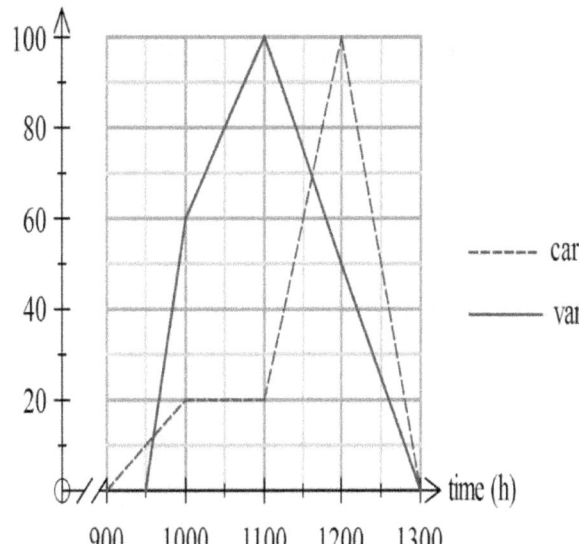

   (c) Estimate the time when the van first overtook the car.

   (d) Determine the average speed of the car over its entire journey.

3. Alex leaves home at 8 am and drove at a constant speed of 60 km/h for 30 minutes. He stopped at his friend's house and had a two hour game playing break. They then drove to the Waterpark 80 km away at a constant speed of 40 km/h. They spent 3 hours at the Waterpark after which Alex drove back reaching home at 5 pm with his friend as they were planning a sleepover.

(a) Complete the distance-time graph on the axes below.

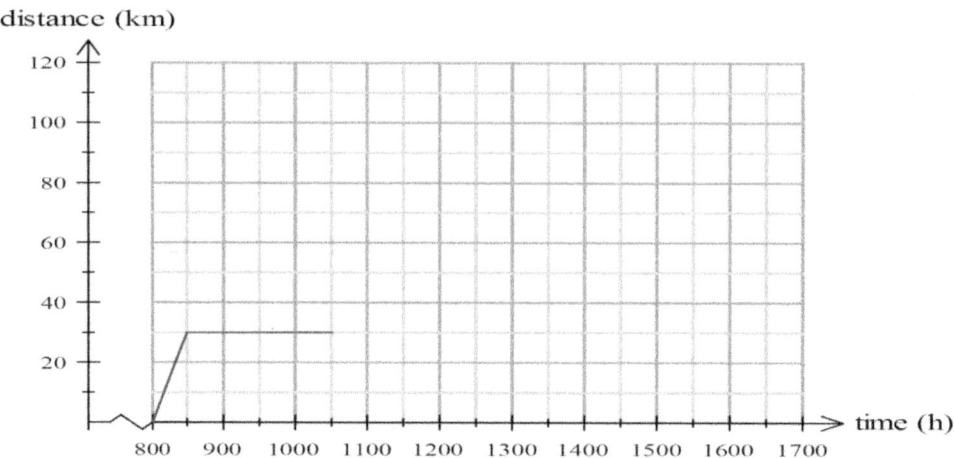

(b) Calculate the return journey's speed.

4. Relax Islands taxes the first $30 000 of an individual's income at a rate of 12%, and all income over $30 000 is taxed at 25%.

(a) Pinocchio makes $18 000 and Geppetto makes $40 000. How much is each taxed?

(b) Write a piecewise function T that specifies the total tax on an income of x dollars.

5. Tina is employed as a sales person and is paid commission as follows:
First $ 20000 → 4 %
Amount exceeding $ 20000 → $800 + 3% of each $1 over $20000
Write a piecewise function C that specifies the total commission on an income of x dollars.

6. The diagram below shows the tax system in an American State. As we can see the tax system is progressive meaning the more you earn the more tax you pay.

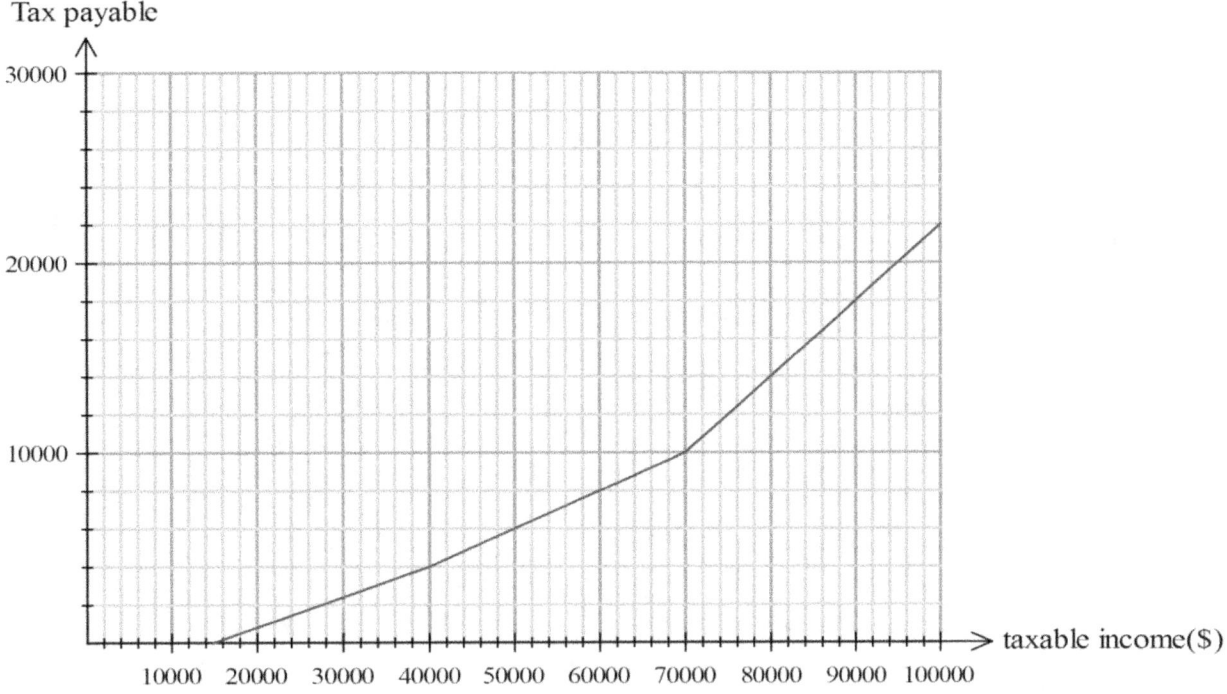

Use the graph to estimate the tax payable by
(a) An accountant with a yearly income of $80000.

(b) A teacher with a yearly income of $64000.

(c) A soccer player with a yearly income of $98000.

Angela paid $12000 in tax last calendar year,

(d) What was her taxable income?

Jimmy is a quantity surveyor and earns $100 000 a year. However, because his wife went on maternity leave and was no longer working, Jimmy decided to declare two incomes of $50 000 each to the tax office instead. How much can he save in tax by doing so?

7. When it comes to selling a property, Kingston Real Estate's agents charge a commission fee. How much fees someone pays depends on the location of the property. The graph below shows the amount of fees paid in two different areas: metro and country.

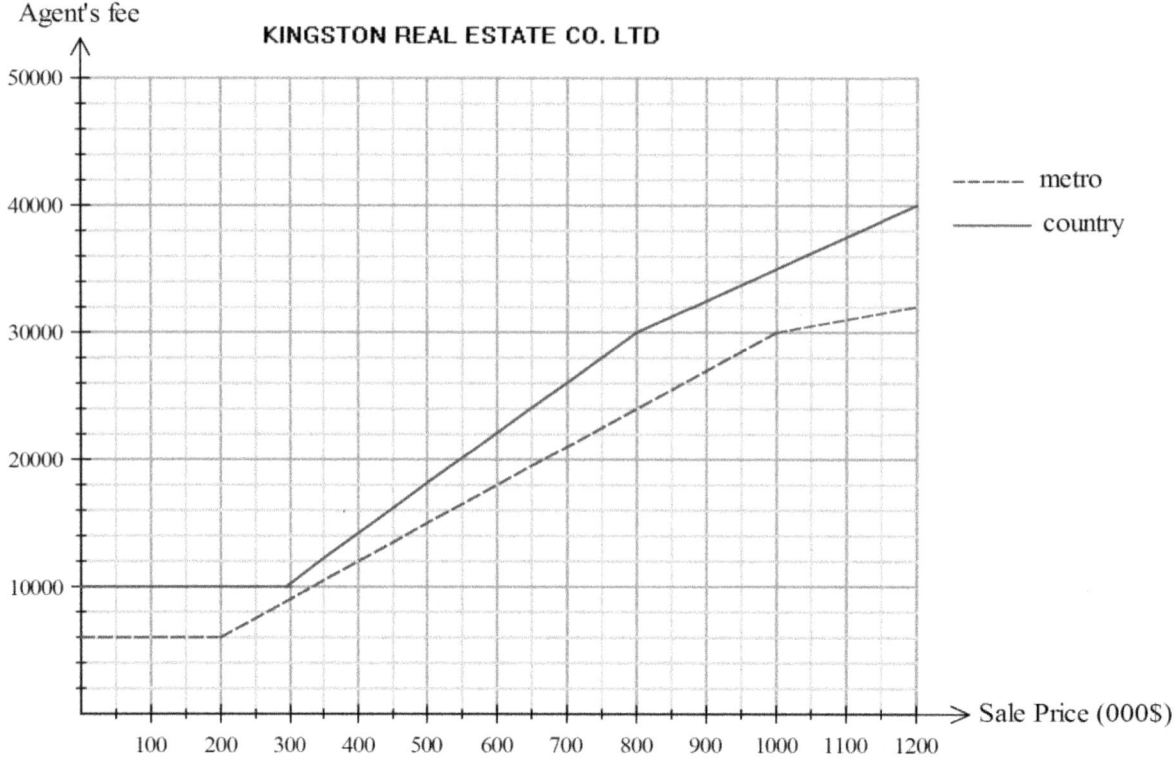

Use the graph to estimate the fees payable on a property sold for
(a) $700 000 in the country.

(b) $600 000 in the metro.

Angela paid an agent fee $22000 for having her house sold by Kingston Real Estate.

(c) Is her house located in the metro or the country?

CHAPTER 10

TRIGONOMETRY FOR RIGHT TRIANGLES

10A HYPOTENUSE, ADJACENT AND OPPOSITE

Trigonometry for right-angled triangles is the study of the relationship between the side lengths and its angles. There are three ratios that are used to compare the different lengths of the sides of a right-angled triangle. These ratios are the sine, cosine and tangent ratios commonly named as the trigonometric ratios. The reference angle, usually θ, determines the name of the sides as shown below.

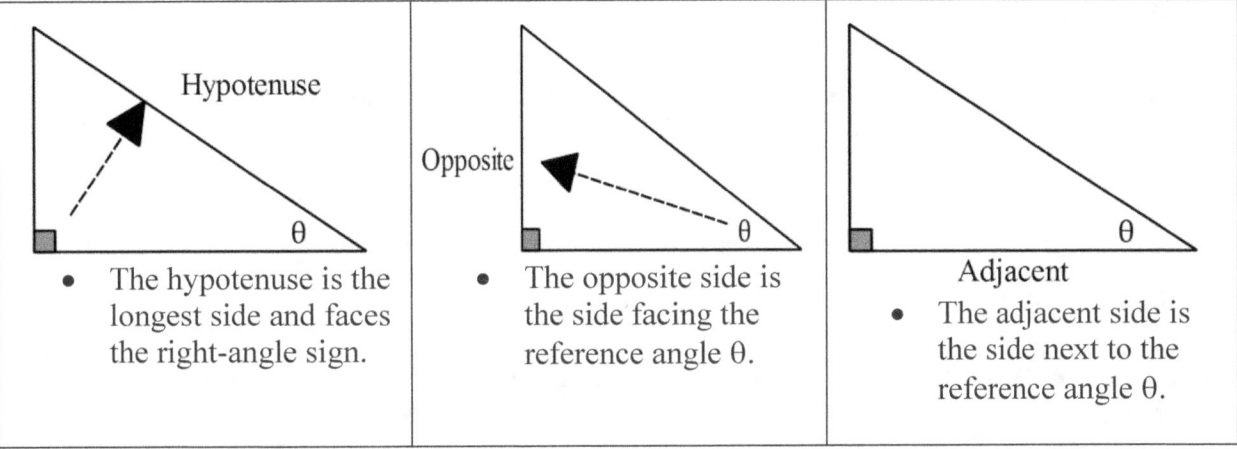

- The hypotenuse is the longest side and faces the right-angle sign.
- The opposite side is the side facing the reference angle θ.
- The adjacent side is the side next to the reference angle θ.

CLASS ACTIVITY 1
In the following triangles label the hypotenuse (H), the opposite (O) and the adjacent (A).

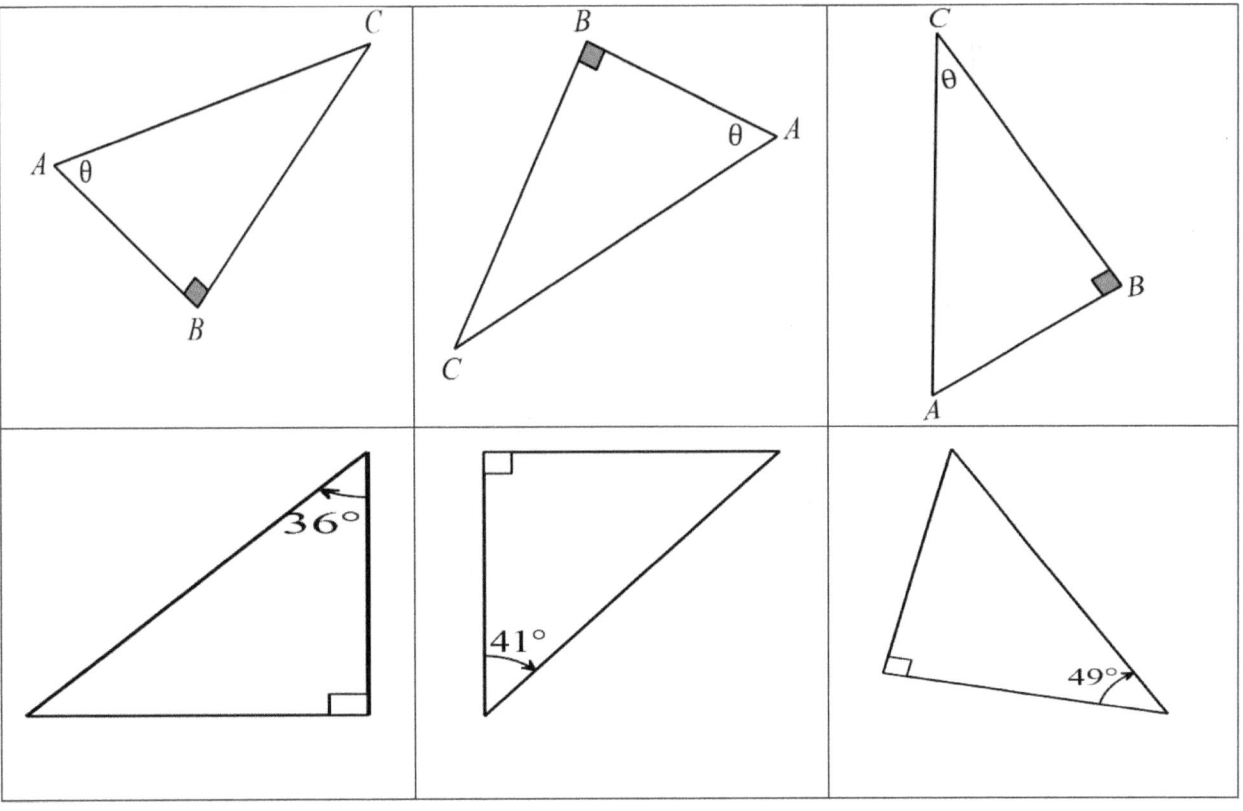

10B TRIGONOMETRIC RATIOS

As seen earlier, there are three trigonometric ratios: sine (sin), cosine (cos) and tangent (tan). A simple way of remembering the trigonometric ratios is by using the acronym SOH CAH TOA.

$$\sin\theta = \frac{Opposite}{Hypotenuse}, \cos\theta = \frac{Adjacent}{Hypotenuse}, \tan\theta = \frac{opposite}{adjacent}$$

EXAMPLES

Find the exact value of sinθ.

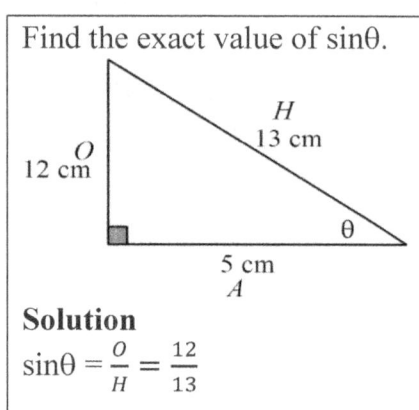

Solution
$\sin\theta = \frac{O}{H} = \frac{12}{13}$

Find the exact value of cosθ.

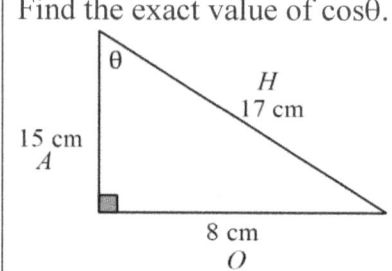

Solution
$\cos\theta = \frac{A}{H} = \frac{15}{17}$

Find the exact value of tanθ.

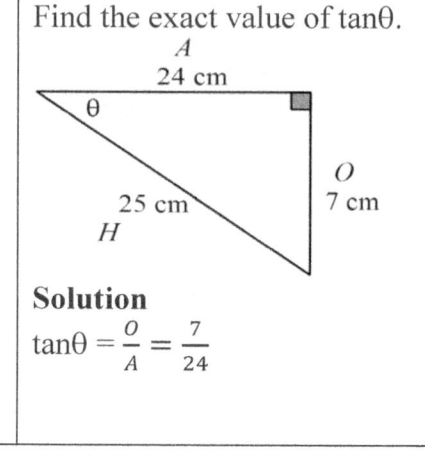

Solution
$\tan\theta = \frac{O}{A} = \frac{7}{24}$

CLASS ACTIVITY

Write down as a fraction the value of sin∠ FDE.

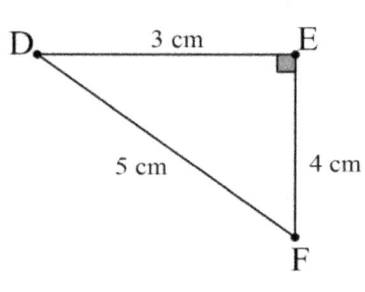

Find the value of cos∠ HGI.

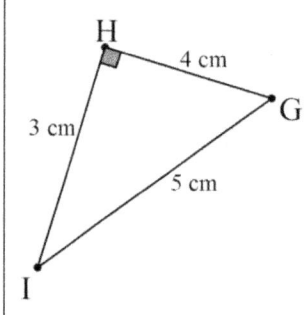

Express as a simplest fraction the value of tan∠ BAC.

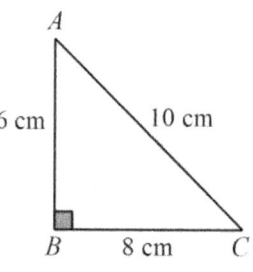

Find the exact value of cos∠ QRP.

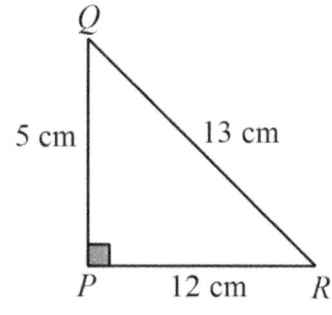

Find the exact value of tan ∠YXZ.

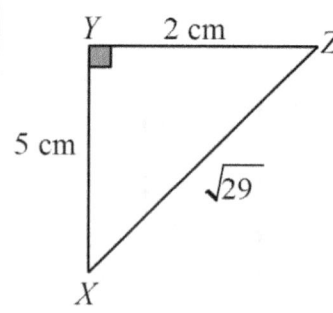

10C USING THE CALCULATOR

CLASS ACTIVITY

Use your calculator to find the value of the pronumerals in each of the following cases, giving your answers to one decimal place.

1. $x = 20 \times \sin 36°$ $ = 11.8$	2. $y = 24 \times \cos 45°$	3. $x = 13 \times \tan 28°$
4. $x = 32 \times \sin 51°$	5. $y = 32 \times \cos 63°$	6. $z = 15 \times \tan 54°$
7. $x = \dfrac{10}{\sin 27°}$	8. $x = \dfrac{12}{\cos 46°}$	9. $y = \dfrac{9}{\tan 30°}$
10. $\sin 30° = \dfrac{x}{10.6}$ (x in the numerator, always multiply) $x = 10.6 \times \sin 30°$ $ = 5.3$	11. $\tan 60° = \dfrac{x}{8.6}$	12. $\cos 45° = \dfrac{x}{16.5}$
13. $\sin 40° = \dfrac{25}{x}$ (x in the denominator, divide) $x = \dfrac{25}{\sin 40°}$ $ = 38.9$	14. $\tan 49° = \dfrac{12}{x}$	15. $\cos 35° = \dfrac{14.7}{y}$
16. $\sin 52° = \dfrac{x}{9.8}$	17. $\tan 32° = \dfrac{25}{x}$	18. $\cos 60° = \dfrac{x}{18}$
19. $\sin 63° = \dfrac{22}{x}$	20. $\tan 78° = \dfrac{x}{10.2}$	21. $\cos 49° = \dfrac{25.4}{y}$

10D FINDING SIDE LENGTHS

To be able to find the lengths of the sides of a right-angled triangle, we have to identify the correct trigonometric ratio to be used. The following steps might be helpful.

- If the triangle is not given, draw one showing all known information.
- Label the sides with the letters O, A and H. (remember to label only two sides, leave the blank side unlabelled)
- Use SOH-CAH-TOA to decide which trigonometric ratio to be used.
- Form an equation and solve for the unknown.

EXAMPLES

1. Find the value of x.

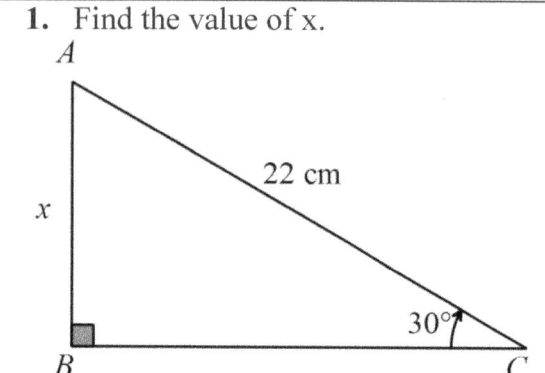

SOLUTION
First we label the sides AB and AC. Leave the side BC blank as it will help to choose the correct trigonometric ratio.

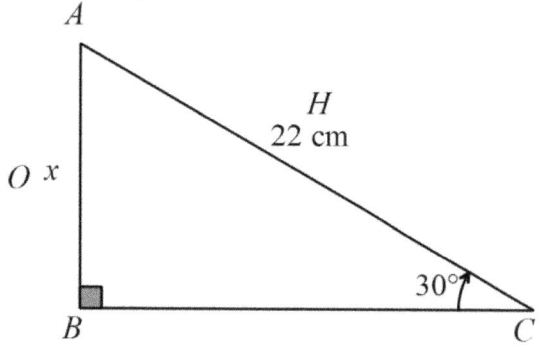

Since we have O and H, we are going to use the sine ratio.

SOH CAH TOA

$$\sin 30 = \frac{x}{22}$$

If the unknown is in the numerator we have to multiply

$$\therefore x = 22 \times \sin 30 = 11 \, cm$$

2. Find the value of y.

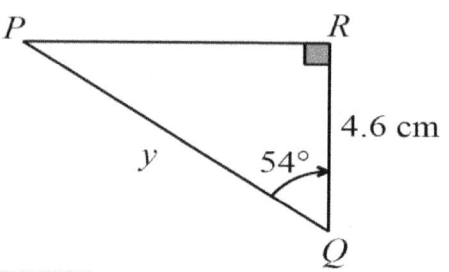

SOLUTION
First we label the sides PQ and QR. Leave the side PR blank

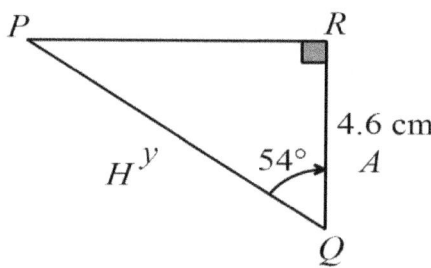

Since we have A and H, we are going to use the cosine ratio.

SOH **CAH** TOA

$$\cos 54 = \frac{4.6}{y}$$

If the unknown is in the denominator we have to divide

$$\therefore y = \frac{4.6}{\cos 54} = 7.83 \, cm$$

EXERCISE 10D

Find the value of each variable, correct to two decimal places where applicable.

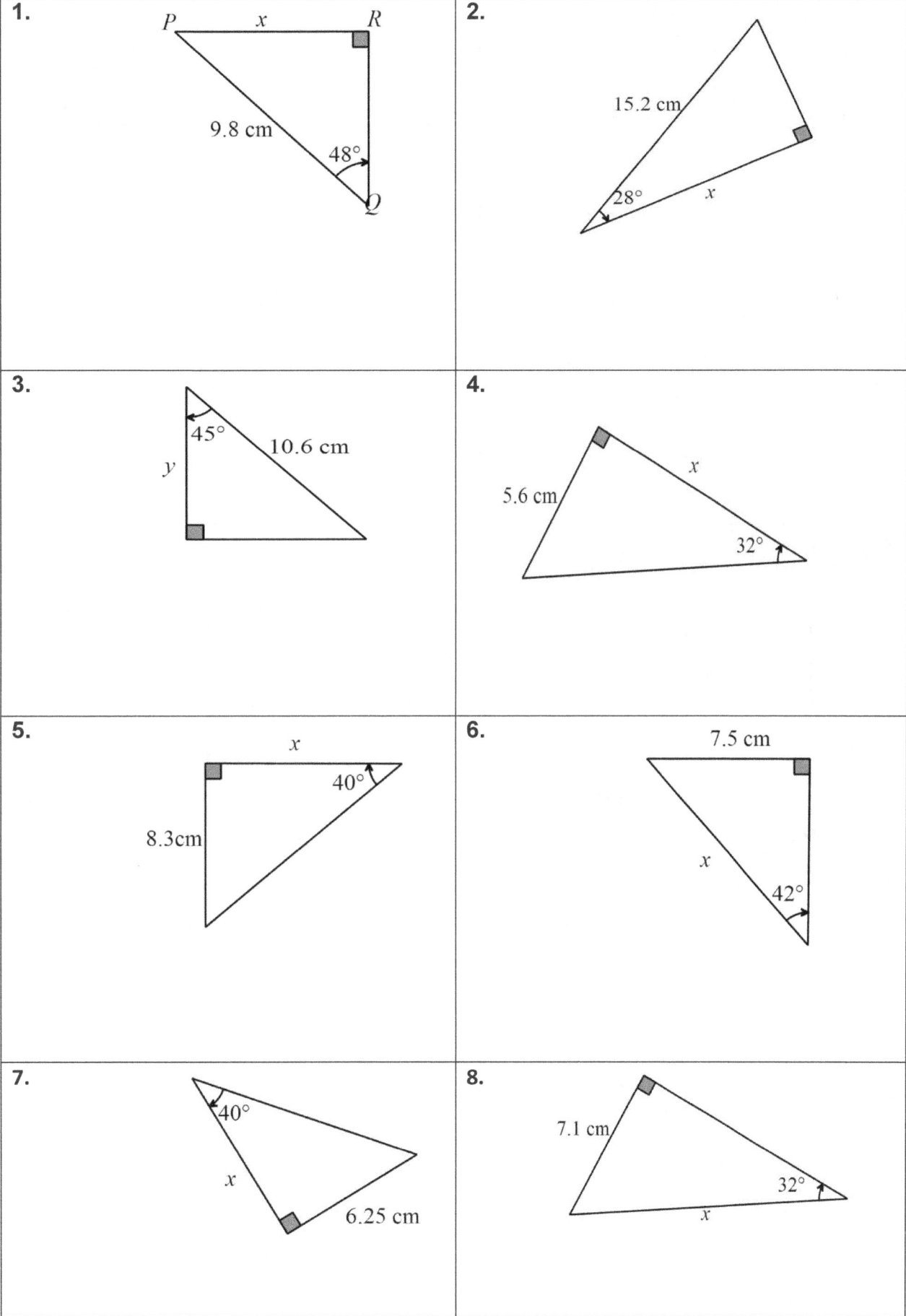

9.

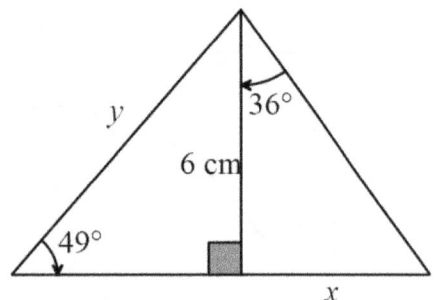

10.

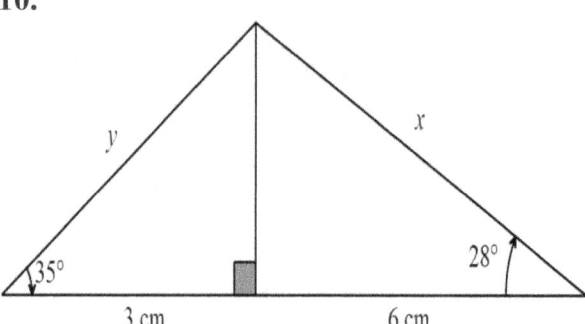

11.

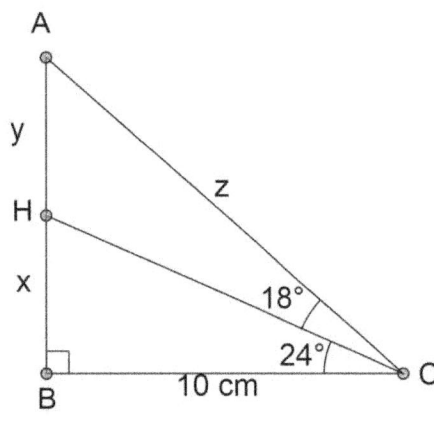

12.

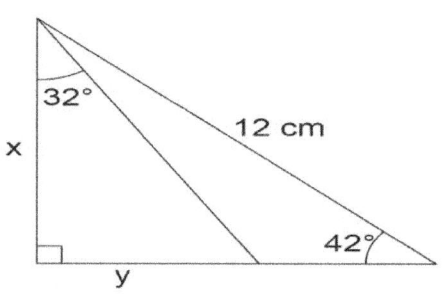

CHAPTER 10 : TRIGONOMETRY FOR RIGHT TRIANGLES

10E FINDING MISSING ANGLES

To be able to find the missing angle in a right angled triangle, at least two of the sides must be known. Consider the triangle ABC on the right. To determine the size of angle BAC (labelled as x in the diagram), we follow similar steps as seen previously in 10D.

- Label the sides with the letters O, A and H.
- Use SOH-CAH-TOA to decide which trigonometric ratio to be used.
- Form an equation and solve for the unknown.
- Use inverse trigonometry to find x. ($sin^{-1}, cos^{-1}, tan^{-1}$)

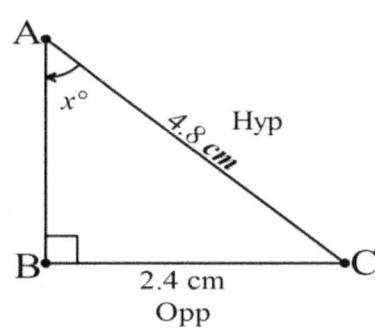

EXAMPLES

1. Find the size of angle BCA (θ) correct to the nearest degree.

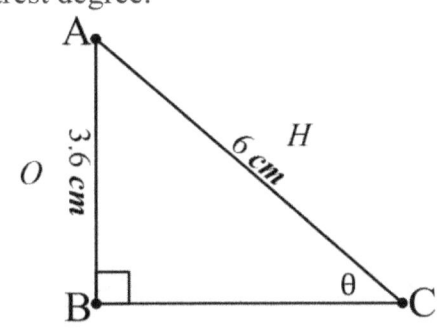

SOLUTION
Using SOH CAH TOA
$$\sin \theta = \frac{3.6}{6}$$
$$\theta = \sin^{-1}\left(\frac{3.6}{6}\right)$$
$$\theta = 37°$$

2. Find the value of θ in the diagram below.

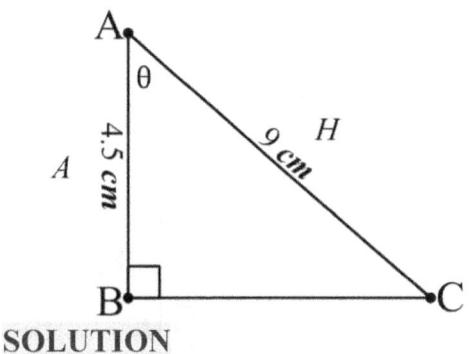

SOLUTION
Using SOH CAH TOA
$$\cos \theta = \frac{4.5}{9}$$
$$\theta = \cos^{-1}\left(\frac{4.5}{9}\right)$$
$$\theta = 60°$$

3. Find the angle that the ladder makes with the wall.

SOLUTION

Using SOH CAH TOA
$$\tan x = \frac{5}{10}$$
$$x = \tan^{-1}\left(\frac{5}{10}\right)$$
$$\theta = 26.6°$$

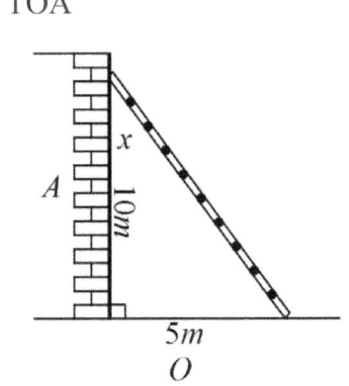

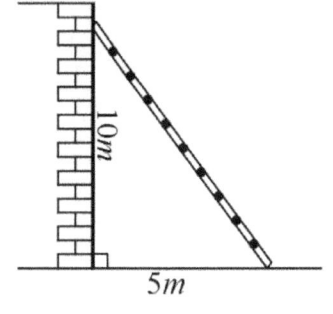

EXERCISE 10E

Find the value of θ in each of the following triangles.

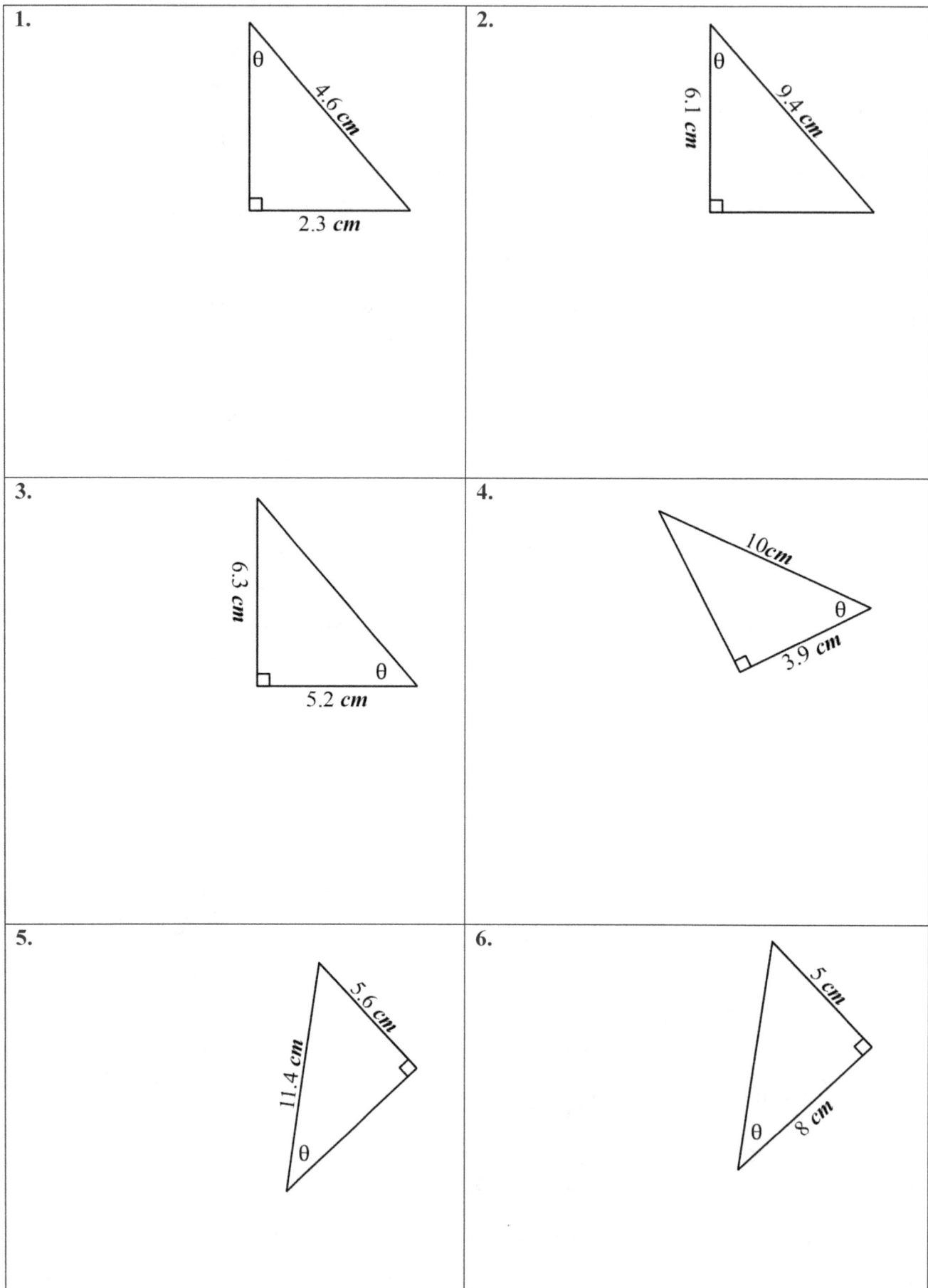

7. In triangle ABC, AB = 10.4cm, AC = 6 cm and ∠ACB = 90°. Find ∠ABC.

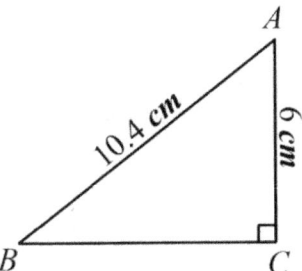

8. In triangle PQR, PQ = 10cm, QR = 6.8 cm and ∠PQR = 90°. Find ∠PRQ.

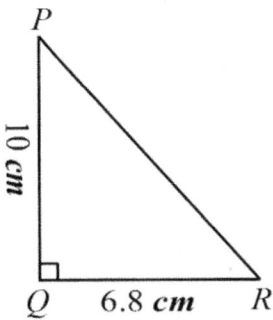

9. Determine the size of angle FDG.

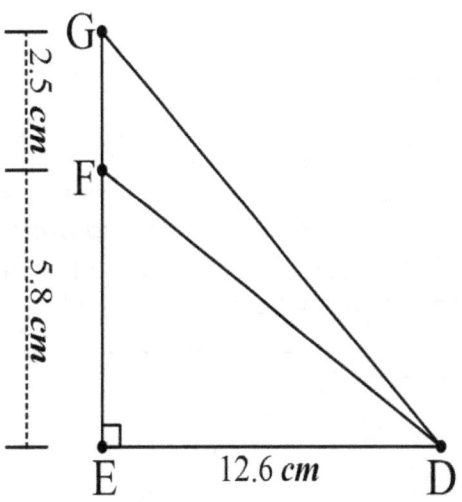

10. In the diagram AB = 14.6 cm, AD = 15 cm and CD = 10.2 cm. Find the difference between the size of angles *x* and *y*.

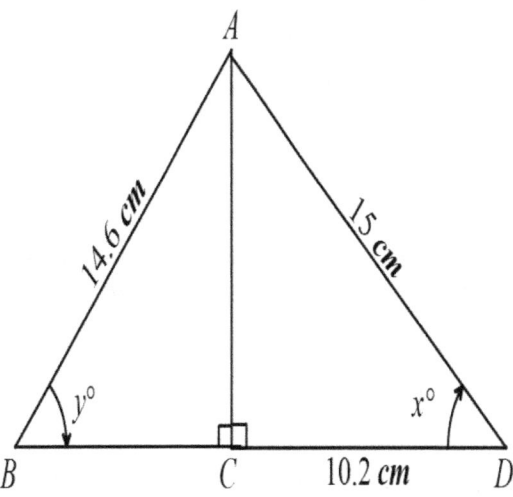

10F APPLICATIONS : ANGLE OF ELEVATION AND DEPRESSION

The **angle of elevation** of an object as seen by an observer is the angle between the horizontal and the line of sight from observer's eye.

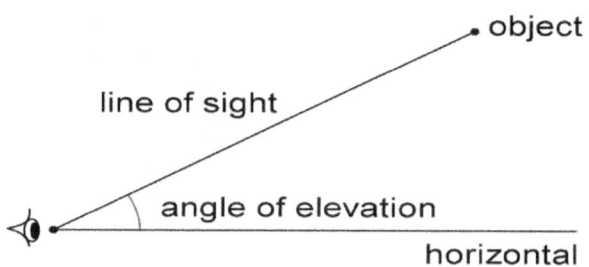

However, if the object is below the level of the observer, then the angle between the horizontal and the observer's line of sight is called the **angle of depression**.

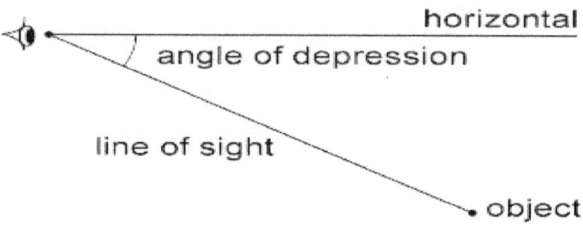

EXAMPLES

1. Tony standing at the top of a cliff sights a ship 650m at sea as shown. The angle of depression of the ship from Toni is 43°. If Tony is 1.82 m, find the height of the cliff, to the nearest metre.

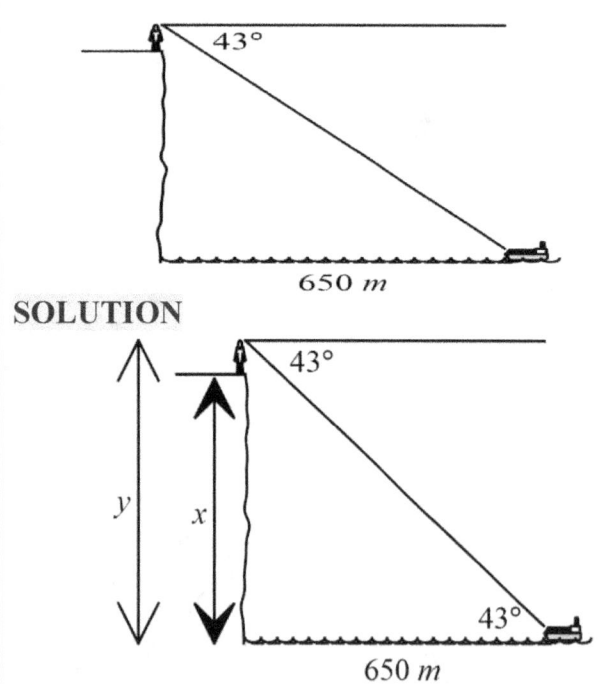

SOLUTION

$$\tan 43 = \frac{y}{650}$$
$$\therefore y = 650 \times \tan 43 = 606.13 \, m$$
height of cliff, $x = 606.13 - 1.82 = 604 \, m$

2. Alex is 1.72m tall. He sights a bird at the top of the tree. The angle of elevation of the top of the tree from Alex's eye-level is 33°. Given that Alex is 32m away from the tree, calculate the height of the tree.

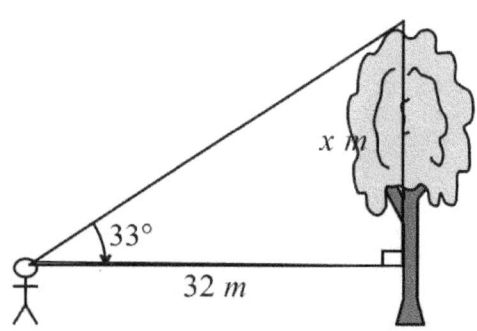

SOLUTION

$$\tan 33 = \frac{x}{32}$$
$$\therefore x = 32 \times \tan 33 = 20.78 \, m$$
height of tree $= 20.78 + 1.72 = 22.50 \, m$

EXERCISE 10F

1. The sun is at an angle of elevation of 51°. A tree casts a shadow 26 metres long on the ground. How tall is the tree?

2. From the top of a lighthouse 82 m high, the angle of depression of a boat is 44°. Find the distance from the boat to the foot of the lighthouse.

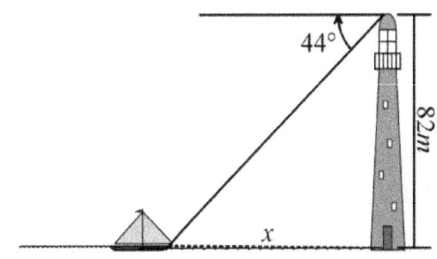

3. At a point on the ground 48 m from the foot of a tree, the angle of elevation to the top of the tree is 35°. Find the height of the tree.

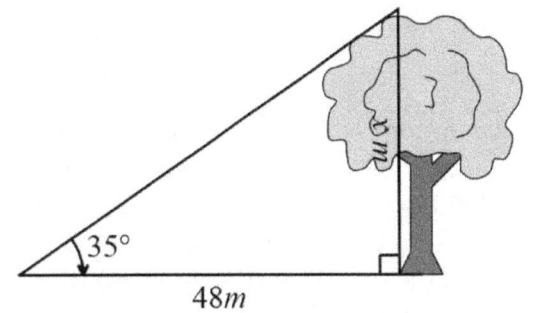

4. A 13 m ladder leans against a wall so that the base of the ladder is 5m from the base of the wall. Label the diagram with correct measurements and find the ladder's angle of elevation?

5. A ship is on the surface of the water, and its radar detects a submarine at a distance of 98m, at an angle of depression of 25°. Determine how deep underwater is the submarine.

6. A dog, who is 15 metres from the base of a light post, spots a bird at the top of the post at an angle of elevation of 39°. Find the direct-line distance between the dog and the bird.

7. The angle of elevation from a car to St George Cathedral is 29°. If the church is 38 m high. Determine the distance of the car from the church.

8. A person living on the 11th floor of a hotel sights the top and bottom of an apartment 140 m away. The angle of elevation for the top of the apartment is 34° and the angle of depression for the base of the building is 45°. Find the height of the apartment?

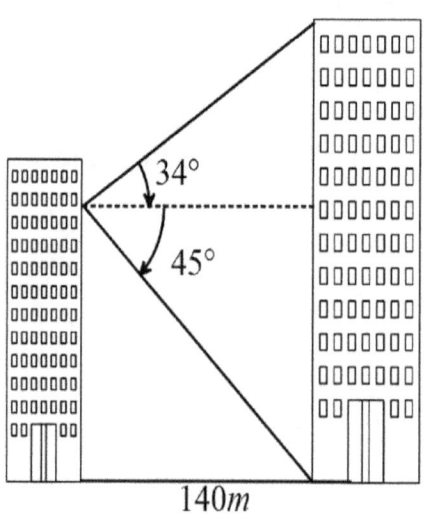

9. An aircraft is travelling horizontally at a contant speed of 500km/h at a height of 10000 m above level ground. At 10 am, Ally sees the aircraft at an angle of elevation 32°. Determine the angle of elevation at 10.30 am.

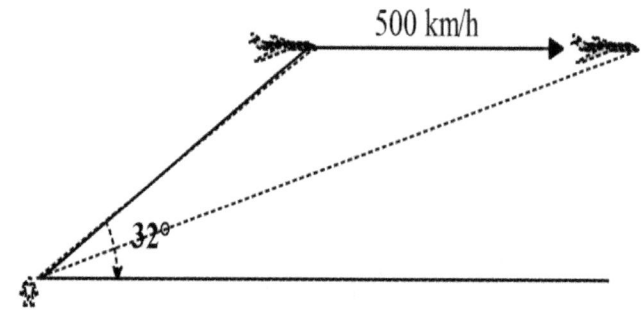

10. The angle of elevation of the top of a tree from a statue's feet is 46° as shown. If the tree casts a shadow of 24m,

   (a) Determine the height of the tree.

   (b) If a statue casts a shadow of 4 m, find the statue's height, assuming the two triangles are similar.

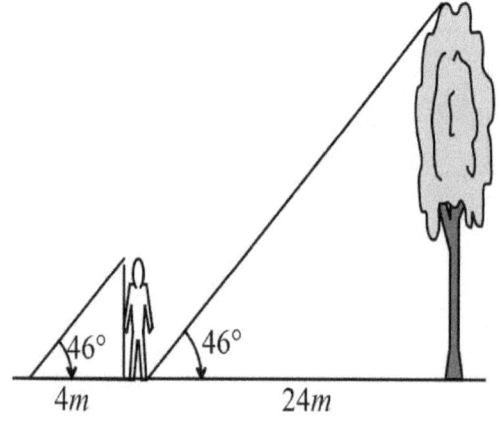

10G BEARINGS

Bearings are a measure of direction, with North taken as being the starting point.
When we travel North, our bearing is 000°, and this is usually represented as a vertical line pointing up.
When we travel in any other direction, our *bearing* is measured **clockwise** from North.
Bearings are also written as 3 digits. Example 059°, 007° and so on.

Consider the diagrams below:

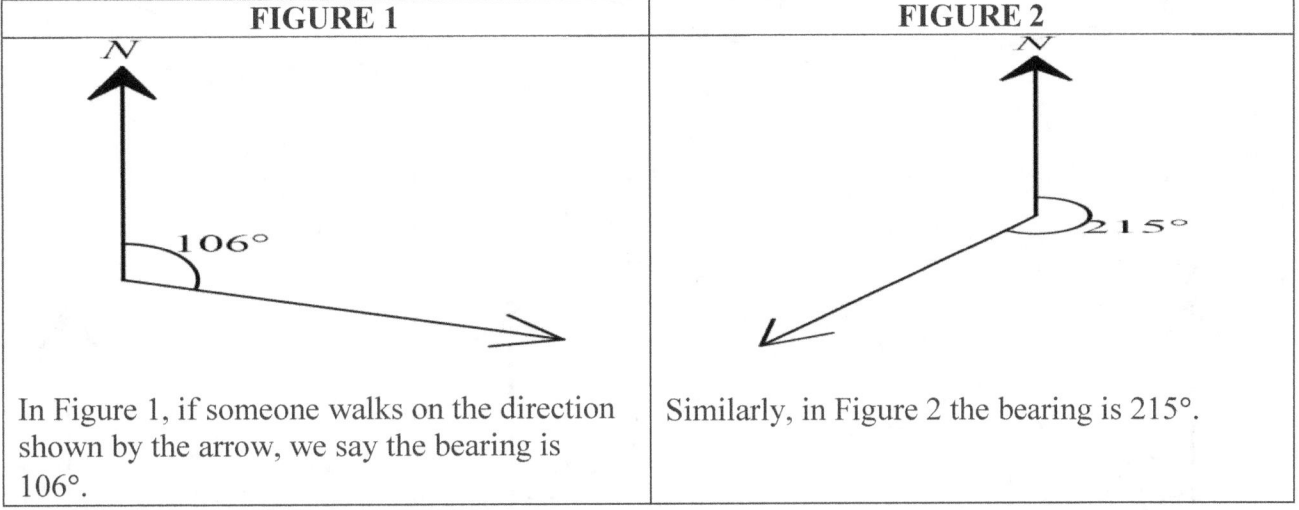

FIGURE 1	FIGURE 2
In Figure 1, if someone walks on the direction shown by the arrow, we say the bearing is 106°.	Similarly, in Figure 2 the bearing is 215°.

EXAMPLE 1
An aeroplane flies from Porto to Rio on a bearing of 065°.
On what bearing should the pilot fly, to return to Porto from Rio?

SOLUTION
First, we find the supplementary angle = 180° − 65° = 115°
Bearing to return to Porto from Rio = 360° − 115° = 245°

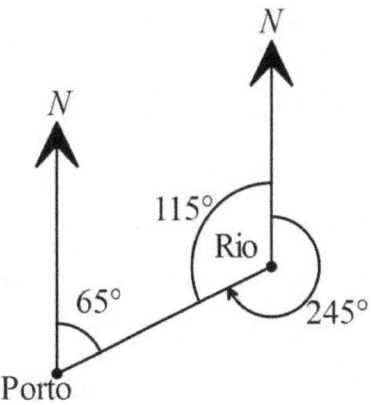

EXAMPLE 2
The diagram shows the positions of three towns R, S and T.
Find the bearing of R from S.

SOLUTION

180° − 57° = 123°
360° − 131° − 123° = 106°

Bearing R from S = 131° + 106° = 237°

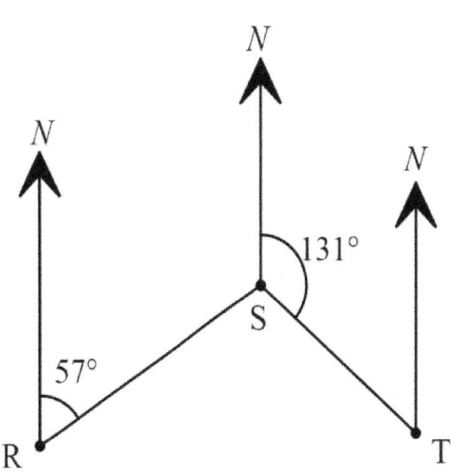

EXERCISE 10G

1. Complete the following by filling the blanks.

112° angle at C, B to southeast	65° angle at A, B to northeast	130° angle at B, A to southwest
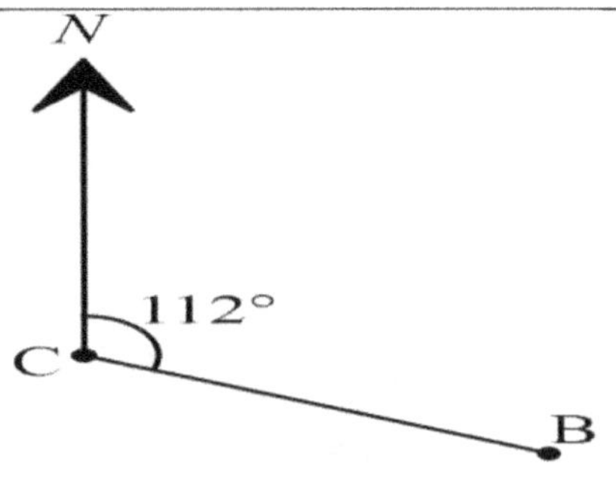 Bearing of B from C = ...	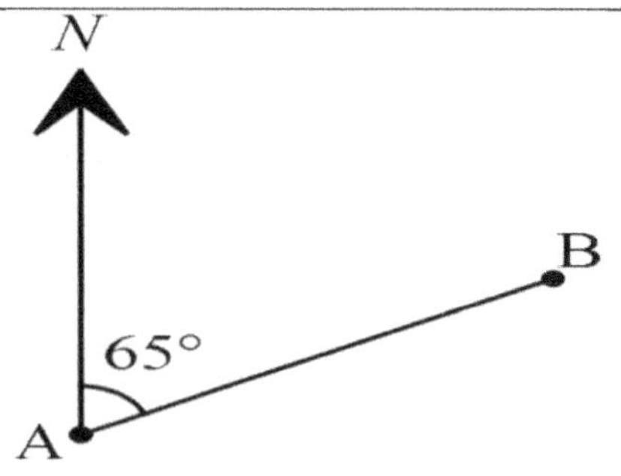 Bearing of B from A = ...	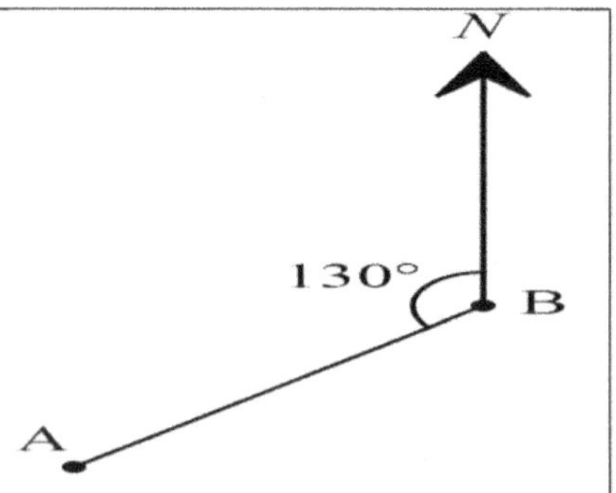 Bearing of A from B = ...
45° angle at O, P to northeast	40° angle at P, Q to northwest	Right angle at Q, P due east
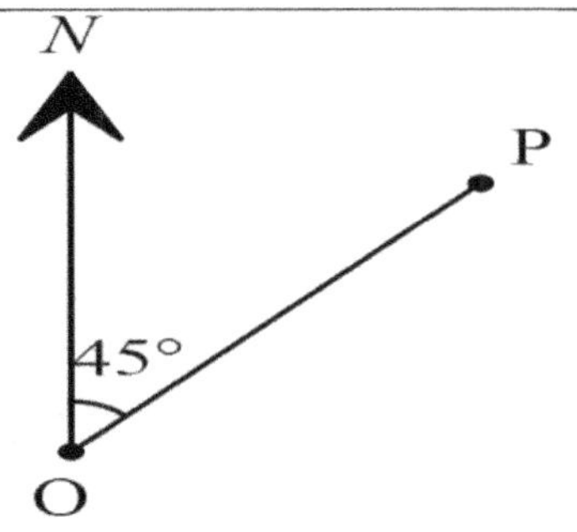 Bearing of P from O = ...	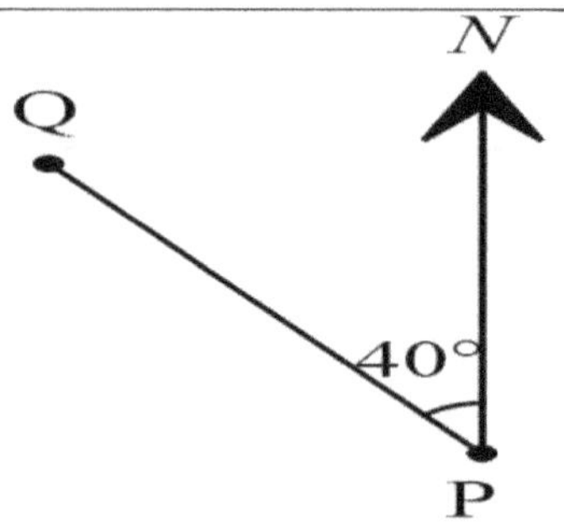 Bearing of Q from P = ...	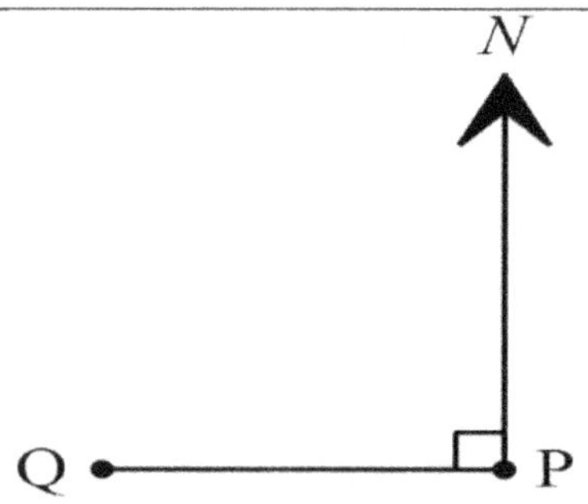 Bearing of Q from P = ...

2. Study the diagram carefully then answer the questions on the right.

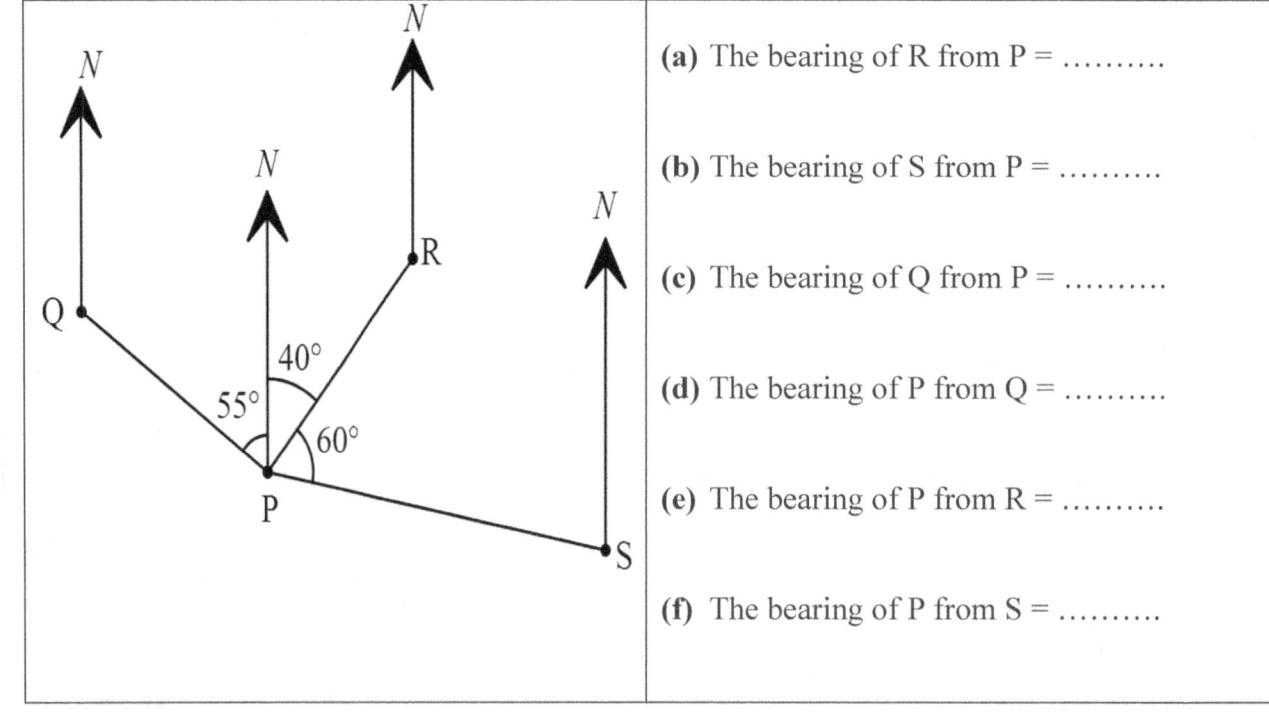

(a) The bearing of R from P = ...

(b) The bearing of S from P = ...

(c) The bearing of Q from P = ...

(d) The bearing of P from Q = ...

(e) The bearing of P from R = ...

(f) The bearing of P from S = ...

3. Study the diagram carefully then answer the questions on the right.

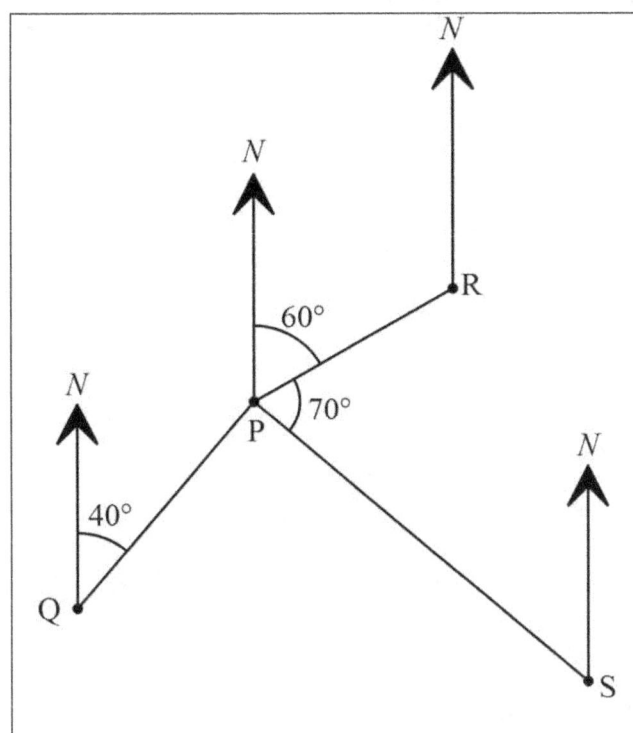

(a) The bearing of R from P = ...

(b) The bearing of S from P = ...

(c) The bearing of Q from P = ...

(d) The bearing of P from Q = ...

(e) The bearing of P from R = ...

(f) The bearing of P from S = ...

4. The map of an holiday island has been shown below.

   What is the bearing of

   (a) Waterfall from the lighthouse,

   (b) Beach from the lighthouse,

   (c) Lighthouse from the church.

   (d) The lighthouse from the village.

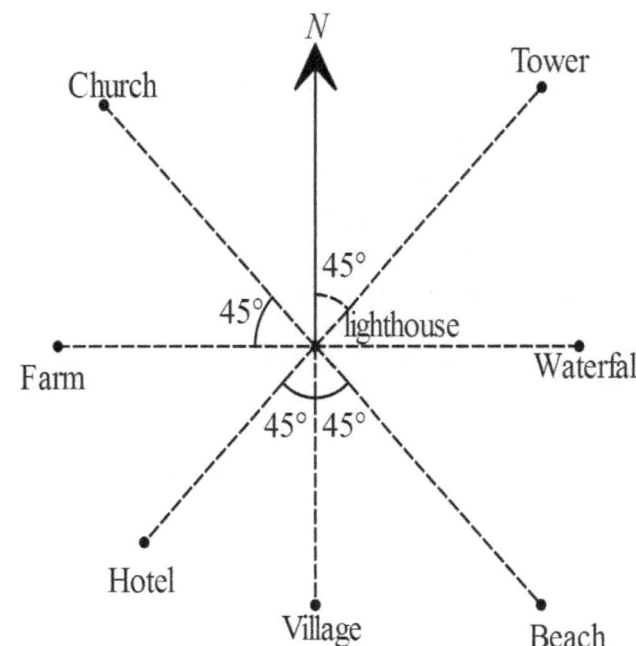

5. A ship sets sail from a harbour A and travels 30 km to a harbour B on a bearing of 060°. It stops for a few hours, then sails to a harbour C on a bearing of 150°, travelling a distance of 40 km.

   (a) Draw a diagram to illustrate the above situation.

   (b) Hence calculate the distance A to C.

   (c) On what bearing must the ship set sail to go back to harbour A?

6. Given that
   A is North of B.
   C is South-East of B.
   C is on a bearing of 160° from A.

   Find the bearing of:

   (a) A from B

   (b) A from C

CHAPTER 11

TRIGONOMETRY FOR NON-RIGHT ANGLED TRIANGLES

11A   AREA OF TRIANGLE : BASE AND HEIGHT KNOWN

If the base and the height of a particular triangle are known, then the area can be calculated by the formula

$$\text{Area} = \frac{1}{2} \times base \times height$$

EXERCISE 11A

Find the area of the following triangles.

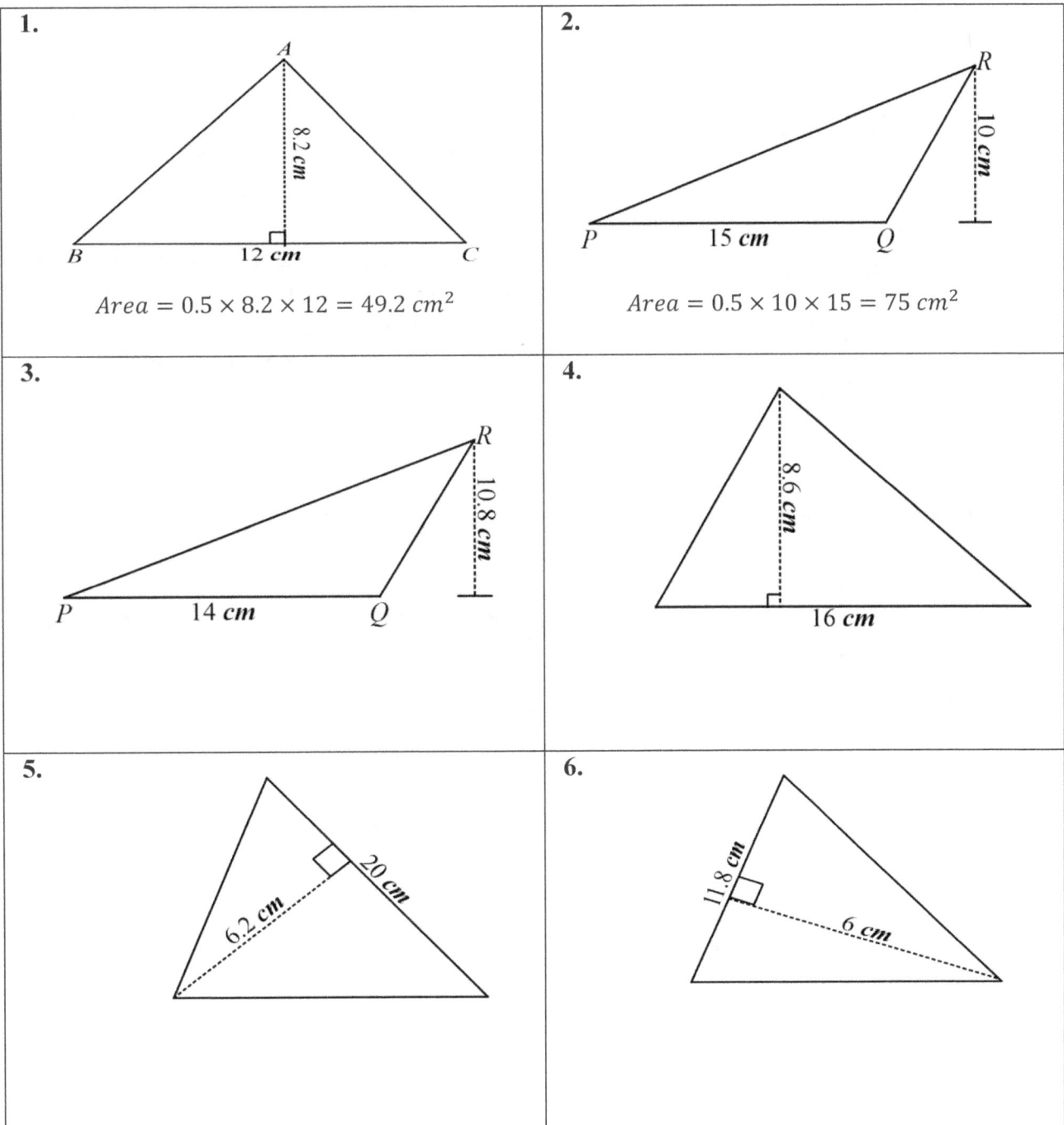

1.
   Area = 0.5 × 8.2 × 12 = 49.2 cm²

2.
   Area = 0.5 × 10 × 15 = 75 cm²

3.

4.

5.

6.

11B AREA OF TRIANGLE : HEIGHT UNKNOWN

If the base and the height of a particular triangle are unknown, then the area can be calculated by the formula

$$\text{Area} = \frac{1}{2} \times a \times b \times \sin C,$$

where the C is the angle included between the two sides being used.

EXAMPLES

Find the area of the following triangles.

1.
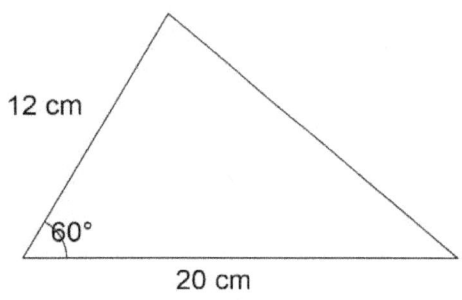

Area $= \frac{1}{2} \times 12 \times 20 \times \sin 60°$
$= 103.92$ cm^2.

2.
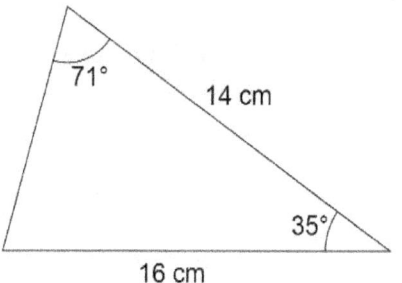

Since the angle lies between the 16 cm and 14 cm sides, we ignore the 71° angle.
Area $= \frac{1}{2} \times 14 \times 16 \times \sin 35°$
$= 64.24$ cm^2.

3. Given that the area of the triangle is 60 cm^2, find the value of x.

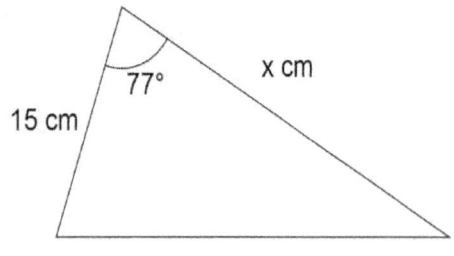

$$solve\left(\frac{1}{2} \times 15 \times x \times \sin 77 = 60, x\right)$$

$$\{x = 8.210432862\}$$

$\frac{1}{2} \times 15 \times x \times \sin 77° = 60$

Use solve facility on calculator as shown on the right.
∴ $x = 8.21 \, cm$.

CHAPTER 11 : TRIGONOMETRY FOR NON-RIGHT ANGLED TRIANGLES

EXERCISE 11B

In questions 1-4, find the area of the triangles.

1.
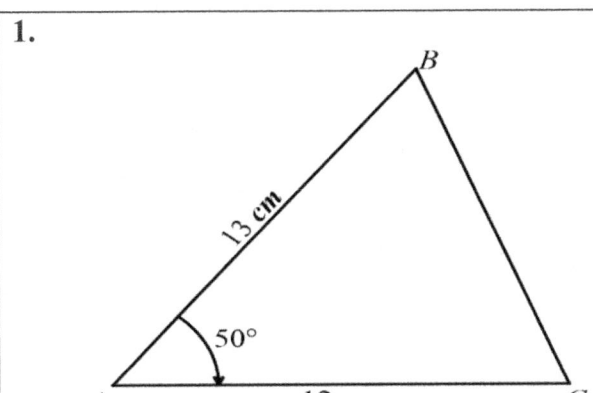

2.
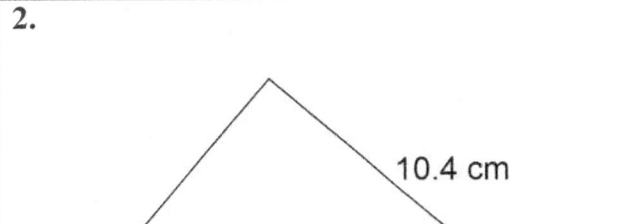
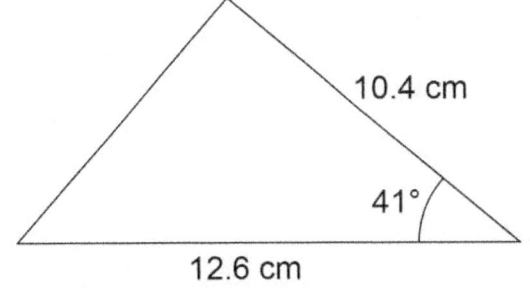

3.
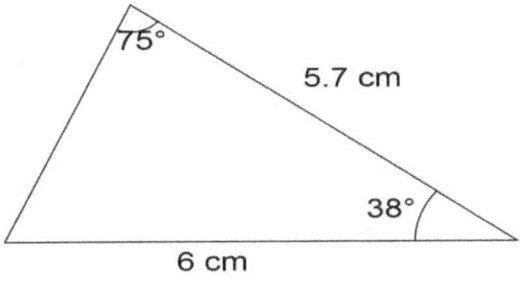

4.
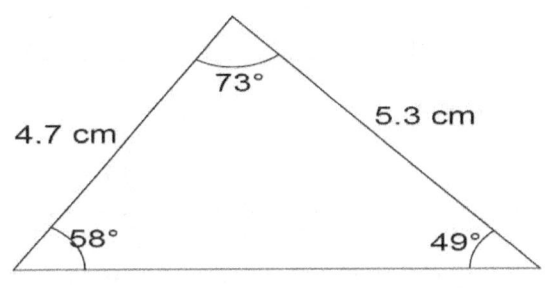

5. Given that the area of the triangle is 12 cm², find the value of x.

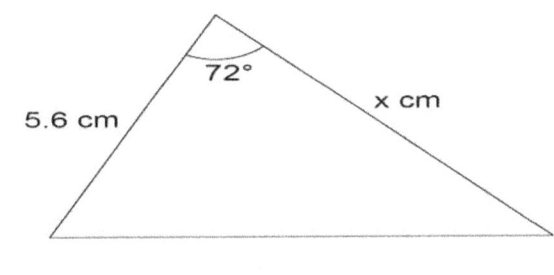

6. If the area of the triangle is 15 cm², find the value of x.

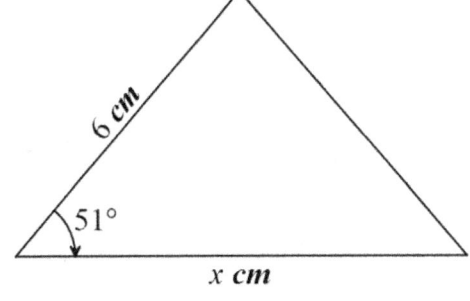

11C AREA OF TRIANGLE: HERON'S FORMULA

If the lengths of the three sides of a triangle are known, a very quick method for calculating the area of the triangle is to use the Heron's formula which states that

$$Area = \sqrt{s(s-a)(s-b)(s-c)},$$

where a, b and c are the lengths of the sides of a triangle and $s = \frac{a+b+c}{2}$.

EXAMPLES

Find the area of the following triangles using the Heron's formula.

1.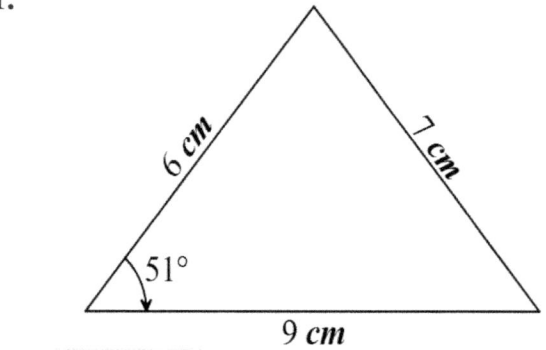

   SOLUTION
   $$s = \frac{6+7+9}{2} = 11$$
   Using $A = \sqrt{s(s-a)(s-b)(s-c)},$
   $$A = \sqrt{11(11-6)(11-7)(11-9)}$$
   $$= 20.98 \ cm^2.$$

2. Use Heron's formula to find the area of a triangle of lengths 9, 10 and 11. Give your answer in exact form.

   SOLUTION
   $$s = \frac{9+10+11}{2} = 15$$
   Using $A = \sqrt{s(s-a)(s-b)(s-c)},$
   $$A = \sqrt{15(15-9)(15-10)(15-11)}$$
   $$= \sqrt{1800} \ cm^2.$$

3. A level block of land is triangular in shape. (Shown as triangle ABC in the diagram below). Because of its shape, The Real Estate Company decides to plant trees on the block instead of selling it. Trees can be obtained locally from a nursery for $4.65 each. It is ideal that one tree be planted every 3.6 m² of land. How many trees are required to fill the area and what will be the total cost to fill the block of land with the trees?

   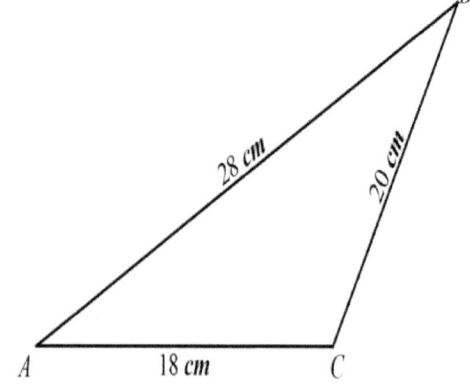

   SOLUTION
   $$\therefore s = \frac{18+28+20}{2} = 33$$

   Using $A = \sqrt{s(s-a)(s-b)(s-c)},$

   $$A = \sqrt{33(33-18)(33-28)(33-20)}$$
   $$= 179.4 \ m^2.$$
   Number of trees = 179.4 ÷ 3.6 ≈ 49 *trees*
   Cost = 49 × 4.65 = $227.85

EXERCISE 11C

1. Use Heron's formula to find the area of the triangle.

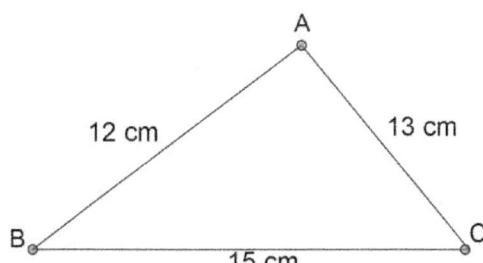

2. Find the area of the triangle.

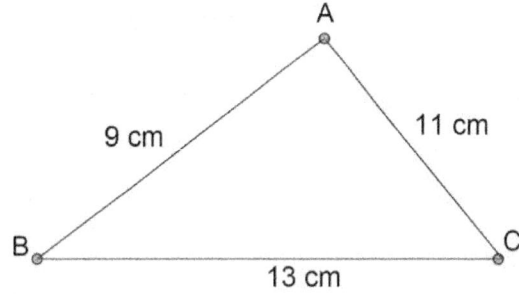

3. Find the area of a triangular playground, to the nearest square metre, with sides of length 8 m, 10 m and 14 m.

4. Find the triangular area, to the nearest square metre, enclosed by three pieces of fencing 50 m, 60 m and 75 m long.

5. Find the area of a triangle with side lengths 11 m, 13 m and 20 m.

6. Use Heron's formula to find the area of a triangle of lengths 3cm, 7cm and 8cm. Give your answer in **exact form**.

11D   THE SINE RULE : CALCULATOR ASSUMED

When do we use the sine rule?

1. For non-right angled triangles
2. At least two angles involved.

The sine rule is given by $\quad \dfrac{\sin A}{a} = \dfrac{\sin B}{b} = \dfrac{\sin C}{c}$

EXAMPLES

1. Find the value of x, correct to 2 decimal places.

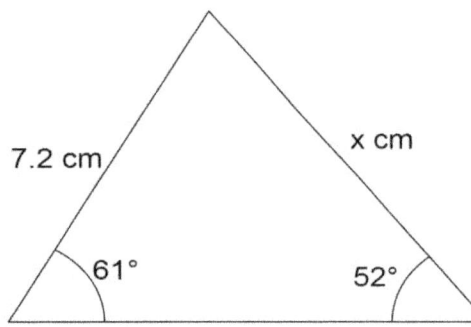

SOLUTION

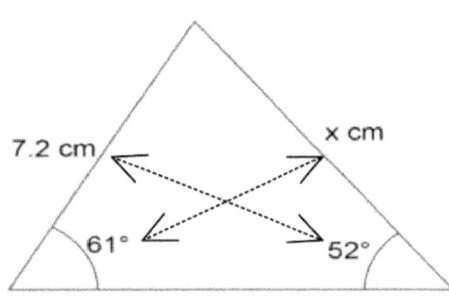

Always use a cross as shown to know which side is divided by which one.

Use the solve facility on your calculator (Main, Action, Advanced , Solve)

$$\text{solve}\left(\dfrac{\sin(61)}{x} = \dfrac{\sin(52)}{7.2}, x\right)$$

$$\{x = 7.991340045\}$$

$$x = 7.99\ cm$$

2. Find the value of x

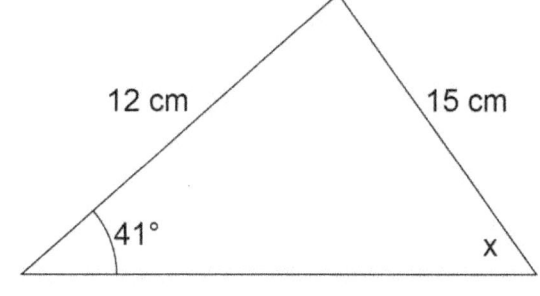

SOLUTION

$$\text{solve}\left(\dfrac{\sin(41)}{15} = \dfrac{\sin(x)}{12}, x\right)$$

$\{x = 360.\,constn(1) + 148.342041,$
$x = 360.\,constn(2) + 31.65795896$

$$x = 31.7° \text{ or } 148.3°$$

3. Find the value of x:

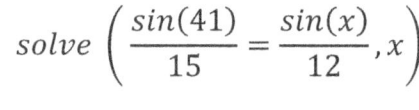

SOLUTION
The angle opposite side x can be found as
$x = 180 - (42 + 63) = 75°$

$$\text{solve}\left(\dfrac{\sin(42)}{11} = \dfrac{\sin(75)}{x}, x\right)$$

$$\{x = 15.87908846\}$$

$$x = 15.9\ cm$$

EXERCISE 11D

Use the solve facility on your calculator to find the value of *x* in each of the following cases.

1. $\dfrac{\sin(60)}{x} = \dfrac{\sin(45)}{7}$

2. $\dfrac{\sin(x)}{9.8} = \dfrac{\sin(61)}{10.4}$

3. $\dfrac{\sin(x)}{28} = \dfrac{\sin(38)}{20}$

4. $\dfrac{\sin(57)}{12} = \dfrac{\sin(49)}{x}$

Find the value of x in each of the following triangles.

5.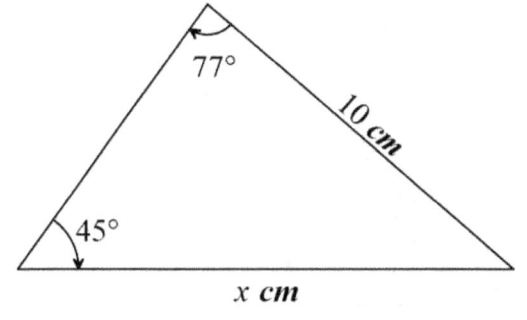

6.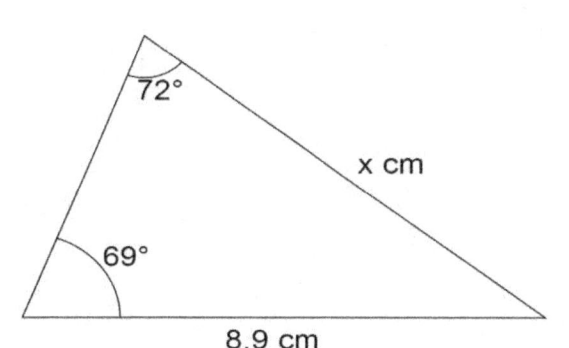

7.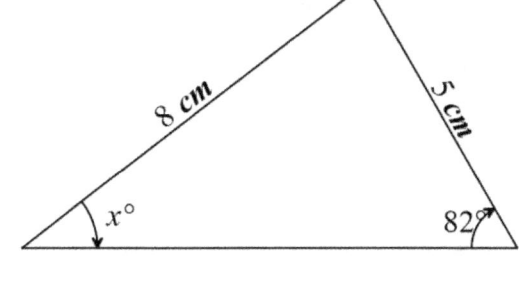

8.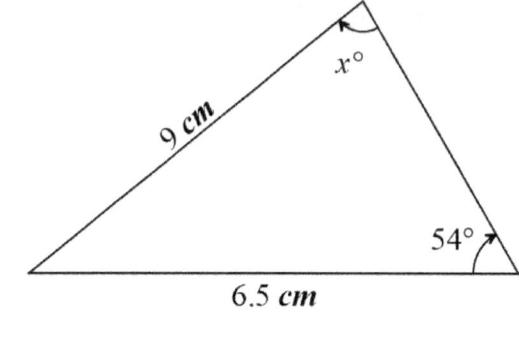

9.

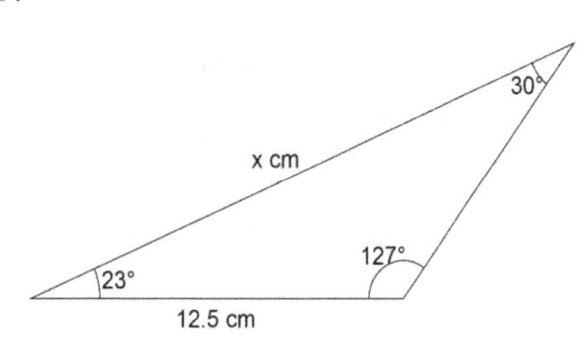

10.

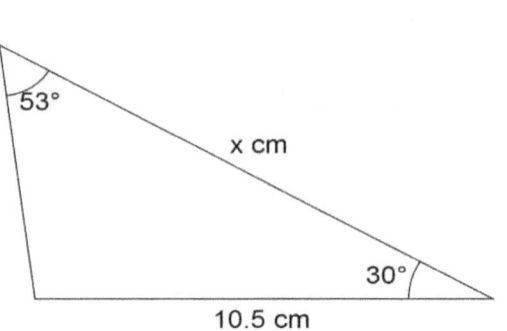

11. In △PTR, TR = 8 cm, ∠PTR = 150° and ∠TPR = 18°. Calculate the length of PT.

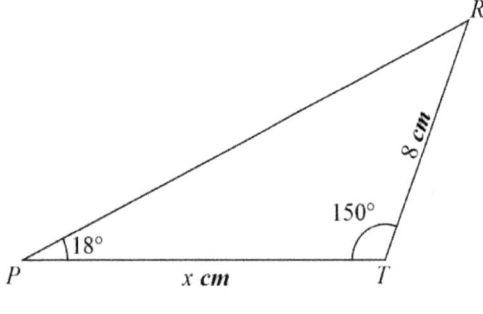

12.

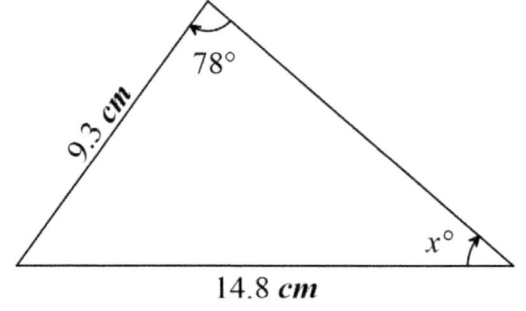

13. In △ADC, AC = 8 cm, AD = 9.6 cm and ∠ACD = 74°. Calculate ∠ADC.

CHAPTER 11 : TRIGONOMETRY FOR NON-RIGHT ANGLED TRIANGLES

11E   THE SINE RULE : NON CALCULATOR

In this part of the chapter, we are still going to use the sine rule to solve triangles but restrict the use of calculators. The examples below will definitely help the reader to understand how such problems can be tackled.

EXAMPLES

1.

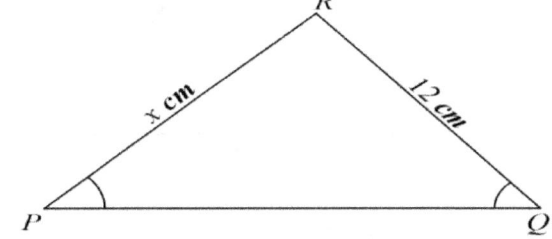

In $\triangle PQR$, $QR = 12$ cm, $sinQ = 0.3$ and $sin\angle P = 0.5$. Calculate the length of PR.

SOLUTION

$$\frac{\sin P}{12} = \frac{sinQ}{x}$$

$$\frac{0.5}{12} = \frac{0.3}{x}$$

Multiply both sides by 10

$$\frac{5}{12} = \frac{3}{x}$$

Cross multiply $5x = 36$

$$\therefore x = \frac{36}{5} = 7.2 \, cm$$

2.

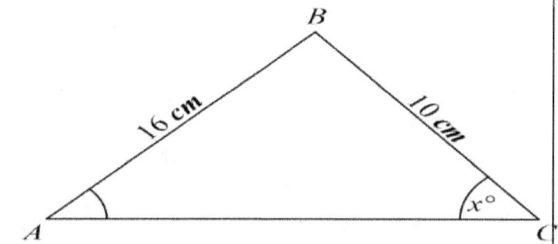

In the above triangle, AB = 16 cm, BC = 10 cm and sinA = 0.25. Calculate the value of sinC.

SOLUTION

$$\frac{\sin A}{10} = \frac{sinC}{16}$$

$$\frac{0.25}{10} = \frac{x}{16}$$

Cross multiply $\quad 10x = 4$

$\therefore x = 0.4$

$\sin C = 0.4$

EXERCISE 11E

1. In $\triangle ABC$ below, BC = 10 cm, sinA = 0.5 and sinB = 0.4. Calculate the length of AC.

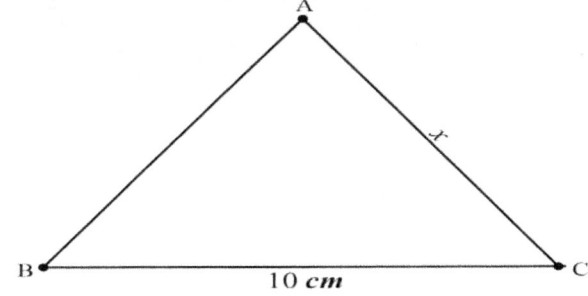

2. In $\triangle DEF$ below, DF = 8 cm, sinE = 0.2 and sinD = 0.5. Calculate the length of EF.

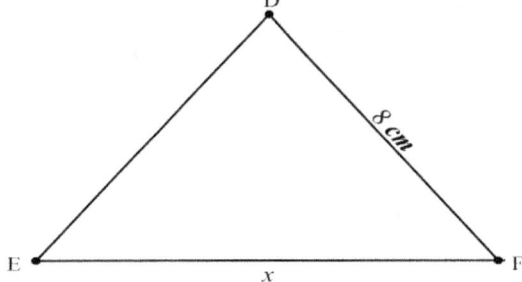

3. In △ABC, BC = 20 cm, sinA = 0.4 and sinB = 0.2. Calculate the length of AC.

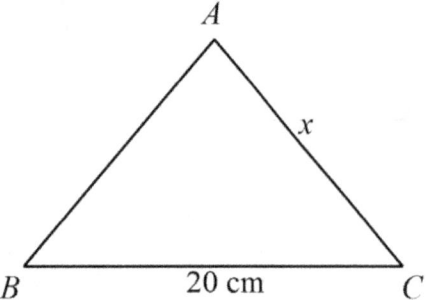

4. In triangle PQR, the length of the side PQ is 6 cm, sin R = 0.6 and sin Q = 0.5. Determine the exact length of the side PR.

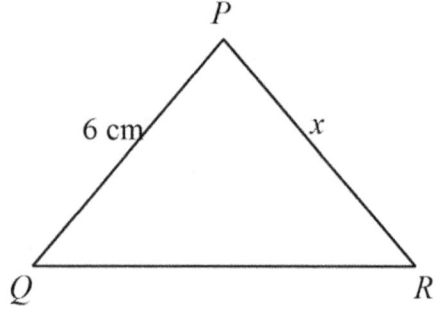

5. In △ABC, BC = 15 cm, sinA = 0.3 and sinB = 0.5. Calculate the length of AC.

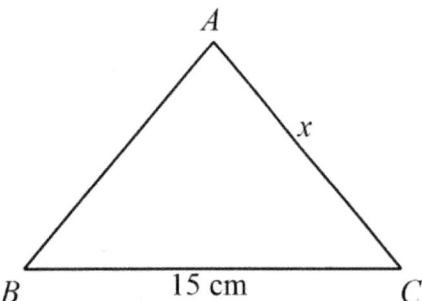

6. In triangle BCD, the length of the side BC is 16 cm, sin D = 0.8 and sin C = 0.4. Determine the exact length of the side BD.

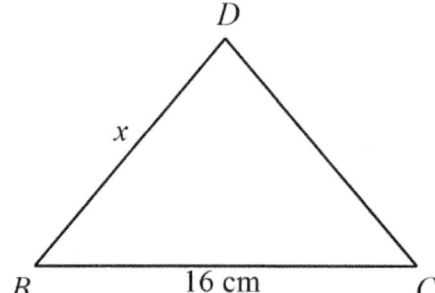

11F THE COSINE RULE (I) – FINDING SIDES

When do we use the cosine rule?

1. For non-right angled triangles
2. At most one angle involved.

The cosine rule is given by $a^2 = b^2 + c^2 - 2bc\, CosA$, where a is the side facing the angle.

EXAMPLES

1. Find the value of x.

SOLUTION

$x^2 = 10^2 + 12^2 - 2 \times 10 \times 12 \times \cos 53°$

Use the solve facility to find x as follows.

(Main, Action, Advanced, Solve)

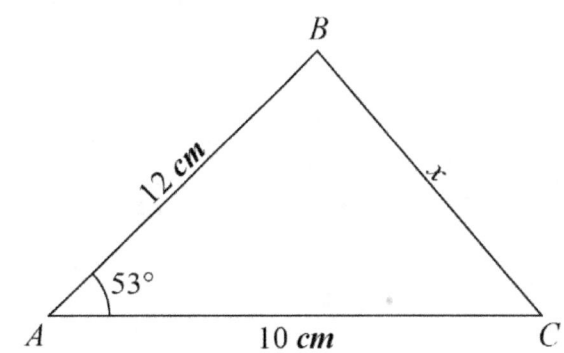

$$\text{solve}\,(x^2 = 10^2 + 12^2 - 2 \times 10 \times 12 \times \cos(53), x)$$

$$\{x = -9.978195951,$$
$$x = 9.978195951\}$$

We reject the negative value of x as the length of a triangle cannot be less than zero.
Hence $x = 9.98\ cm$

2. Find the value of x in the triangle ABC.

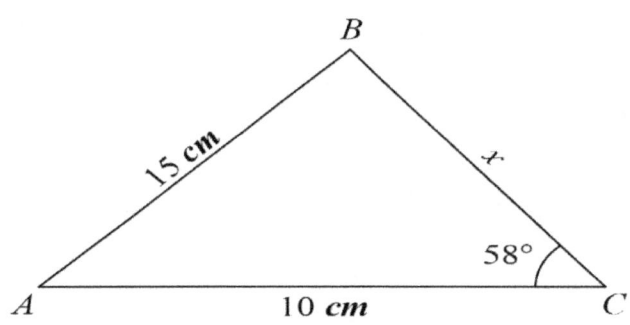

SOLUTION
Use the solve facility on your calculator,

$$\text{solve}\,(15^2 = x^2 + 10^2 - 2 \times 10 \times x \times Cos58, x)$$

$$(x = -7.073415921, x = 17.67180121\}$$

Here again, we reject the negative answer.

$x = 17.7\ cm$

MATHEMATICS APPLICATIONS UNIT 2

EXERCISE 11F

Find the value of x in each of the following using the solve facility in your calculator.

1. $x^2 = 8^2 + 9^2 - 2 \times 8 \times 9 \times \cos 65°$

2. $x^2 = 10^2 + 13^2 - 2 \times 10 \times 13 \times \cos 48°$

3. $x^2 = 7^2 + 6^2 - 2 \times 7 \times 6 \times \cos 125°$

4.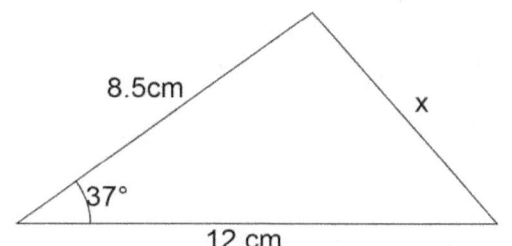

5.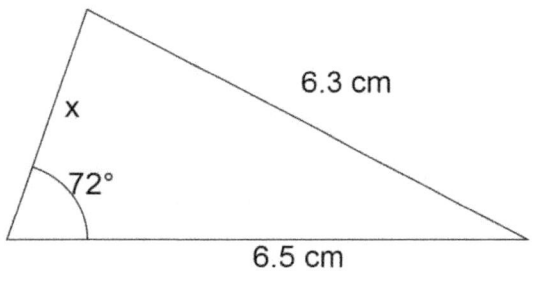

6.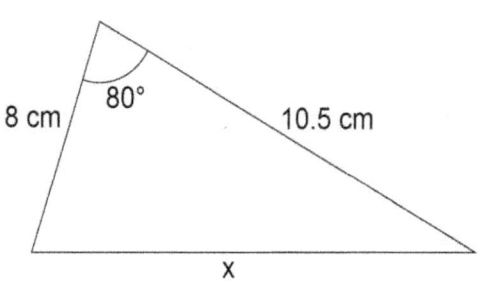

7.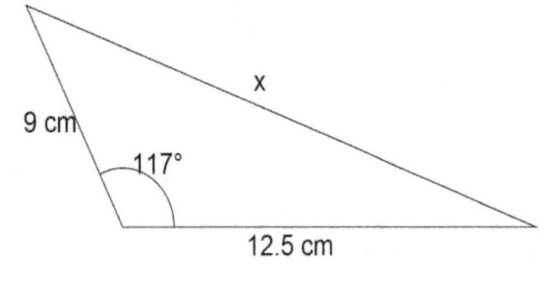

8.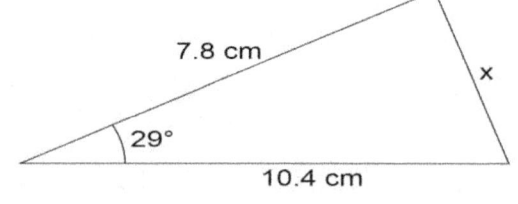

11G THE COSINE RULE (II) – finding angles

To find angles, using the cosine rule make use of the formula

$$\cos A = \frac{b^2 + c^2 - a^2}{2bc}, \text{ where } a \text{ is the side facing the angle.}$$

EXAMPLE

Find the value of x.

SOLUTION

$Solve\left(\cos x = \dfrac{8^2 + 13^2 - 10^2}{2 \times 8 \times 13}, x\right)$

$x = 50.3°$.

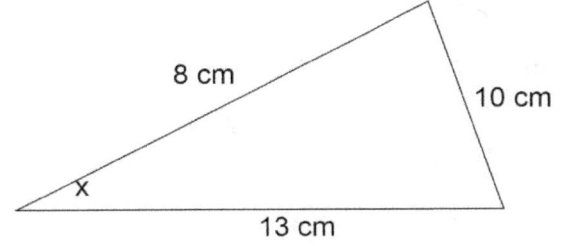

EXERCISE 11G

Find the value of x in each case.

1.
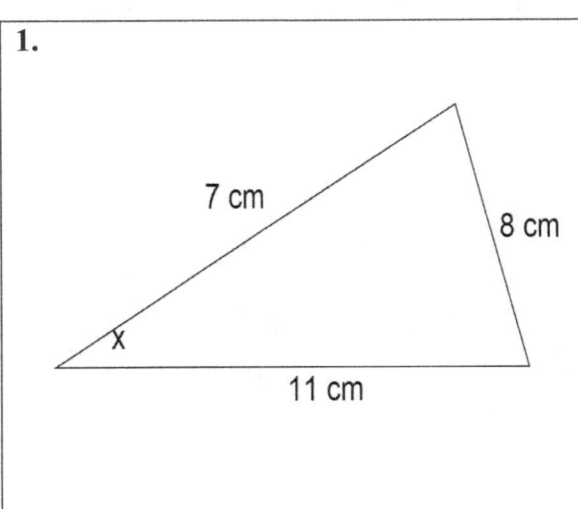

2.
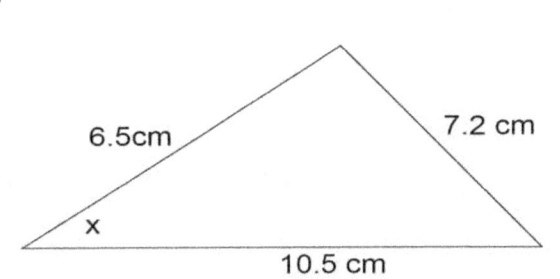

3.
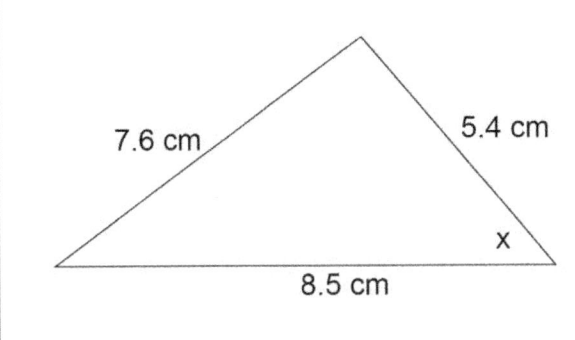

4.
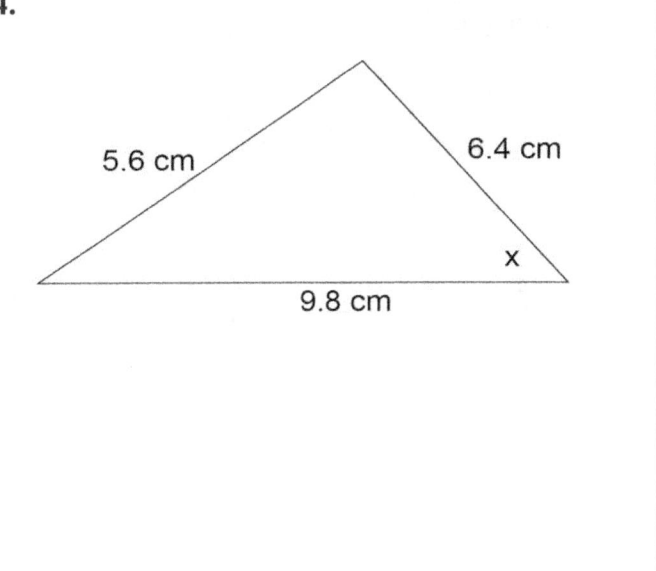

11H THE COSINE RULE (III) – finding the smallest and largest angle

- The smallest angle faces the smallest side.
- The largest angle faces the longest side

EXAMPLE

Find the size of the smallest angle in the given triangle.

SOLUTION

The smallest side being 6 cm, the smallest angle is angle C.

$$Solve\left(\cos C = \frac{7^2 + 9^2 - 6^2}{2 \times 7 \times 9}, C\right)$$
$$\therefore \angle C = 41.8°$$

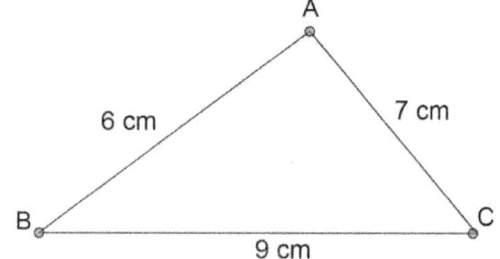

EXERCISE 11H

1. Find the size of the largest angle from the triangle.

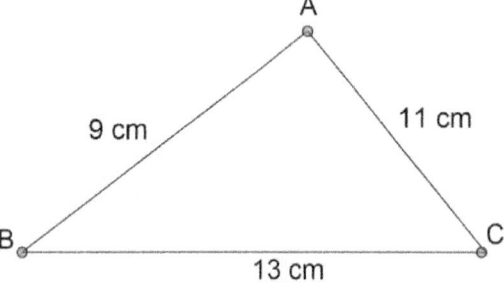

2. Find the size of the smallest angle in the given triangle.

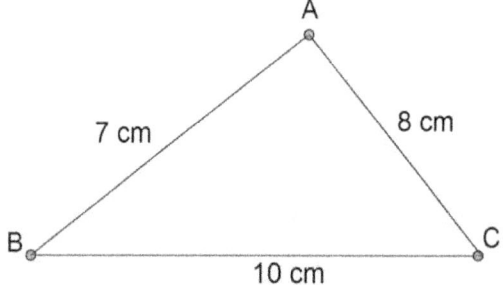

3. Find the size of the largest angle from the triangle.

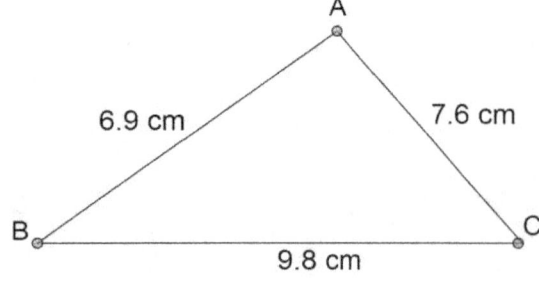

4. Find the size of the largest angle from the triangle.

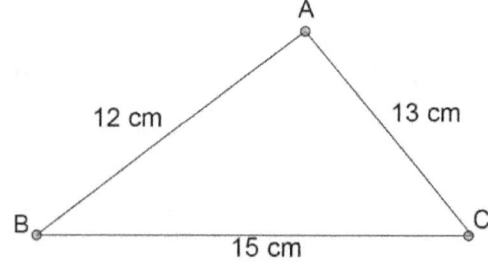

CHAPTER 11 : TRIGONOMETRY FOR NON-RIGHT ANGLED TRIANGLES

11 I   SINE RULE, COSINE RULE AND AREA APPLICATIONS

EXAMPLE

The diagram shows a farmer's block of land having the shape of a quadrilateral labelled ABCD. Given that AD = 63m, DC = 75m and BC = 81m. Also $\angle ADC = 56°$ and $\angle ABC = 47°$.

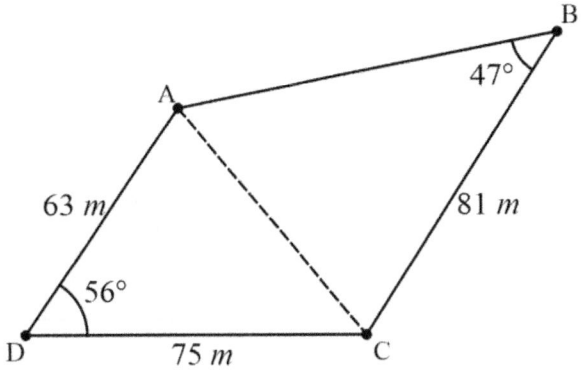

(a) Find the length of AC, correct to the nearest m.

$$AC^2 = 63^2 + 75^2 - 2 \times 63 \times 75 \times cos56 = 65.65 \, m$$

$$\therefore AC \approx 66 \, m$$

(b) Find the area of triangle ADC.

$$A = \frac{1}{2} \times 63 \times 75 \times sin56° = 1958.6 \, m^2$$

(c) Use the sine rule to determine the size of the acute angle BAC, correct to the nearest degree.

$$\frac{sin\angle BAC}{81} = \frac{sin\, 47}{65.65}$$

$$\angle BAC = 64.47° \approx 64°$$

(d) Find the area of the block ABCD.

$$\angle ACB = 180 - 47 - 64.47 = 68.53°$$

$$Area\ of\ \triangle ABC = \frac{1}{2} \times 81 \times 65.65 \times sin68.53° = 2474.3 \, m^2$$

Hence area of quadrilateral ABCD = 1958.6 + 2474.3 = 4432.9 m^2

EXERCISE 11 I

1. The diagram below (not drawn to scale) shows a school oval consisting of three walls AB, BC and AC.

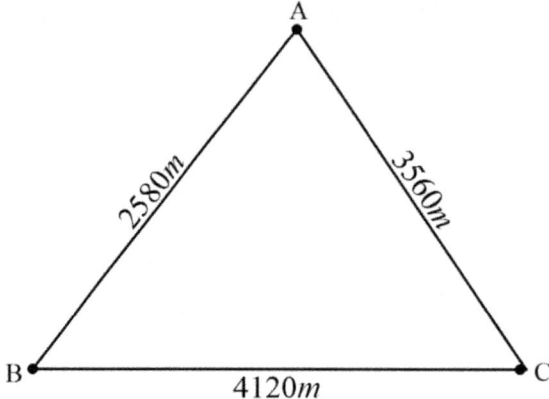

(a) Use trigonometry to determine the size of the angle BAC to the nearest degree.

(b) Use the sine rule to determine the size of the angle ABC.

(c) The section ABC needs to be covered with artificial lawn. The cost of the material is $48 per square metre. Determine the total cost of installing the lawn.

2. The diagram below (not drawn to scale) is a survey plan of a new industrial site land *ABCD*.

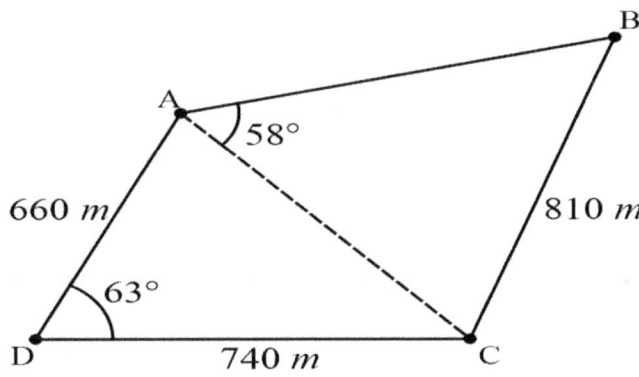

(a) To develop the site a road needs to be constructed along the line segment *AC*. Using trigonometry, calculate the length of this road to the nearest metre.

(b) Alpha Road Resurfacing Co Ltd has obtained the contract of constructing the road AC at a rate of $1450 per metre. What is the total cost of constructing the road?

(c) Using trigonometry, determine the size of ∠ABC.

(d) Using trigonometry, determine the area of the site ABCD.

3.

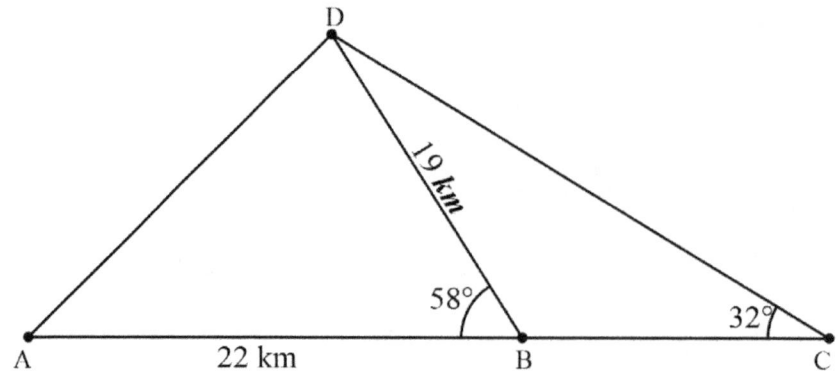

The diagram shows four towns A, B, C and D. ABC is a straight line ABC.

AB = 22 km, BD = 19 km, ∠ABD = 58° and ∠BCD = 32°. Calculate

(a) The length of BC,

A bridge connects towns A and D.
(b) Determine the length of the bridge AD.

(c) The area of triangle ABD,

(d) The bridge being too busy with the commuters, the government has decided to construct a road from Town B connecting to the bridge AD. Find the shortest distance from B to AD.

4. The Dexter's family is planning the front yard of their new house. The diagram below (not drawn to scale) shows the area.

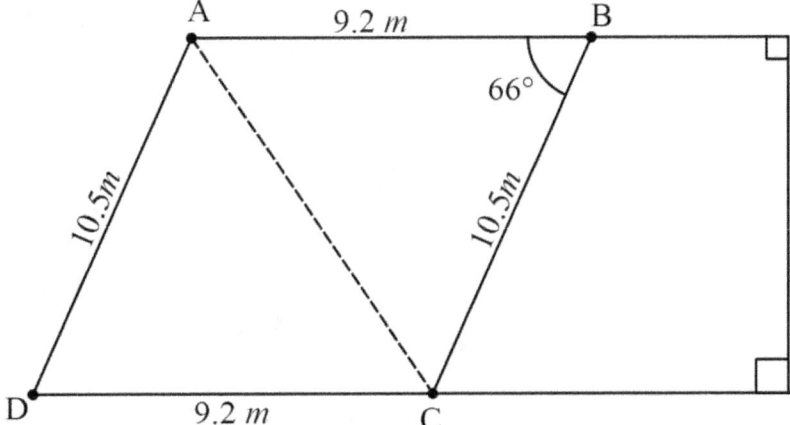

(i) (a) The Dexter's decide to build a limestone wall (one block high) from A to C to split the available space into two tiny little gardens.

(ii) Using trigonometry, calculate the length of this wall.

(iii) Limestone blocks come in 400mm lengths. How many blocks will the Dexter's need to buy?

(c) The playground area ACD is to be covered by sand. Using trigonometry, determine

(i) the size of the angle CAD.

(ii) the area of the playground ACD.

5. The diagram below shows a sketch of a block of land Lot 91. Find BOTH the **area** and **perimeter** of the block.

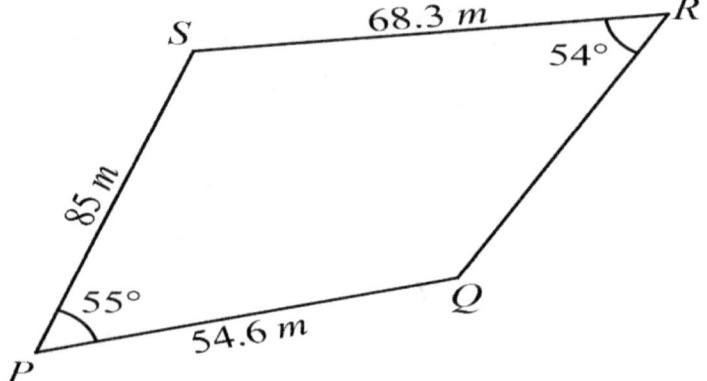

CHAPTER 12

SIMULTANEOUS LINEAR EQUATIONS

Simultaneous equations in this chapter are two equations, each containing two unknown letters. We have to use both equations to find the value of the unknown letters through different methods such as elimination, substitution or graphical. Skills needed to tackle this chapter are

- Solving linear equations
- Algebra rules
- Directed numbers

12A   ELIMINATION METHOD

In elimination method, we have to make the coefficients of one the two variables the same by multiplying by a scalar, if applicable.

❖ If the coefficients have the same sign, either positive or negative, we need to subtract one equation from the other.
❖ If the coefficients have the different signs, one positive and one negative, we need to add both equations

EXAMPLES
Solve the following pairs of simultaneous equations using the elimination method.

EXAMPLE 1	EXAMPLE 2
$4x + y = 10$ $2x + y = 4$ **SOLUTION** First we label both equations $4x + y = 10$ .... equation 1 $2x + y = 4$ .... equation 2 Clearly the coefficient of y is same in both equations. They both have the same sign, so we subtract to eliminate y. $\quad 4x + y = 10$ $\quad 2x + y = 4$ Subtracting we have, $\quad 2x = 6$ $\quad x = 3$ To find the value of y, substitute $x = 3$ in either equation Using $\quad 2x + y = 4$ $\quad\quad\quad 2(3) + y = 4$ $\quad\quad\quad 6 + y = 4$ $\quad\quad\quad y = -2$	$2x + 3y = 9$ $3x + y = 10$ **SOLUTION** First we label both equations $2x + 3y = 9$ .... equation 1 $3x + y = 10$ .... equation 2 ($\times 3$) Multiply the second equation by 3 to make the coefficient of y same as in equation 1. $\quad 2x + 3y = 9$ $\quad 9x + 3y = 30$ Subtracting we have, $\quad -7x = -21$ $\quad x = 3$ To find the value of y, substitute $x = 3$ in equation 2. (We can also use equation 1.) $\quad 3x + y = 10$ $\quad 3(3) + y = 10$ $\quad 9 + y = 10$ $\quad y = 1$

EXAMPLE 3	EXAMPLE 4
$$2a + b = 7$$ $$5a - 2b = 22$$ **SOLUTION** First we label both equations $2a + b = 7 \ldots\ldots$ equation 1 ($\times 2$) $5a - 2b = 22 \ldots\ldots$ equation 2 We can make the coefficients of b the same by multiplying equation 1 by 2. $4a + 2b = 14$ $5a - 2b = 22$ The signs are different, so we add to eliminate b Adding we have, $$9a = 36$$ $$a = 4$$ To find the value of b, substitute $a = 4$ in either equation Using $2a + b = 7$ $$2(4) + b = 7$$ $$8 + b = 7$$ $$b = -1$$ Hence $a = 4, b = -1$	$$3x + 4y = 23$$ $$2x + 3y = 16$$ **SOLUTION** First we label both equations $3x + 4y = 23 \ldots\ldots$ equation 1 ($\times 3$) $2x + 3y = 16 \ldots\ldots$ equation 2 ($\times 4$) In this case neither the coefficients of x are same nor the coefficient of y. We can choose to eliminate one of them, it is a personal choice. To eliminate x multiply equation 1 by 2 and equation 2 by 3 so that both become $6x$ Or To eliminate y multiply equation 1 by 3 and equation 2 by 4 so that both become $12y$. Say we want to eliminate y. $$9x + 12y = 69$$ $$8x + 12y = 64$$ Subtracting we have, $$x = 5$$ To find the value of y, substitute $x = 5$ in either equation Using $2x + 3y = 16$ $$2(5) + 3y = 16$$ $$10 + 3y = 16$$ $$3y = 6$$ $$\therefore y = 2$$

EXERCISE 12A

Solve the following pairs of simultaneous equations using the elimination method.

1. $$5x + y = 9$$ $$2x + y = 3$$	2. $$x + 2y = 11$$ $$x - y = -1$$

3.	$3x + y = 5$ $5x + 2y = 8$	4.	$x + 2y = 11$ $2x - y = -3$
5.	$x + 3y = 13$ $2x - y = -2$	6.	$2x + y = 9$ $4x + 3y = 17$
7.	$3x + 4y = 17$ $2x - 5y = 19$	8.	$2x + y = 7$ $4x - 3y = 19$

9.
$$3x + 2y = 11$$
$$2x + 5y = 0$$

10.
$$2x + 3y = 11$$
$$3x + 5y = 17$$

11.
$$4x + 5y = 17$$
$$2x + 3y = 10$$

12.
$$7x + 2y = 8$$
$$4x + 5y = -34$$

13.
$$2x + 3y = 10$$
$$5x + 4y = 11$$

14.
$$3x + 2y = 5$$
$$x + 2y = -3$$

12B SUBSTITUTION METHOD

The **substitution method** is more valid in case of a pair of simultaneous equations having two unknown and when it is possible to make one of the unknown the subject of the formula. It simply involves putting one of the equations into the other. We can also use the substitution method even if both equations of the linear system are in standard form. We can start by solving one of the equations for one of its variables as illustrated below.

Substitution method can be applied in four simple steps:

Step 1:
Solve one of the equations for either x or y as the subject of formula.

Step 2:
Substitute the solution from step 1 into the other equation.

Step 3:
Solve this new equation.

Step 4:
Solve for the second variable.

EXAMPLES

Solve the following simultaneous equations using the substitution method.

1.
$$y = 2x + 1$$
$$2x + 3y = 27$$

SOLUTION
First we label both equations
$y = 2x + 1$ .... equation 1
$2x + 3y = 27$ ... equation 2

From equation 1, we can see that y is already the subject of the formula
So we replace equation 1 in equation 2 and solve for x first
$$2x + 3(2x + 1) = 27$$
$$2x + 6x + 3 = 27$$
$$8x + 3 = 27$$
$$8x = 24$$
$$\therefore x = 3$$
To find y, replace $x = 3$ in equation 1
$$y = 2(3) + 1 = 7$$
So $\quad x = 3$ and $y = 7$.

2.
$$x - 3y = 8$$
$$2x + 5y = 5$$

SOLUTION
First we label both equations
$x - 3y = 8$ .... equation 1
$2x + 5y = 5$ .... equation 2

Rearrange equation 1 to make x the subject
$x = 3y + 8$ .... equation 3
$\quad$ Replace this in equation 2, we have
$$2(3y + 8) + 5y = 5$$
$$6y + 16 + 5y = 5$$
$$11y + 16 = 5$$
$$11y = -11$$
$$\therefore y = -1$$
To find x replace $y = -1$ in equation 3
$$x = 3(-1) + 8$$
$$x = 5$$
Hence $\quad x = 5, y = -1$

EXERCISE 12B

Solve the following simultaneous equations using the substitution method.

1.
$$y = 3x + 2$$
$$x + 2y = 11$$

2.
$$y = x - 4$$
$$5x + y = 32$$

3.
$$x = 3y + 2$$
$$2x + y = 18$$

4.
$$x = y - 3$$
$$2x + y = 9$$

5.
$$x + 3y = 13$$
$$y = 2x + 2$$

6.
$$x = 5y - 3$$
$$3x + y = 23$$

Solve the following simultaneous equations using the substitution method.

7.	$3x + y = 11$ $5x + 2y = 19$	8.	$x + 3y = 3$ $2x - 5y = 17$
9.	$x = 4y - 3$ $2x + y = 12$	10.	$y = 6 - 3x$ $2x + 3y = 18$
11.	$x - 2y = 3$ $2x + 3y = 20$	12.	$y = 10 - 4x$ $2x + 3y = 0$

12C SIMULATNEOUS EQUATIONS : USING SOLVE CAPACITY

To solve a pair of simultaneous equations on CAS, make use of the following steps:

- Main
- Keyboard
- 2D (skip this step for latest CAS)
- Select the symbol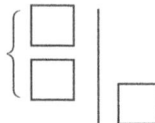
- The following will appear on your calculator
- Insert the 1st equation in the first box, the 2nd equation in the second box and x,y in the third box.
- Press EXE and the answer will appear as shown below.

Note that most graphic calculators have similar functions to solve simultaneous equations.

EXAMPLE
Use the solve facility on your calculator to solve the simultaneous equations
$$x + y = 50$$
$$2x + 3y = 120$$

$$\begin{cases} x + y = 50 \\ 2x + 3y = 120 \end{cases} \bigg| x, y$$

$$\{x = 30, y = 20\}$$

EXERCISE 12C

Use the solve facility on your calculator to solve the following simultaneous equations.

1. $y = 2x + 5$ and $y = 3x - 1$	2. $3x - y = 10$ and $2x + 5y = 1$
3. $y = x + 4$ and $y = 4x - 8$	4. $4x + 3y = 11$ and $5x - y = 9$
5. $x + y = 10$ and $2x - y = 8$	6. $x + y = 9$ and $2x - y = 6$

12D USING GRAPHICAL FACILITY ON CALCULATOR

The next couple of examples will demonstrate how we can make use of technology to solve a pair of simultaneous equations by plotting both graphs and finding the coordinates of their point of intersection. Being straight lines obviously there will be only one point of intersection thereby only one value of x and one value of y. This method is slightly harder than the previous section as it requires some mathematical skills of re-arranging and making y the subject of the formulae each time.

Use the following steps on your calculator:

**MENU → GRAPH &TABLE →
INSERT BOTH EQUATIONS →
TICK THE BOX →
CLICK ON THE GRAPH ICON
          (1ST ON TOP LEFT)**

EXAMPLES

Use the graphical facility on your calculator to solve the following pairs of simultaneous equations.

(a) $y = 2x + 5$
    $x + y = 11$ ....(2)

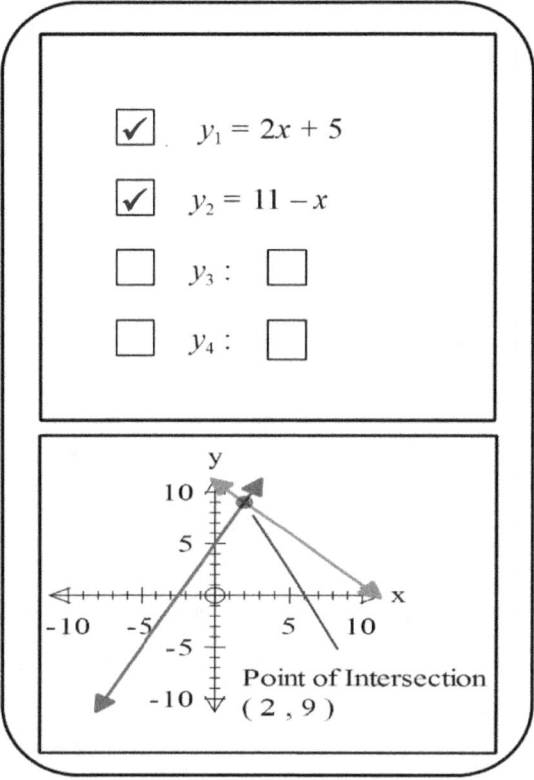

Note that for equation (2), we have to make y the subject of the formula in order to be able to insert $11 - x$ in the required box.

As we can see from the graph, the point of intersection is (2,9).

Hence $x = 2$ and $y = 9$.

(b) $y = 3x - 1$
    $2x + 3y = 19$ ....(2)

Again, we have to re-arrange equation (2) making y the subject of the formula as shown below.
$$2x + 3y = 19$$
$$3y = 19 - 2x$$
$$\therefore y = \frac{1}{3}(19 - 2x)$$

The point of intersection being (2,5)
Hence $x = 2$ and $y = 5$.

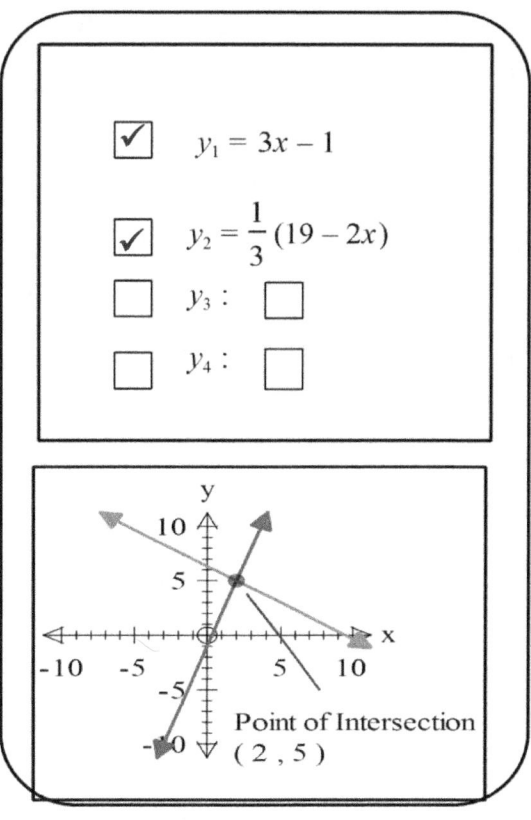

EXERCISE 12D

Using the graphical facility on your calculator solve the following simultaneous equations.

1. $x + 2y = 11$
   $x - y = -1$

2. $2x + 5y = -15$
   $y = x + 4$

3. $2x + 3y = 12$
   $y = x - 6$

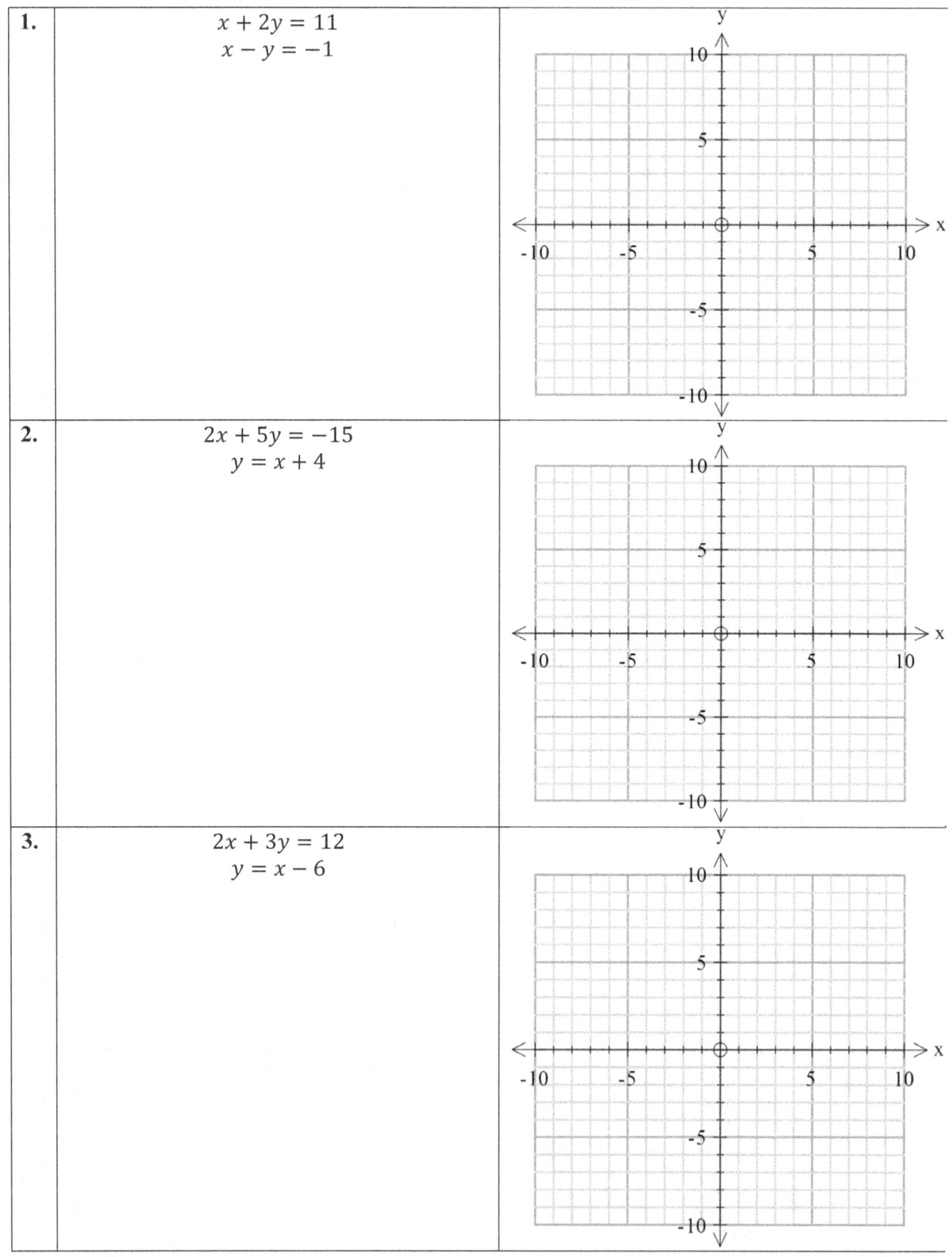

CHAPTER 12 : SIMULTANEOUS LINEAR EQUATIONS

12E   THE GRAPHICAL METHOD STEP BY STEP APPROACH

Simultaneous equations can also be solved graphically. We have to graph both lines on the same set of axes and the solution is given by the coordinates of the point of intersection of the two lines.

It is definitely not the best method to solve a pair of simultaneous equations as it is often hard to graph the lines accurately. Furthermore, reading the point of intersection off the graph is sometimes difficult when the solutions are not whole numbers.
Nevertheless, the graphical method remains a very useful tool for solving simultaneous equations.

CLASS ACTIVITY
State the solution in each of the following

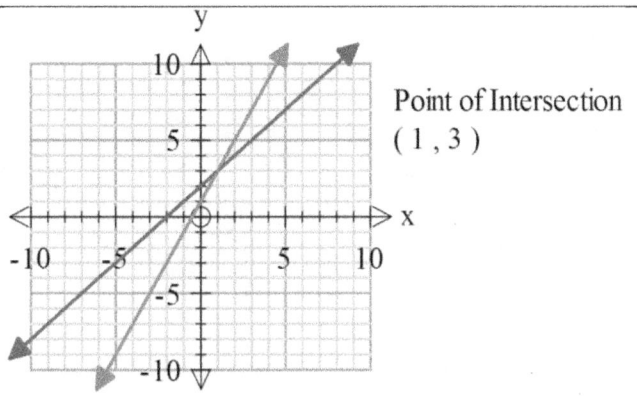

Point of Intersection (1 , 3)

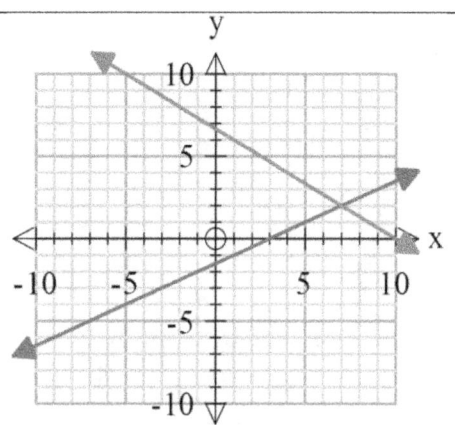

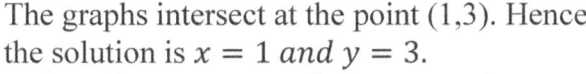

The graphs intersect at the point (1,3). Hence the solution is $x = 1$ and $y = 3$.

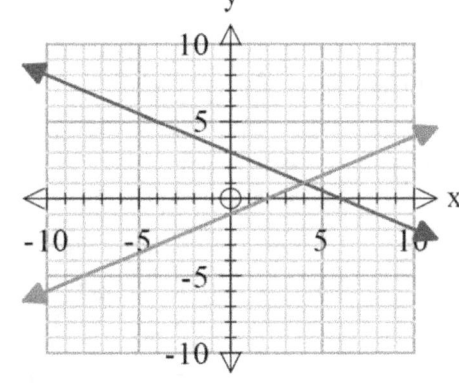

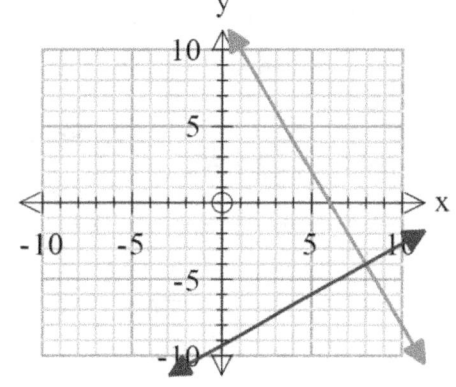

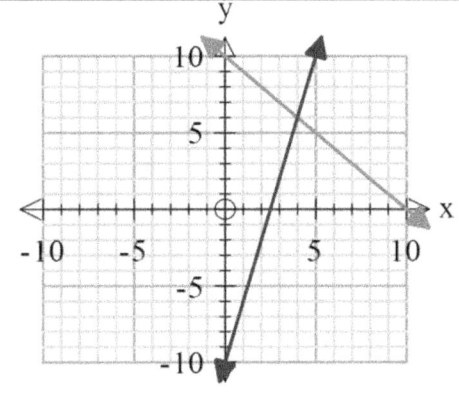

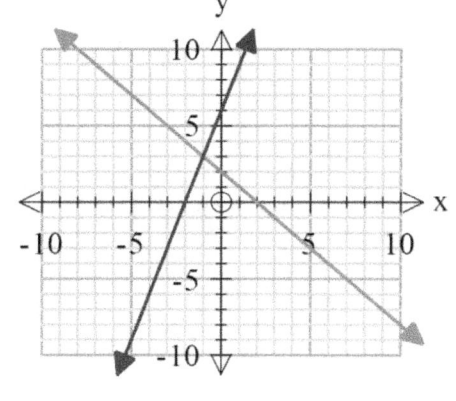

EXAMPLES

Solve these simultaneous equations by using the graphical method.

1.
$$2x + y = 10$$
$$x + y = 7$$

SOLUTION
For example, to draw the line $2x + y = 10$ pick two easy numbers to plot. One when $x = 0$ and one where $y = 0$.

When $x = 0$ in the equation $2x + y = 10$
This means $y = 10$
So one point on the line is (0, 10)
When $y = 0$
$2x = 10$ so $x = 5$
So another point on the line is (5, 0)
Similarly, the line $x + y = 7$ crosses the axes at (0,7) and (7,0).
The graphs intersect at the point (3,4)
$$\therefore x = 3, y = 4$$

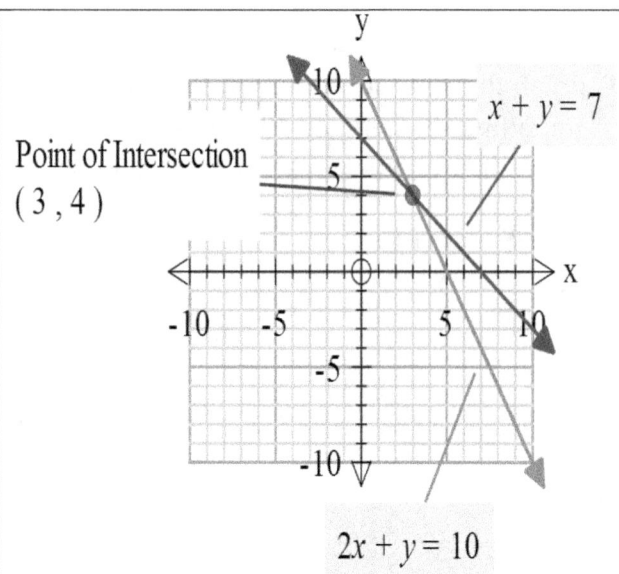

Point of Intersection (3, 4)

2.
$$y = x + 3$$
$$2x + y = 9$$

SOLUTION
To draw the line $y = x + 3$, we can construct a table of values as shown

x	0	1	2
y	3	4	5

We can then plot the pairs of values to draw the line $y = x + 3$.
To draw the line $2x + y = 9$, we can use the intercept method as shown in example 1 above.
When $x = 0, y = 9$
So one point on the line is (0, 9)
When $y = 0$
$2x = 9$ so $x = 4.5$
So another point on the line is (4.5, 0)
The graphs intersect at the point (2,5)
$$\therefore x = 2, y = 5$$

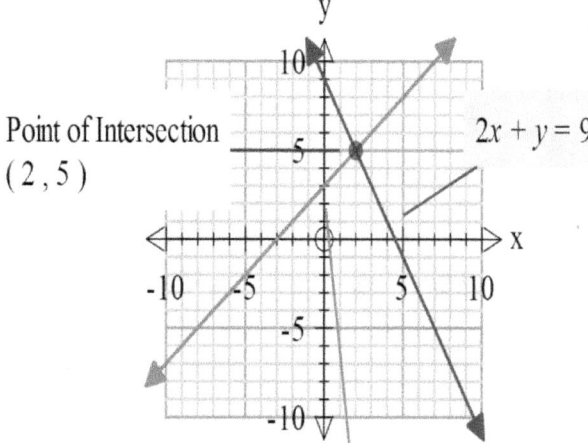

Point of Intersection (2, 5)

EXERCISE 12E

Solve these simultaneous equations by using the graphical method. Check your answer using CAS.

1.	$x - 2y = 10$ $y = x - 5$	
2.	$2x - y = 6$ $y = x - 4$	
3.	$4x + y = 8$ $y = x - 7$	

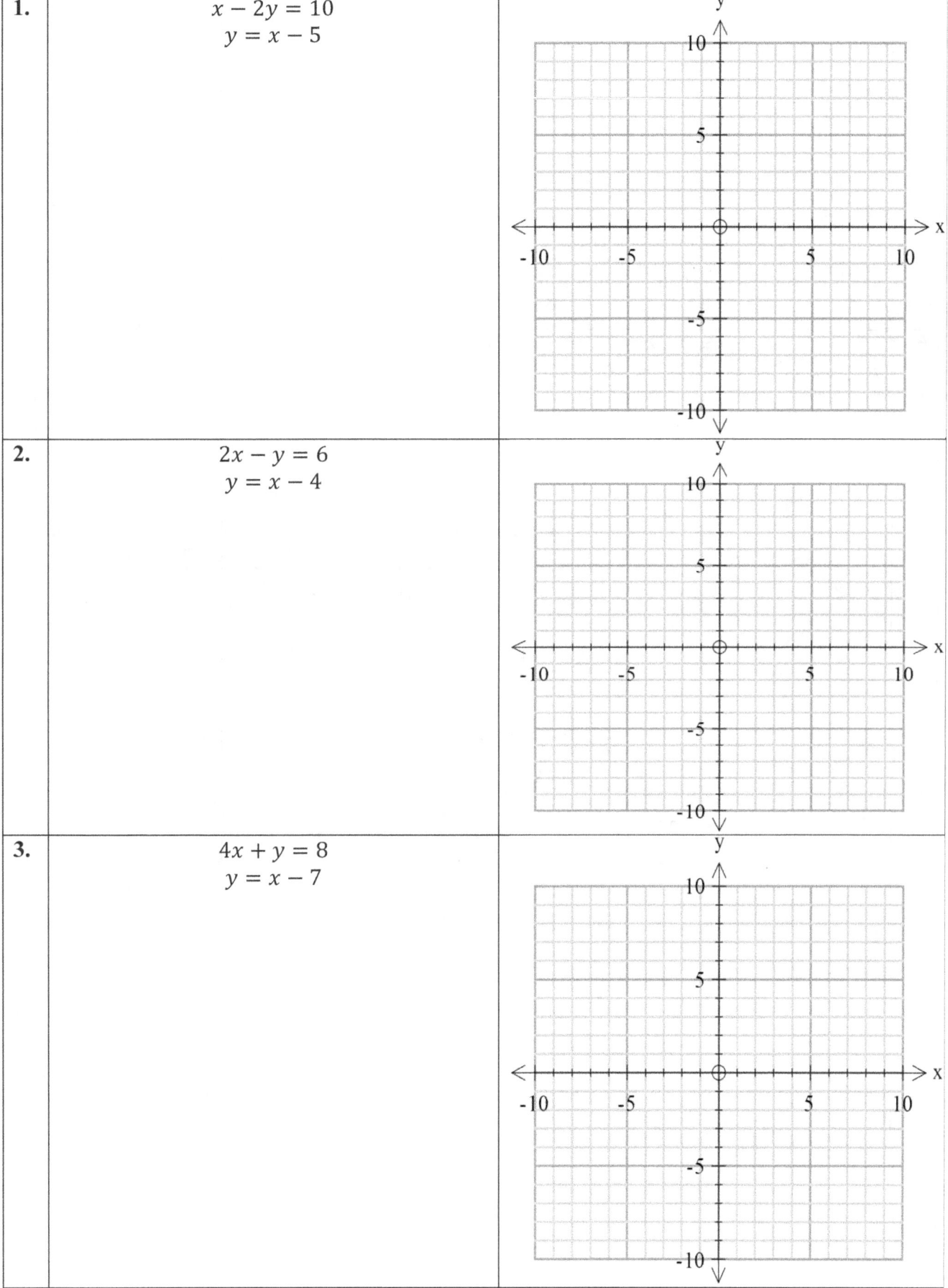

4.

$$y = -2x + 6$$
$$x + 2y = 6$$

5.

$$y = -2x + 7$$
$$3x + 2y = 10$$

6.

$$y = 2x + 4$$
$$x + y = 7$$

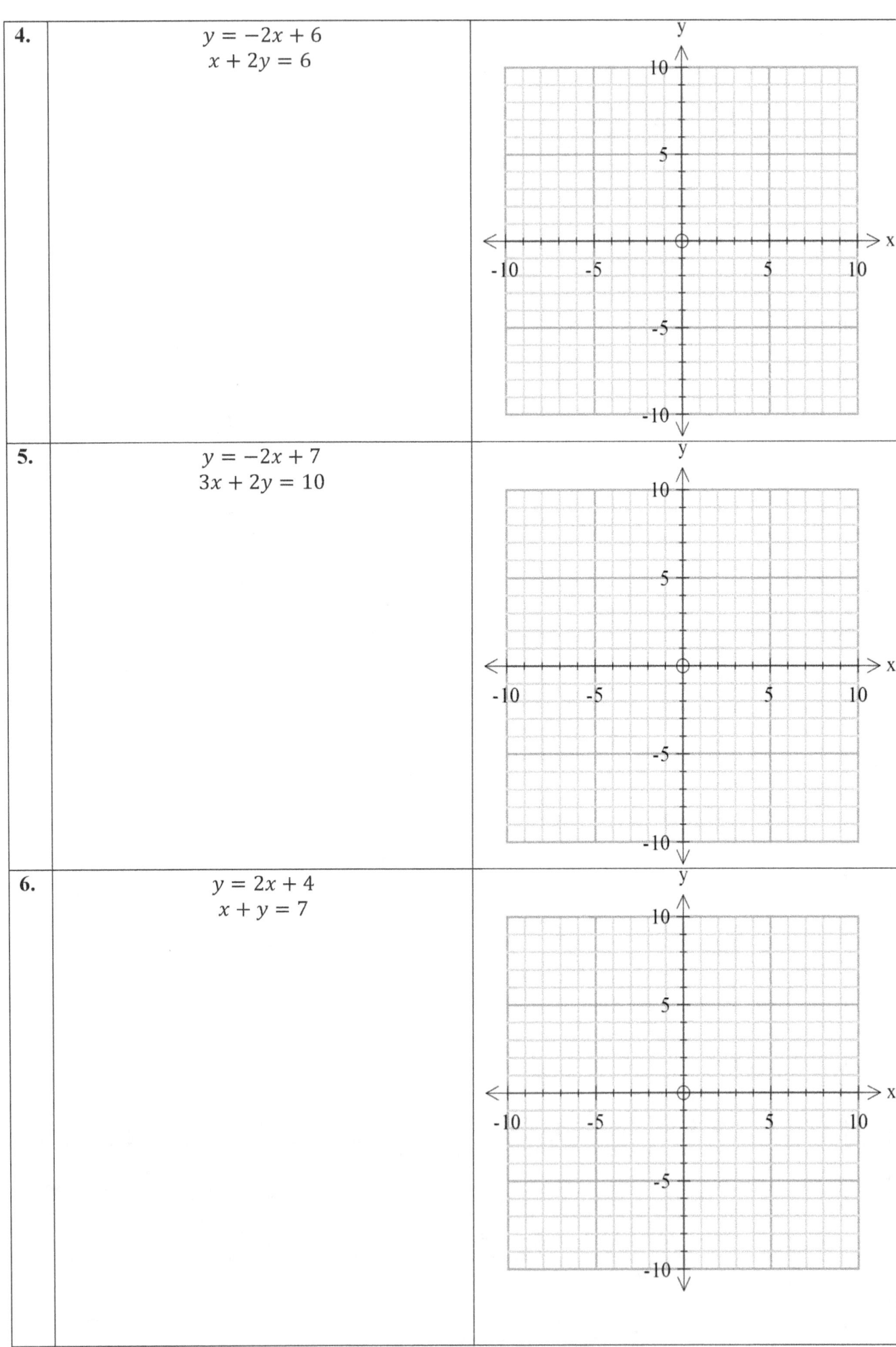

12F APPLICATIONS

EXAMPLES

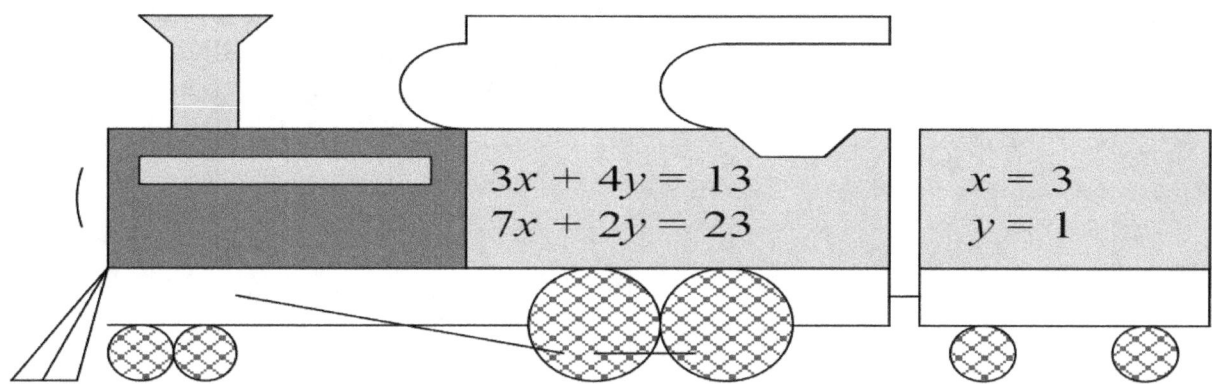

1. The admission fee at the Royal Show is $3.50 for children and $5 for adults. On a certain day, 2200 people enter the fair and $8750 is collected. Two simultaneous equations can be written from this information.
   One of the equation is $3.5x + 5y = 8750$

   (a) Explain clearly what the term $5y$ represents in this situation.

   Answer : money raised from adults admission fees

   (b) Write down the second equation.

   Answer : $x + y = 2200$

   (c) Solve the pair of equations to determine the number of children and the number of adults who attended the show.

   $$\begin{array}{l} 3.5x + 5y = 8750 \\ x + y = 2200 \end{array} \bigg| \, x, y$$

   $$\{x = 1500, y = 700\}$$

   Therefore 1500 children and 700 adults attended the show.

2. Ten years ago, Alex was 12 times as old as Tim and in ten years' time, Alex will be twice as old as Tim. Find their present ages.
   Solution
   Let Alex's present age be x.
   Let Tim's present age be y.
   Ten years ago, Alex was $(x - 10)$ years old and Tim was $(y - 10)$ years old.
   Since Alex was 12 times as old as Tim, we have
   $$x - 10 = 12(y - 10)$$
   $$x - 10 = 12y - 120$$
   $$x = 12y - 110 \; ...... equation \; (1)$$
   Also in ten years' time
   Alex will be $(x + 10)$ years old and Tim was $(y + 10)$ years old.
   As Alex will be twice as old as Tim,
   $$x + 10 = 2(y + 10)$$
   $$x + 10 = 2y + 20$$
   $$x = 2y + 10 \; ...... equation \; (2)$$
   Solving equations (1) and (2) on CAS, we have

   $$\begin{array}{l} x = 12y - 110 \\ x = 2y + 10 \end{array} \bigg| \, x, y$$

   $$\{x = 34, y = 12\}$$

   Hence Alex is 34 years old and Tim is 12 years old currently.

EXERCISE 12F

1. Two numbers x and y are such that the sum of 2 numbers is 23 and their difference is 3. Write a pair of equations and use any appropriate method to find the 2 numbers.

2. Seven footballs (x) and three soccer balls (y) cost a total of $314. Four footballs and five soccer balls cost a total of $255. Write down a pair of simultaneous equations involving x and y. Hence find the cost of each football and each soccer ball.

3. The length of a rectangle (y) is 5cm longer than its width (x). The perimeter of the rectangle is 54 cm.

    (a) Write two equations that connect width (x) and length (y).

    (b) Solve the equations using substitution and so state the **_length_** and **_width_** of the rectangle.

4. A car travels for x hours at 60 km/h and then travels for y hours at 80 km/h. If it has travelled for 8 hours and covered a total distance of 540 km, find x and y.

5. A rectangle has a perimeter of 52 cm while the difference between the length and the width is 5 cm. Find the length and the width.

6. A piggy bank contains 40 coins, all of them are either 5-cents coins or 20-cent coins. If the value of the coins in the piggy bank is $6.50, find the number of each kind of coin.

7. Alisha has 25 coins in her purse, consisting of 50-cent coins and 20-cent coins, which total $10.40. How many of each does she have?

8. At the Perth arena 1000 tickets were sold during a concert. Adult tickets cost $8.50, children's ticket cost $4.50, and a total of $7300 was collected. How many tickets of each kind were sold?

9. Anna is x years old and Bob is y years old. Last year, Bob was 6 times as old as Anna.

(i) Form an equation in x and y and show that it simplifies to $y = 6x - 5$.

(ii) In 19 years' time, Bob will be twice as old as Anna.
Form another equation in x and y and show that it simplifies to $y = 2x + 19$.

(iii) Hence find the present ages of Anna and Bob.

10. Mc Café charges $2.50 for a cup of tea and $3.25 for a cup of coffee. One morning the café sold 80 cups of tea and coffee altogether, and charged $245 in total. Two simultaneous equations can be written from this information.
One of the equation is
$2.5x + 3.25y = 245$.

(a) Explain clearly what the term $2.5x$ represents in this situation.

(b) Write down the second equation.

(c) Solve the pair of equations to determine the number of cups of each type of drink sold.

CHAPTER 13

THE NORMAL DISTRIBUTION

A continuous random variable X has a normal distribution if it has a p.d.f

$$f(x) = \frac{1}{\sigma\sqrt{2\pi}} e^{-\frac{(x-\mu)^2}{2\sigma^2}}$$

The graph of f(x) is given below.

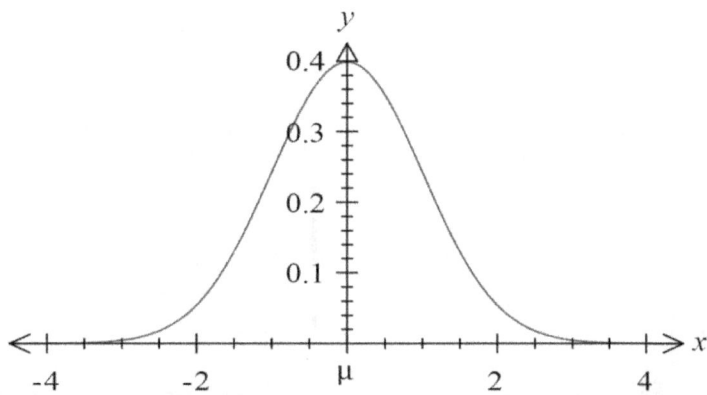

From this graph we can deduce some of the important properties of the normal distribution.

- ❖ The distribution is symmetrical about the mean μ.
- ❖ The mean, mode and median coincide and are equal due to symmetry.
- ❖ The domain of the function is $-\infty < x < \infty$
- ❖ The horizontal axis is an asymptote as $x \to -\infty$ and $x \to +\infty$.
- ❖ Area under the curve is 1.

From the p.d.f above, we can see that the probability distribution of X depends only on μ and σ. Hence instead of remembering the formula it is sufficient to refer to the random variable X as having a normal distribution by using the notation

$$X \sim N(\mu, \sigma^2)$$

CLASS ACTIVITY

Complete the following table.

	Mean μ	Standard deviation σ		Mean μ	Standard deviation σ
$X \sim N(50, 5^2)$	50	5	$X \sim N(65, 7^2)$		
$X \sim N(42, 3^2)$			$X \sim N(35, 10)$		
$X \sim N(20, 6^2)$			$X \sim N(100, 9^2)$		
$X \sim N(30, 25)$			$X \sim N(80, 10^2)$		

13A USING TECHNOLOGY

To solve problems in normal distribution, the use of technology is very important as in old days we use to make use of tables and it was really time consuming. Make use of the following steps on your calculator to determine probabilities in normal distribution questions.

- Menu
- Statistics
- Calc
- Distribution
- Normal CD
- Next
- Insert the values of μ, σ, lower and upper to obtain the answer.

EXAMPLES

1. If $X \sim N(50, 5^2)$, determine $P(X > 54)$.
   SOLUTION

   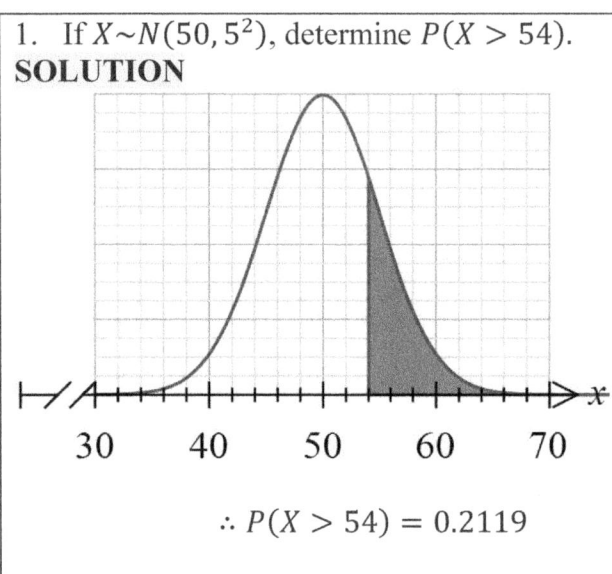

Lower	54
Upper	∞
σ	5
μ	50

Prob	0.2118554
Z Low	0.8
Z Up	2E+998
σ	5
μ	50

   $\therefore P(X > 54) = 0.2119$

2. The length of rods in a large batch is normally distributed with mean 120 mm and standard deviation 1.5 mm. What percentage of rods would you expect to measure between 116 mm and 124 mm?
   SOLUTION

   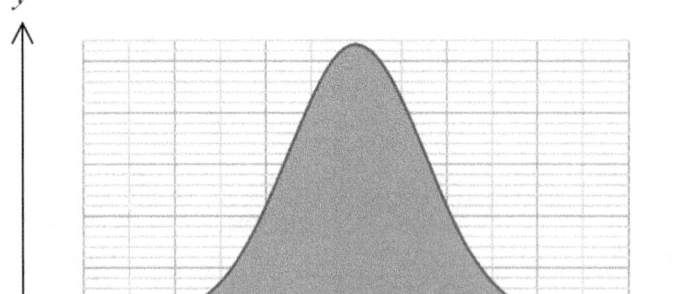

   $$P(116 < X < 124) = 0.9923$$

   Hence 99.23% of the rods measure between 116 mm and 124 mm.

EXERCISE 13A

Use technology to answer the following questions.

1. If $X \sim N(56, 10^2)$, determine $P(X < 66)$.	2. If $X \sim N(28, 3^2)$, determine $P(X > 36)$.
3. If $X \sim N(100, 25)$, determine $P(X > 100)$	4. If $X \sim N(60, 4^2)$, find $P(56 < X < 64)$.
5. If $X \sim N(-5, 9)$, determine $P(X > 0)$	6. If $X \sim N(12, 2^2)$, determine $P(9 < X < 13)$
7. The height of girls at a particular age follows a normal distribution with mean 130 cm and standard deviation 3 cm. Find the probability that a girl picked up at random from this age group has a height (a) Less than 134 cm (b) Between 131 cm and 133 cm.	8. A certain type of vegetable has a mass which is normally distributed with mean 2 kg and standard deviation 0.25 kg. In a lorry load of 600 of these vegetables, estimate how many will have a mass greater than 2.1 kg.
9. The number of hours of the life of a torch battery is normally distributed with mean 120 hours and a standard deviation of 16 hours. Find the probability that a troch battery has a life of (a) More than 140 hours (b) Between 110 and 128 hours.	10. The number of marks of 1000 candidates in an examination is normally distributed with a mean of 58 marks and a standard deviation of 10 marks. Given that the pass mark is 50 marks, estimate the number of candidates who pass the examination.
11. A company packing spices knows that the weight of 500 packets form a normal distribution with a mean weight of 16 grams and a standard deviation of 0.2 gram. How many of these 500 packets are expected to weigh less than 15.6 grams?	12. The masses of tablets of chocolates produced by a certain machine are found to be normally distributed with a mean of 140g and a standard deviation of 5g. Estimate the number of tablets in a batch of 200 whose masses are greater than 145g.

13B THE 68%, 95% AND THE 99.7% RULE

For continuous random variables, taking a large number of measurements and analysing the results usually give rise to a normal curve as shown below.

Given a normal distribution, 68% of the scores lie within 1 standard deviation either side of the mean as shown on the right. Also, 1 standard deviation above or below the mean accounts for 34% of the area.	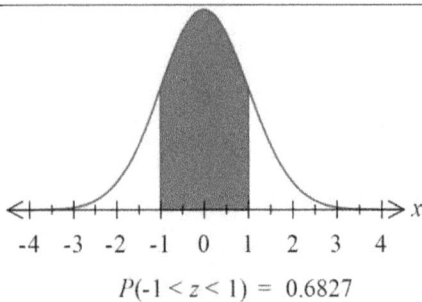 $P(-1 < z < 1) = 0.6827$	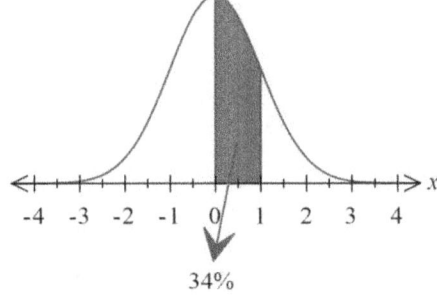 34%
Similarly, the probability of a randomly selected score from a normally distributed population to lie within 2 standard deviations either side of the mean is 95%. Likewise, 47.5% of scores lie 2 standard deviations from the mean.	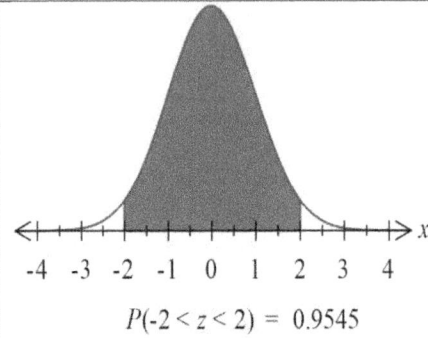 $P(-2 < z < 2) = 0.9545$	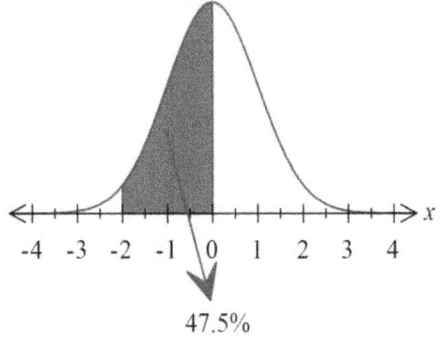 47.5%
As seen from the graph on the right 99.7% of scores lie within 3 standard deviations either side of the mean.	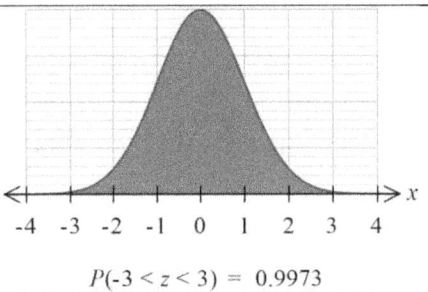 $P(-3 < z < 3) = 0.9973$	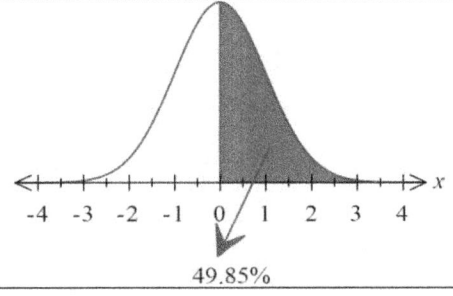 49.85%

EXAMPLES

1. If $X \sim N(26, 4^2)$, without using a calculator determine $P(22 < X < 30)$.

SOLUTION

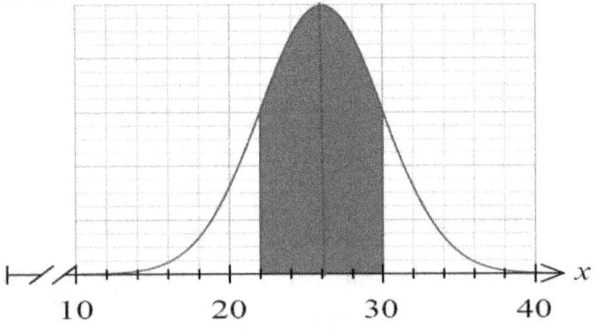

Since the scores lie within 1 standard deviation of the mean,
$$P(22 < X < 30) = 0.68$$

2. If $X \sim N(40, 5^2)$, without using a calculator determine $P(30 < X < 45)$

SOLUTION

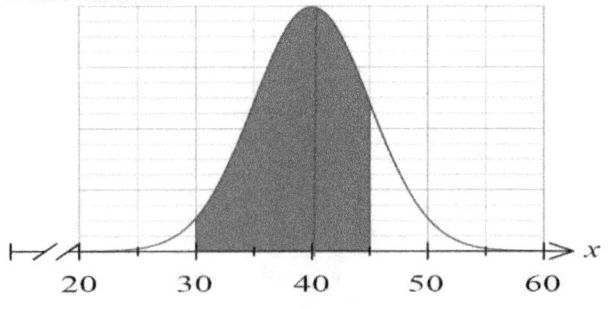

Since the scores lie 2 standard deviations below the mean and 1 standard deviation above the mean,
$$P(30 < X < 45) = 0.5 \times 0.95 + 0.5 \times 0.68$$
$$= 0.815$$

EXERCISE 13B

1. If $X \sim N(30, 5^2)$, without using a calculator determine $P(25 < X < 35)$

2. If $X \sim N(40, 3^2)$, without using a calculator determine $P(34 < X < 46)$

3. If $X \sim N(18, 2^2)$, without using a calculator determine $P(16 < X < 20)$

4. If $X \sim N(50, 4^2)$, without using a calculator determine $P(X > 50)$

5. If $X \sim N(30, 2^2)$, without using a calculator determine $P(24 < X < 36)$

6. If $X \sim N(60, 4^2)$, without using a calculator determine $P(56 < X < 68)$

7. If $X \sim N(40, 5^2)$, without using a calculator determine $P(X < 45)$

8. If $X \sim N(20, 5^2)$, without using a calculator determine $P(X > 15)$

9. If $X \sim N(25, 6^2)$, without using a calculator determine $P(X < 31)$

10. If $X \sim N(20, 5^2)$, without using a calculator determine $P(X > 10)$

11. After extensive testing, it was found that the lifetimes of Power bulbs had a mean of 2500 hours and a standard deviation of 100 hours. Assuming that the lifetime of a bulb is modelled by a normal distribution, find
    (a) the probability that a Power bulb will have a lifetime between 2400 and 2600 hours.

    (b) The probability that a Power bulb has a lifetime exceeding 2700 hours.

12. The random variable X has a normal distribution with mean 6 and standard deviation 1.5. Calculate
    (a) $P(X < 6)$

    (b) $P(4.5 < X < 9)$

13. The weights of oranges in a supermarket shipment are normally distributed with a mean of 160 g and a standard deviation of 20 g. The distribution is such that 68%, 95% and 99.7% of the oranges have weights within one, two and three standard deviations from the mean respectively.

    Determine the probability that a randomly chosen orange from the supermarket

    (i)   weighs between 120 g and 200 g.

    (ii) weighs more than 160 g.

    (iii) weighs exactly 100 g.

    (iv) weighs between 140 g and 160 g.

14. At a hardware store, the lengths of a large number of wooden rods marked as 2 m long, were actually normally distributed with a mean of 202 cm and a standard deviation of 3 cm.

    (i)   State the median length of the wooden rods.

    (ii) Find the probability that the length of a randomly chosen plank is between 199 cm and 205 cm.

    (iii) Find the probability that the length of a randomly chosen plank is less than 1.99 m.

CHAPTER 13: THE NORMAL DISTRIBUTION

13C   INVERSE NORMAL DISTRIBUTION

In this section, we are going to make use of the capabilities of our calculators to use inverse normal distribution to determine unknown quantities.

Make use of the following steps on your calculator to achieve your goal.

- ❖ Menu
- ❖ Statistics
- ❖ Calc
- ❖ Inv. Distribution
- ❖ Inverse Normal CD
- ❖ Next
- ❖ Choose the correct tail setting and insert the values of μ, σ and the given probability to obtain the answer.
- ❖ Next

EXAMPLES

1. Determine the value of k below.

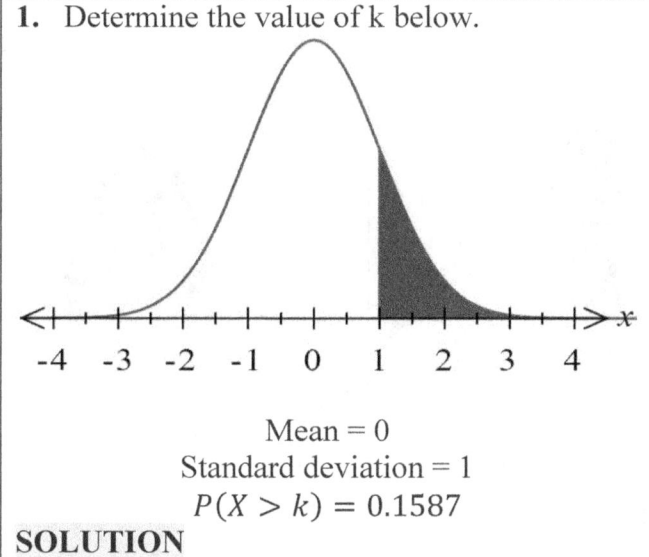

Mean = 0
Standard deviation = 1
$P(X > k) = 0.1587$

Tail setting	Right
Prob	0.1587
σ	1
μ	0

X_1InvN	0.9998151
Prob	0.1587
σ	1
μ	0

SOLUTION
Hence $k = 1$

2. The masses of apples sold at a fruit and vegetable shop are normally distributed with a mean mass 150 g and standard deviation 10 g.
Determine   (a) the mass exceeded by 5% of the apples.
              (b) the range of the masses of the central 50% of the apples (i.e IQR)

SOLUTION
(a) Mean = 150
Standard deviation = 10
$P(X > k) = 0.05$
Hence $k = 166\ g$

(b) range = 157 − 143 = 14 g.

Tail setting	Centre
Prob	0.5
σ	10
μ	150

X_1InvN	143.2551
X_2InvN	156.7449
Prob	0.5
σ	10
μ	150

EXERCISE 13C

Determine the value of k in each of the following.

1. 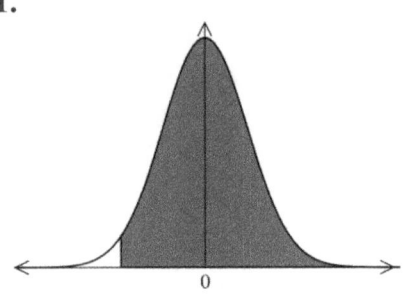 Mean = 0 Standard deviation = 1 $P(X > k) = 0.9772$	2. 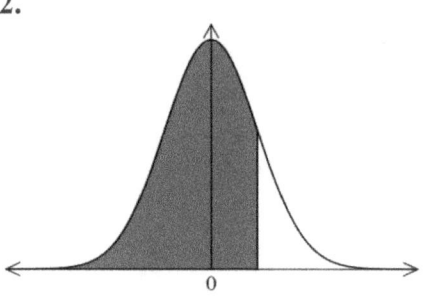 Mean = 0 Standard deviation = 1 $P(X < k) = 0.8413$	3. 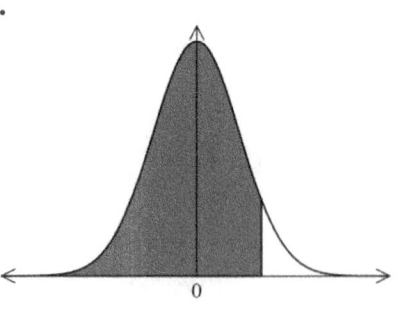 Mean = 0 Standard deviation = 1 $P(X < k) = 0.9332$
4. 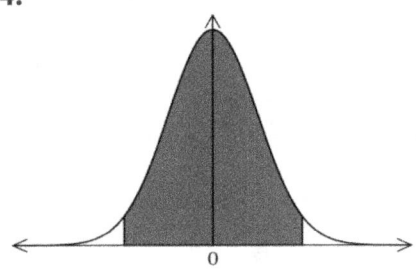 Mean = 0 Standard deviation = 1 $P(-k < X < k) = 0.9545$	5. 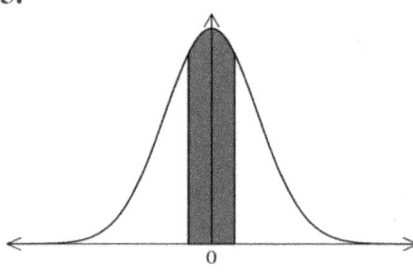 Mean = 0 Standard deviation = 1 $P(-k < X < k) = 0.3829$	6. 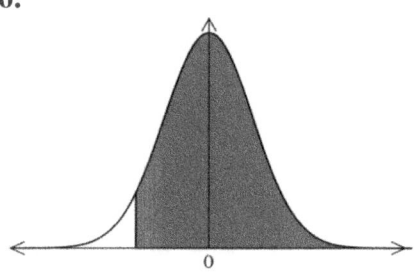 Mean = 0 Standard deviation = 1 $P(X > k) = 0.95$
7. 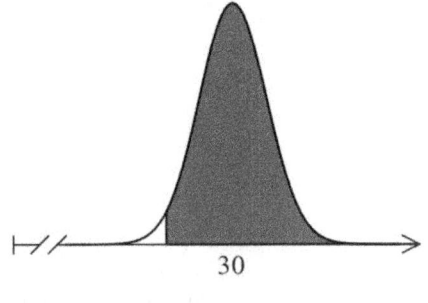 Mean = 30 Standard deviation = 2 $P(X > k) = 0.9772$	8. 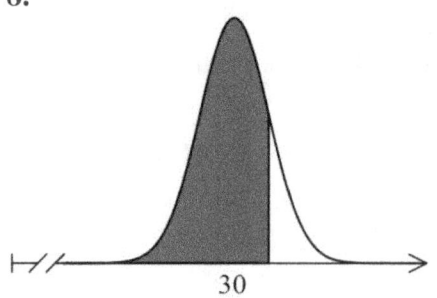 Mean = 30 Standard deviation = 2 $P(X > k) = 0.8413$	9. 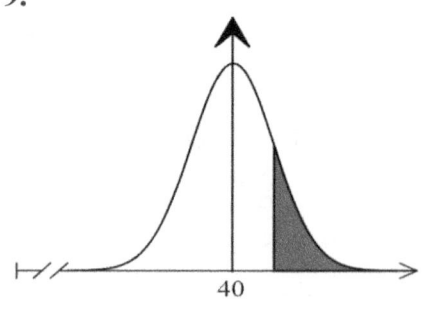 Mean = 40 Standard deviation = 5 $P(X > k) = 0.1587$

10. The life span of a species of insects is modelled by a normal distribution with a mean of 360 hours and a standard deviation of 20 hours. Determine the life span exceeded by 7% of the insects.	**11.** The weight of a consignment of sacks of sugar is normally distributed with a mean of 30 kg and a standard deviation of 2 kg. Determine to the nearest kg, the weight below which 15% of the sacks fall.
12. The marks for a mathematics examination at a school are normally distributed with a mean of 59% and a standard deviation of 12%. (a) State the median examination score. (b) Determine the interquartile range of the examination scores. (c) The top 10% students were awarded an A grade. Determine the minimum cut off for an A grade.	**13.** The masses of cabbages sold at a supermarket are normally distributed with a mean mass 850 g and standard deviation 50 g. (a) If a cabbage is chosen at random, determine the probability that its mass exceeds 900g. (b) Determine the mass exceeded by 15% of the cabbages correct to three (3) significant figures.
14. Top Jewellery Ltd purchased fresh water pearls for the production of its necklaces. The diameters of the pearls were found to be normally distributed, with a mean of 1.2 cm and a standard deviation of 0.15 cm. (a) What proportion of the pearls will have a diameter exceeding 1.05 cm? (b) Below what size will the diameter of 5% of the pearls fall?	**15.** The marks for a Mathematics Applications Unit 2 examination at a High School are normally distributed with a mean of 61% and a standard deviation of 8%. (a) State the modal examination score. (b) Determine the interquartile range of the examination scores. (c) The bottom 10% students were awarded a D grade. Determine the maximum cut off for a D grade.

13D QUANTILES OR PERCENTILES

Consider the diagram on the right.
90% of the scores lie below 50. We say 50 is the 90th percentile.

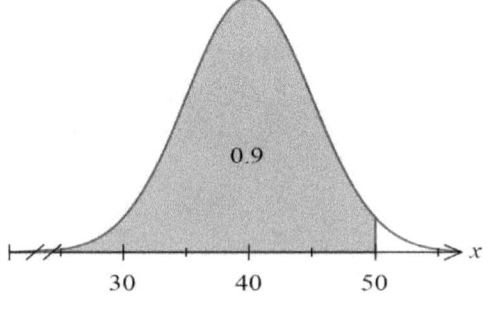

In general, the kth percentile of a set of values divides them so that k % of the values lie below and (100 – k)% of the values lie above.

- ❖ The 25th percentile is referred to as the lower quartile Q_1.
- ❖ The 50th percentile is referred to as the median Q_2.
- ❖ The 75th percentile is referred to as the upper quartile Q_3.

If we are given the percentage or statistical probability of being at or below a certain x-value, to find the percentile, we have to find the x-value that corresponds to it.

EXAMPLES

| 1. If $X \sim N(40, 5^2)$, determine the 60th percentile.
SOLUTION
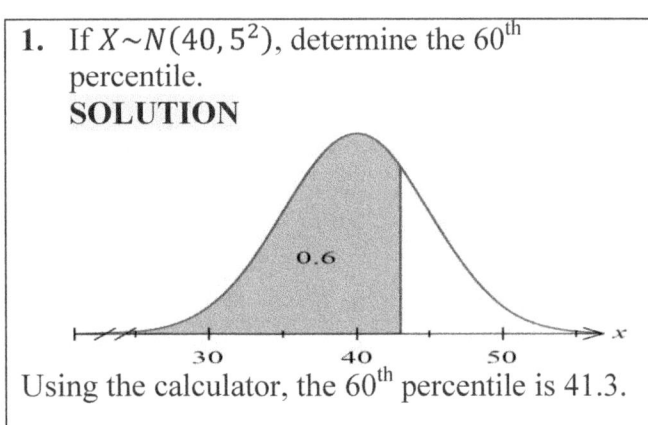
Using the calculator, the 60th percentile is 41.3. | 2. If $X \sim N(60, 3^2)$, determine the 0.85 quantile.
SOLUTION
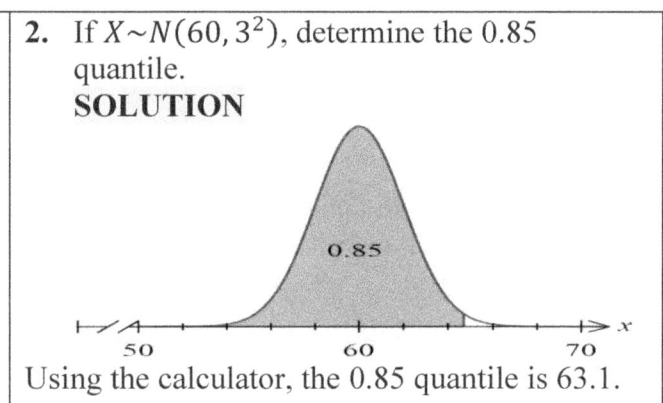
Using the calculator, the 0.85 quantile is 63.1. |

EXERCISE 13D

1. If $X \sim N(50, 4^2)$, determine the 65th percentile.	2. If $X \sim N(60, 5^2)$, determine the 0.42 quantile.
3. If $X \sim N(35, 3^2)$, determine the 78th percentile.	4. If $X \sim N(20, 2^2)$, determine the 0.55 quantile.
5. If $X \sim N(6, 16)$, determine the 52nd percentile.	6. If $X \sim N(10, 25)$, determine the 33rd percentile.
7. If $X \sim N(36, 3^2)$, determine the 88th percentile.	8. If $X \sim N(12, 2^2)$, determine the 0.5 quantile.

13E THE Z-SCORE OR STANDARD SCORE

A normal distribution that is standardized has a mean of 0 and a standard deviation of 1. It is also referred to as the standard normal distribution. If we know the population mean µ and the standard deviation σ of a set of normally distributed scores, we can standardize each "raw" score, x, by converting it into a z score by using the following formula on each individual score:

$$Z = \frac{x - \mu}{\sigma} = \frac{raw\ score - mean}{standard\ deviation}$$

A z score reflects how many standard deviations above or below the population mean a raw score is.

EXAMPLE 1
Jack sat two different tests in Semester One. His results in each test, the class average as well as the standard deviation of each test are given in the table below. Express each as a standard score and comment on your result.

	Jack's score	Class Average (mean)	Standard deviation
Test 1	34	36	4
Test 2	45	40	5

SOLUTION
In Test 1, Jack's z-score is $Z = \frac{34-36}{4} = -0.5$
This indicates that Jack scored half a standard deviation below the mean.

In Test 2, Jack's z-score is $Z = \frac{45-40}{5} = 1$
This indicates that Jack scored one standard deviation above the mean.

EXAMPLE 2
The average score on a mathematics test was 68 marks with an standard deviation of 6 marks. If Adrian's z-score is -1.5, how many marks did he score?
SOLUTION

$$Z = \frac{x - 68}{6} = -1.5$$

Solve on CAS, $x = 59$. Adrian scored 59 marks.

EXAMPLE 3
In a normally distributed data set, the mean is 96 and the z-score for a raw value of 104 is 2. Find the value of the standard deviation.
SOLUTION

$$Z = \frac{104 - 96}{\sigma} = 2$$

Solve on CAS, $\sigma = 4$.

EXERCISE 13E

1. A set of data has a normal distribution with a mean of 20 and standard deviation 4. Find the z-scores of the measurements 16 and 28.	2. A set of data has a normal distribution with a mean of 500 and standard deviation 50. Find the z-scores of the measurements 450 and 525.
3. Cars currently sold by Cheap Cars Ltd have an average of 145 horsepower with a standard deviation of 25 horsepower. What is the z-score for a car with 180 horsepower?	4. Aisha's score on an Investigation Test was 72%. The class average was 75 and the standard deviation was 8%. What was her z-score?
5. Suppose a data set is normally distributed with a mean of 120 and a standard deviation of 10. (a) What data value is 2 standard deviations above the mean? (b) What data value is 1.5 standard deviations below the mean?	6. In a normally distributed data set, find the value of the standard deviation if the following additional information is given. (a) The mean is 240 and the z-score for a data value of 225 is -0.5. (b) The mean is 24 and a z-score for the data value of 18 is -2.

7. Harold sat two different tests in Semester Two. His results in each test, the class average as well as the standard deviation of each test are given in the table below. Express each as a standard score and comment in which test Harold's performance was better.

	Harold's score	Class Average (mean)	Standard deviation
Test 1	42	50	8
Test 2	38	42	5

MATHEMATICS APPLICATIONS UNIT 2 WORKED SOLUTIONS

CHAPTER 1

UNIVARIATE DATA : CLASSIFY, ORGANISE AND DISPLAY

1A TYPES OF DATA

Data can be defined as 'a series of observation, measurement or facts.' Thus to obtain data we need to observe, measure and collect facts about a variable. This variable can take the form of swimming level, country of birth, favourite sport, height, number of rooms in a house, temperature in December and so on.

Group A	Group B
• swimming level • country of birth • favourite sport	• height • number of rooms in a house • Temperature in December

If we study the above table carefully, we can notice important differences between the variables and the way they are measured or observed. In Group A, for instance, we can see that the data are mostly non-numerical and cannot be measured. They are usually termed as **categorical data**. For Group B, on the other hand, all the variables have numbers assigned to them and can be measured. These types of variables are termed as **numerical data**.

Categorical variables classified into two types data : nominal and ordinal. Each of these will be visited in detail with examples below. Numerical data likewise can be subdivided into two types: discrete data and continuous data. These two types of data will be explained with examples as well.

CATEGORICAL VARIABLES

Categorical variables are values or observations that can be sorted into groups or categories. Examples of categorical variables are eye colour, sex, age group, and educational level. Categorical variables can further be divided into two groups: **nominal data** and **ordinal data.**

NOMINAL VARIABLES

Nominal data are values or observations that can be assigned a code in the form of a number where the numbers are simply labels. Nominal data can be counted but not ordered or measured. For example, in a data set of students boys could be coded as 0, girls as 1. The following are examples of nominal variables.

What is your gender?	Which state were you born?	What is the colour of your car?
▲ Male (0) ▲ Female (1)	▲ New South Wales (1) ▲ Western Australia (2) ▲ South Australia (3) ▲ Queensland (4) ▲ Victoria (5)	▲ Red ▲ White ▲ Brown ▲ Green ▲ Grey

ORDINAL VARIABLES

Ordinal variables are values or observations that can have numerical values attached to them; they can be counted and ordered **but not measured**. In the table below, there are some examples of ordinal variables.

Your income level	How do rate the new shops?	Karate level belts
▲ high ▲ medium ▲ low	▲ Unsatisfactory ▲ Satisfactory ▲ Good ▲ Very good ▲ Excellent	▲ White ▲ Yellow ▲ Orange ▲ Green ▲ Blue

NUMERICAL VARIABLES

Numerical data are values or observations that involve measurement or count. We can make measurements such as a student's weight, a tree's height, the temperature in different suburbs at 8 am on Sunday. As a count, we can have examples such as number of classrooms in a school, shoe size at the show store or number of pages of homework done.
Numerical data can further be split into two groups: discrete and variable.

DISCRETE AND CONTINUOUS VARIABLES

Discrete variables represent data that can be counted; it takes particular values and changes in steps. Shoe size is a concrete example of a discrete variable. When we ask our friends their shoe size we expect answers such as 7 or 8.5. The sizes are either a whole number or half a size. Similarly, the number of people attending the Sunday prayer during the month of January is a perfect example of discrete data. A teacher can help 8, 9 or 10 students but not 10.2 students. Continuous variables are measurements that can take any value within a certain range. Continuous data can be counted, ordered and measured. For example, height (1.35m), weight (75.45 kg), temperature (21.45°C), the amount of vitamin C in an orange (35 cl), time to do one lap of the oval (52.546 s). When we measure someone's weight we do not obtain exact values such as 75 kg. The person's weight can be close to 75 kg. As a result we might say that the person's weight lies between 75 kg and 76 kg. Similarly, the time to finish a race is not an exact measurement as it might involve fractions of seconds as well.
The diagram below is a summary of the different types of data.

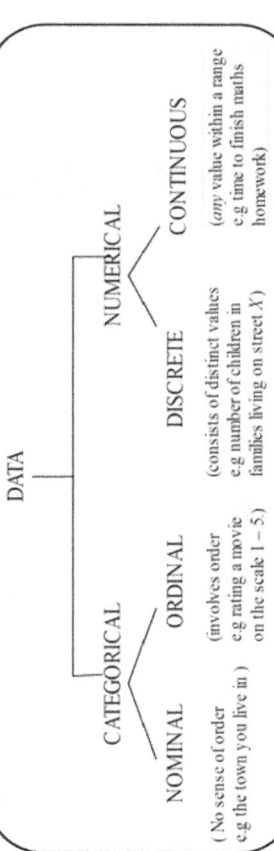

CHAPTER 1 : UNIVARIATE DATA – CLASSIFY, ORGANISE & DISPLAY SOLUTIONS

EXERCISE 1A

For each of the following categorical data sets, tick whether it is nominal or ordinal.

	Nominal	Ordinal
1. The eye colour of members of your family	✓	
2. Your favourite sport	✓	
3. Language spoken at home	✓	
4. Capital cities in Europe	✓	
5. Grade descriptors in Mathematics		✓
6. Eyesight test results of tennis players at US Open		✓

For each of the following numerical data sets, tick whether it is discrete or continuous.

	Discrete	Continuous
7. The number of oranges in each 1 kg bag	✓	
8. The height of students in Year 11		✓
9. The average time to run a 100m race		✓
10. The number of children in each family in your street	✓	
11. Number of admissions in Year 1	✓	
12. Shoe size of players in Man Utd team	✓	
13. Size of a car's gas tank		✓

For each of the following data sets, tick the appropriate box.

	CATEGORICAL		NUMERICAL	
	Nominal	Ordinal	Discrete	Continuous
14. Year level at school		✓		
15. Travel time to work				✓
16. Colour of cars in a parking lot	✓			
17. Blood type (A, B, O, AB)		✓		
18. Number of pets in a family			✓	
19. Age of patients in a nursing home				✓
20. Blood pressure of patients				✓
21. Stage of Cancer (I, II, III or IV)		✓		
22. The number of staff in schools			✓	
23. The cholesterol level in teenagers		✓		
24. The number of lollies in 50g packets			✓	
25. Number of goals scored in a match			✓	

1B DISPLAYING CATEGORICAL DATA

Categorical data are usually displayed with tables, bar charts and pie graphs. An example of each has been given below.

EXAMPLE 1

The table below shows the subject selection of 80 students in Year 11 Mathematics ATAR Courses at Fluency Senior High.

Courses	Number enrolled
Mathematics Essentials	30
Mathematics Applications	40
Mathematics Methods	10

Display the information as a

(a) Bar chart
(b) Pie chart

SOLUTION

(a)

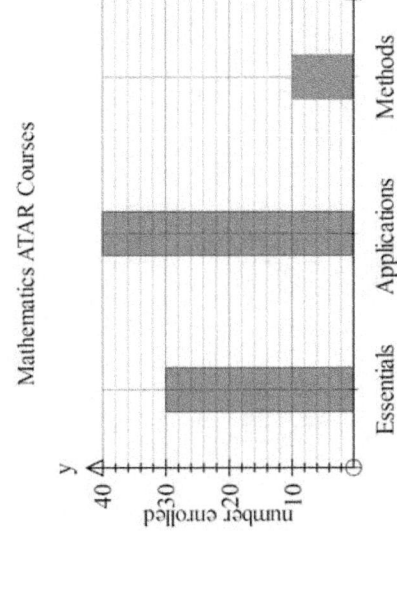

(b)

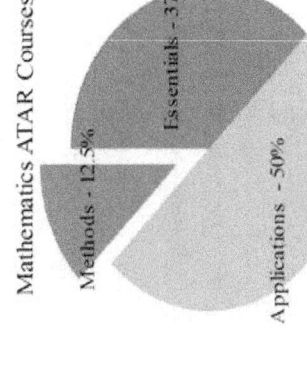

CHAPTER 1 : UNIVARIATE DATA – CLASSIFY, ORGANISE & DISPLAY SOLUTIONS

EXERCISE 1B

1. Little Johnny has a huge collection of toy cars in his playroom. The table below shows the colours and the number of each car.

Colour	Number of cars
Red	8
Yellow	2
Green	4
Blue	6

Display the above information as a bar chart and pie chart. Use the axes and circle given below.

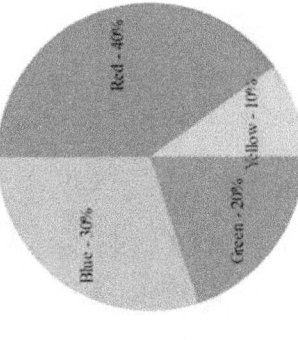

2. Mr Bates did a survey about the eye colour of his students in Year 11 Application Mathematics. He recorded his results in the table below.

Eye Colour	Number of students
Brown	12
Blue	8
Other	5

Display the information on a bar chart below.

3. Are the recreational facilities at the new sport complex adequate? The responses produced the following results.

	Males	Females
Agree	8	5
Neutral	10	15
Disagree	12	10

Display the above table on a side-by-side bar graph below.

4. The table below shows the results of a survey relating to the BMI of 120 staff members of a company.

	BODY MASS INDEX			
Gender		Underweight	About Right	Overweight	Total
	Male	10	50	20	80
	Female	18	20	2	40
	Total	28	70	22	120

Display the above table on a side-by-side percentage bar graph for each gender below.

	Males	Females
Underweight	12.5%	45%
About Right	62.5%	50%
Overweight	25%	5%

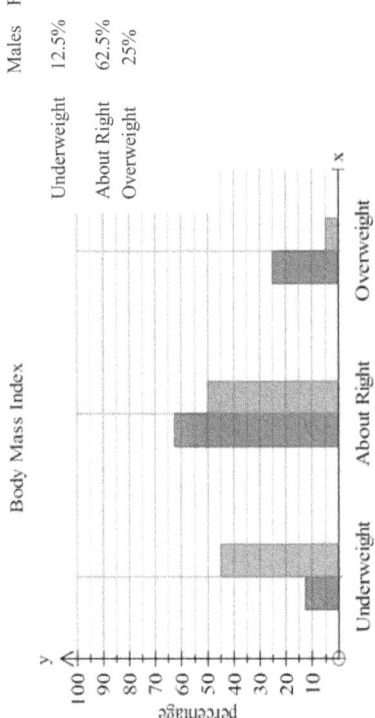

CHAPTER 1: UNIVARIATE DATA – CLASSIFY, ORGANISE & DISPLAY SOLUTIONS

1C DISPLAYING DISCRETE VARIABLES

The Natural Valley Pre-School Playgroup has 30 students in its kindergarten class. In an attempt to find out how many books each child has read, the following data was collected.

Number of books	23	24	25	26	27	28	29	30
Frequency	2	0	4	2	5	1	6	2

Clearly, the variable here which is number of books can be counted and as a result is a discrete variable. We can display the above table on a **dot frequency diagram** as shown.

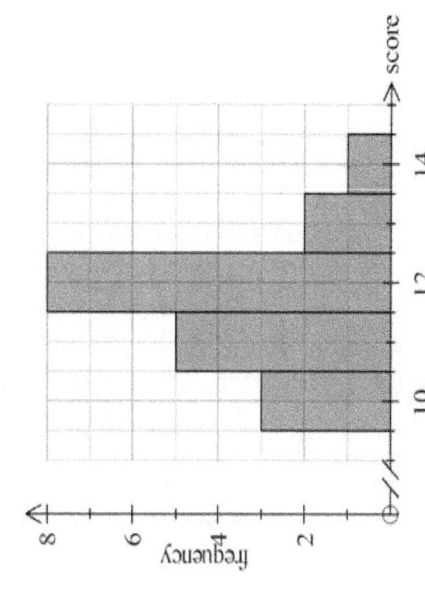

The above table can also be represented by a frequency histogram as shown below.

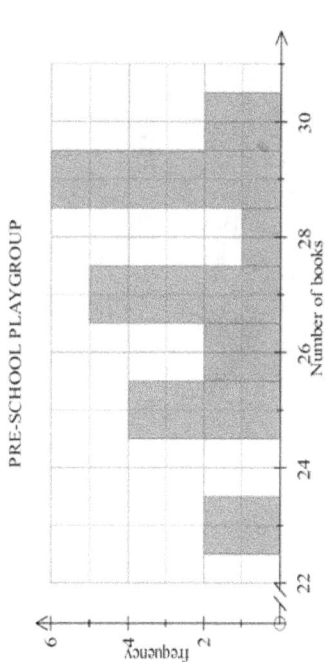

Note that the scores must be written half-way in between the rectangles. The score of 23 lies between 22.5 and 23.5. The number 22.5 is called the **lower boundary** whereas 23.5 is termed as the **upper boundary**. Similarly the score 30 lies between a lower boundary of 29.5 and an upper boundary of 30.5.

As we can see a histogram can be a popular graphing tool in statistics. It can be used to summarise discrete data as well as continuous data. The latter will be dealt in the next part of the chapter. A histogram has an appearance similar to a vertical bar graph as seen earlier, but there are no gaps between the bars. (Unless we have a frequency of zero). In the above example no child read 24 books accounting for the gap between 23.5 and 24.5 on the horizontal axis.

EXERCISE 1C

Draw a frequency histogram for the following:

1.

Score	5	6	7	8	9
frequency	2	7	5	1	4

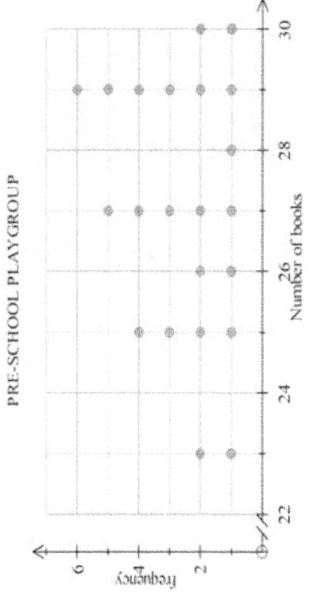

2.

Score	10	11	12	13	14
frequency	3	5	8	2	1

CHAPTER 1 : UNIVARIATE DATA – CLASSIFY, ORGANISE & DISPLAY SOLUTIONS

3. An ordinary die was rolled 40 times and the results listed below.

2	5	4	6	1	1	3	3
3	4	1	6	2	2	1	3
6	1	1	2	1	5	6	6
3	1	5	3	4	3	3	6
5	1	2	5	6	2	3	

Complete the table below and hence draw a dot frequency diagram on the given set of axes.

score	Frequency	Relative frequency $\left(\dfrac{frequency}{total\ frequency}\right)$	Percentage frequency
1	10	$\dfrac{10}{40}$	25%
2	6	$\dfrac{6}{40}$	15%
3	9	$\dfrac{9}{40}$	22.5%
4	3	$\dfrac{3}{40}$	7.5%
5	5	$\dfrac{5}{40}$	12.5%
6	7	$\dfrac{7}{40}$	17.5%
Total	40	1	100%

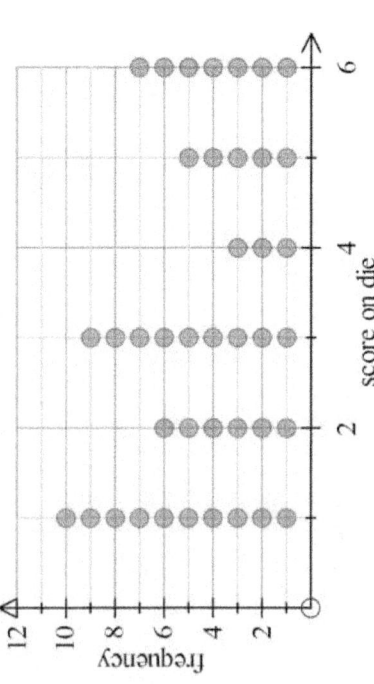

ROLL A DIE

4. The dot frequency diagram shows the amount of pocket money received on a daily basis by a group of 25 students while attending school.

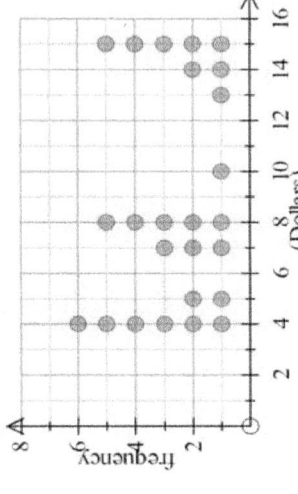

DAILY POCKET MONEY

Use the diagram to complete the following table.

Pocket money ($)	Frequency	Relative frequency	Percentage frequency
4	6	$\dfrac{6}{25}$	24%
5	2	$\dfrac{2}{25}$	8%
7	3	$\dfrac{3}{25}$	12%
8	5	$\dfrac{5}{25}$	20%
10	1	$\dfrac{1}{25}$	4%
13	1	$\dfrac{1}{25}$	4%
14	2	$\dfrac{2}{25}$	8%
15	5	$\dfrac{5}{25}$	20%

5. List two advantages and one disadvantage of using categorical data.

Advantage : Very practical and useful for small data sets

Frequency can be quickly recorded using tallies

Disadvantage : The responses are sometimes too narrow.

CHAPTER 1 : UNIVARIATE DATA – CLASSIFY, ORGANISE & DISPLAY SOLUTIONS

1D DISPLAYING CONTINUOUS VARIABLES

Imagine there is a large spread for the amount of pocket money received by a group of students. Then to display the information on a dot frequency diagram might be very time consuming and tiresome. In such a case we can resort to grouping the data into classes and display the information as a frequency histogram as illustrated in the example below.

EXAMPLE 1

25 students travel the following distances in km to attend school every week days.

3	5	4	8	7
4	11	15	2	23
13	10	9	8	29
20	32	37	11	6
7	8	19	52	25

(a) Tabulate the above results in a grouped frequency table.

Distance travelled (km)	frequency
$0 \leq d < 10$	12
$10 \leq d < 20$	6
$20 \leq d < 30$	4
$30 \leq d < 40$	2
$40 \leq d < 50$	0
$50 \leq d < 60$	1

Note that the class $0 \leq d < 10$ includes numbers ranging from 0 to 9 and the class $10 \leq d < 20$ contains numbers from 10 to 19 inclusive.

(b) Hence display the information as a frequency histogram.

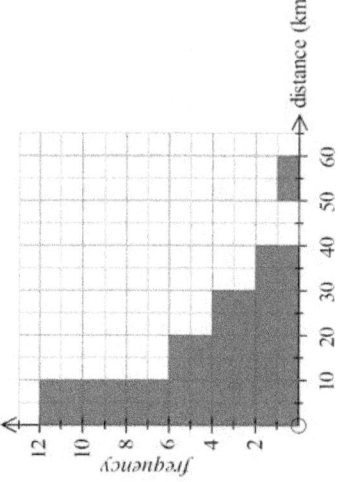

MATHEMATICS APPLICATIONS UNIT 2

EXAMPLE 2

A group of people carried out a survey to find out how far workers in a particular factory travel to get to work on a particular day. The travelling distance were recorded, to the nearest kilometre, for each of the 30 workers. The results are shown in the table.

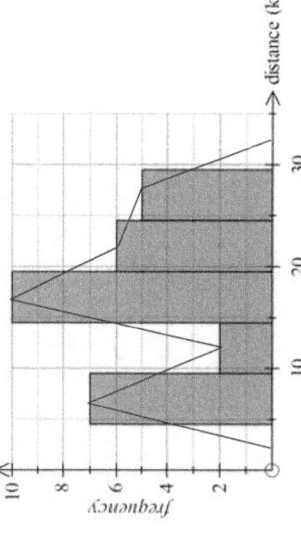

Distance to work (minutes)	Number of workers
$5 < x \leq 9$	7
$10 < x \leq 14$	2
$15 < x \leq 19$	10
$20 < x \leq 24$	6
$25 < x \leq 29$	5

Display the above table in a frequency histogram and include the frequency histogram.

SOLUTION

As we can observe, the classes are all of equal width but are discontinuous. The task of the reader is to make the classes continuous implying that each rectangle starts where the previous one leaves off. This can be done as shown in the table.

Distance to work (km)	CONTINUITY (subtract 0.5 from the lower class boundary and add 0.5 to the upper class boundary)	Number of workers
$5 < x \leq 9$	$4.5 < x \leq 9.5$	7
$10 < x \leq 14$	$9.5 < x \leq 14.5$	2
$15 < x \leq 19$	$14.5 < x \leq 19.5$	10
$20 < x \leq 24$	$19.5 < x \leq 24.5$	6
$25 < x \leq 29$	$24.5 < x \leq 29.5$	5

The frequency histogram can be drawn by joining the midpoint of each rectangle by means of straight lines a shown.

CHAPTER 1: UNIVARIATE DATA – CLASSIFY, ORGANISE & DISPLAY SOLUTIONS

EXERCISE 1D

1. The table below shows the age distribution of 16 people in a holiday home during Christmas. Display the information as a frequency histogram.

Age (x years)	$0 < x \leq 10$	$10 < x \leq 20$	$20 < x \leq 30$	$30 < x \leq 40$
Number of people	8	3	4	1

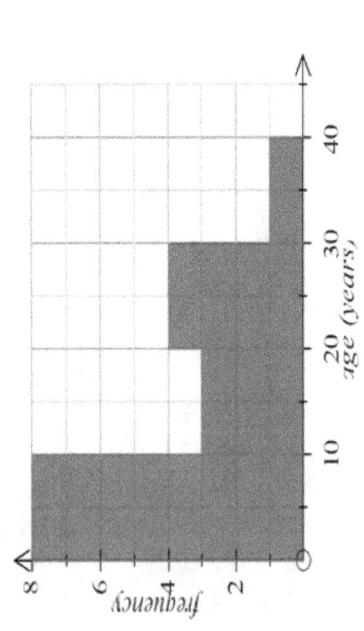

2. Auto Repairs Ltd advertises the company as one of the quickest and most efficient. The table below shows the time taken, in minutes, to get a car repair fixed.

Time in minutes	$0 < x \leq 15$	$15 < x \leq 30$	$30 < x \leq 45$	$45 < x \leq 60$
Number of repairs	2	5	3	6

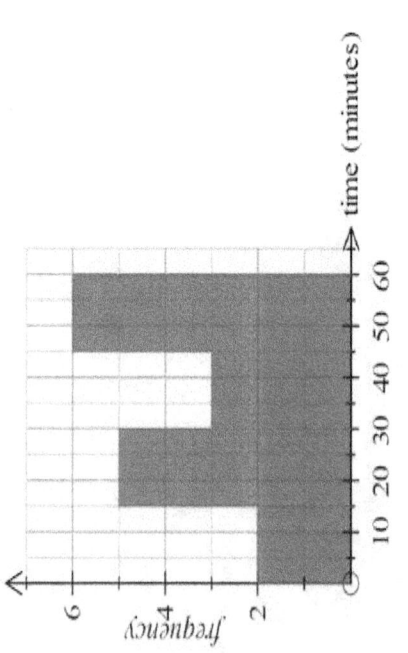

3. The table below shows the speed of cars recorded from 7.30 am to 9 am in a 40 km/h school zone in a suburb. Display the information as a frequency histogram.

Speed km/h	CONTINUITY	Number of cars
$30 < x \leq 39$	$29.5 < x \leq 39.5$	6
$40 < x \leq 49$	$39.5 < x \leq 49.5$	10
$50 < x \leq 59$	$49.5 < x \leq 59.5$	7
$60 < x \leq 69$	$59.5 < x \leq 69.5$	0
$70 < x \leq 79$	$69.5 < x \leq 79.5$	1

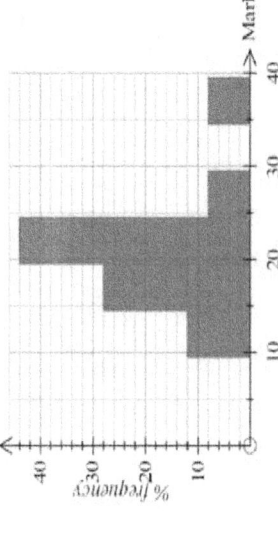

4. The table below shows the marks obtained by a group of 25 Year 11 Mathematics Applications students in a recently held assessment. Draw a percentage frequency histogram.

Marks	CONTINUITY	Number of students (f)	Percentage frequency
$10 < x \leq 14$	$9.5 < x \leq 14.5$	3	12%
$15 < x \leq 19$	$14.5 < x \leq 19.5$	7	28%
$20 < x \leq 24$	$19.5 < x \leq 24.5$	11	44%
$25 < x \leq 29$	$24.5 < x \leq 29.5$	2	8%
$30 < x \leq 34$	$29.5 < x \leq 34.5$	0	0%
$35 < x \leq 39$	$34.5 < x \leq 39.5$	2	8%

5. Thomas is fond of his dice. He has loads of them and likes to experiment. During his first experiment he rolled a tetrahedral die 40 times having sides labelled 1,2,3 and 4. Which of the following histogram is most likely to be obtained from his results?

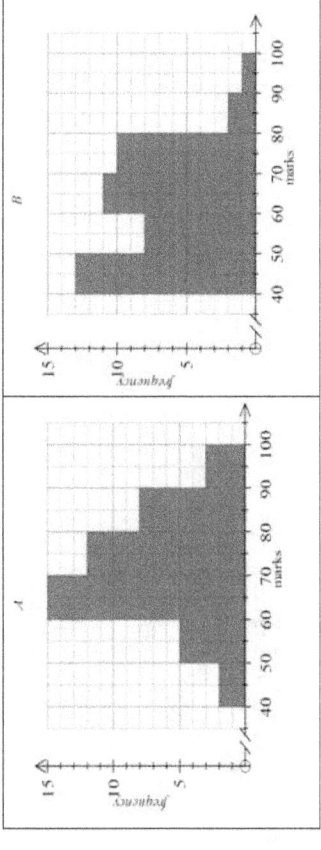

The frequency in histogram A adds up to only 39, therefore cannot be the one. Histogram B, has a frequency total of 40 and the scores are more or less equally distributed. Hence histogram B is the likely histogram.

6. In his next experiment he took 5 ordinary dice and rolled them 12 times. Which of the following histogram is more likely to be displayed?

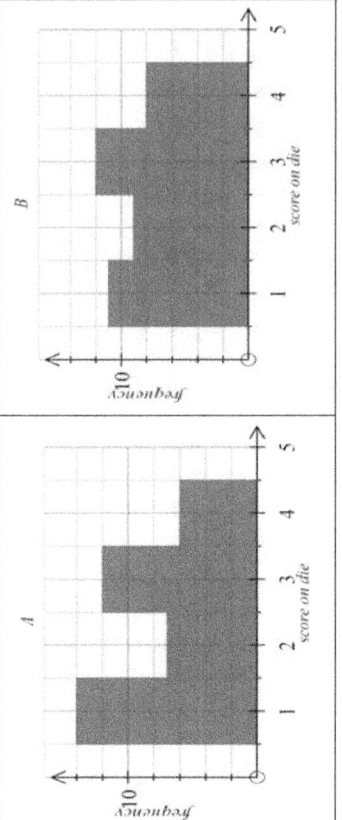

Assuming it is an unbiased die, we would expect to obtain 10 of each of the scores on the die. The most probable histogram will be A as the outcomes are more or less equally distributed between the numbers on the die.

Histogram B, on the other hand seems to be biased against scores 1 and 2.

7. In his last experiment, Thomas recorded the marks obtained by all his classmates in their recent semester one exams in mathematics. He grouped the scores and came up with a histogram. Which of the following can be his class performance given that his teacher is lenient and set easy exams?

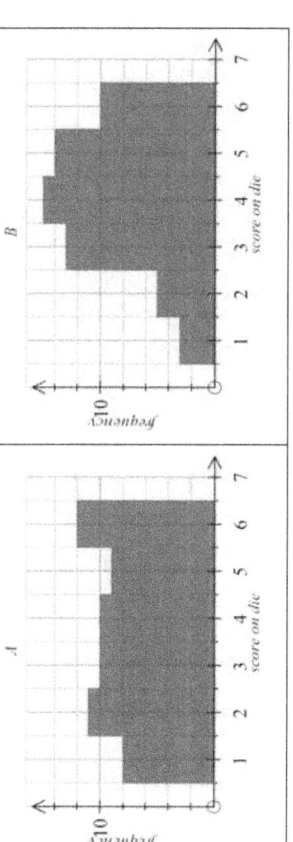

Since the teacher marks leniently, we would expect more students in the higher end. Clearly, the histogram more appropriate in this case is A. Histogram B has a modal score of 40-50 which does not indicate leniency.

CHAPTER 2

SUMMARISING DATA

2A MEASURES OF CENTRAL TENDENCY

Also known as measures of location, a measure of central tendency is a single value that attempts to describe a set of data by identifying the central position within that set of data. The mean, median and mode are all very useful statistical measures, but under different situations, some measures of central tendency become more appropriate to use than others. In this part of the chapter, we will look at the mean, mode and median, and learn how to calculate them and under what conditions they are most appropriate to be used.

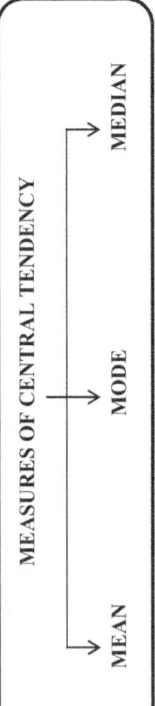

MEASURES OF CENTRAL TENDENCY

MEAN   MODE   MEDIAN

MEAN

The mean or simply known as average is the most popular measure of central tendency. To calculate the mean we divide the sum of all the values in the data set by the number of values in the data set. Mathematically, we denote the mean by the symbol $\bar{x}$ (pronounced x bar), and calculated as

$$\bar{x} = \frac{\Sigma x}{n} \text{ or } \bar{x} = \frac{\Sigma x f}{\Sigma f}$$

where Σ, the Greek letter, pronounced "sigma", which means "the sum of...".

MODE

The mode is the most frequent score in a data set. As such, it is the number with the highest frequency.

MEDIAN

The median is the middle score for a set of data that has been arranged in order. To calculate the median we first need to rearrange that data into ascending order and find the middle score. The position of the median value can be calculated using the formula

median location $= \frac{n+1}{2}$, where n is the number of values in the data set.

RANGE

Range is the difference between the highest and the lowest data.

Range is not a measure of central tendency as it measures a spread of a set of data. However, since it is very easy to compute and it will be seen quite a few times during the course of this chapter, it has been included in some of the exercises to familiarise students with this measure of dispersion or spread.

In the table an attempt has been made to compare the three measures of central tendency by looking at each other's advantages and disadvantages.

	Advantages	Disadvantages
Mean	➢ Most popular in fields such as engineering and business ➢ all the data used to find the answer ➢ useful for comparing data sets ➢ It is unique as it has only one answer	➢ Affected by extreme values (outliers) ➢ Cannot be calculated for nominal data
Mode	➢ Is an actual value of the data ➢ Quick and easy to determine ➢ Not affected by outliers ➢ Only average that can be found for nominal data	➢ Often more than answer or none ➢ Not used for anything else ➢ Can change from sample to sample
Median	➢ Works on ordered data as well as numerical data ➢ Not affected by extreme scores ➢ Useful in calculating percentiles ➢ It is unique as there is only one answer	➢ May not be representative of a sample ➢ Not appropriate for nominal data

EXERCISE 2A

Tick the most appropriate measure of central tendency in each case.

		mean	mode	median
1.	Wages of staff at a high school	✓		
2.	Most common form of transport		✓	
3.	Performance in a recent assessment	✓		
4.	House prices in a suburb			✓
5.	Most frequent burger purchased		✓	
6.	Yummy chocolate bars sold		✓	
7.	Daily high temperature during a week in December	✓		
8.	Hours spent playing online games			✓
9.	Allowance received by retired politicians			✓
10.	Favourite cereal for breakfast		✓	
11.	Age of Year 11 students at a school	✓		
12.	Distance travelled to school			✓
13.	Favourite movie of college students		✓	
14.	Preferred sports channel		✓	
15.	Age of people at a family get-together			✓

2B CALCULATING MEAN, MODE, MEDIAN AND RANGE FOR RAW DATA

EXAMPLES

| 1. Find the mean, mode and median of 2,3,5,8,3,4,10
Also state the range.
SOLUTION
mean, $\bar{x} = \frac{2+3+5+8+3+4+10}{7} = \frac{35}{7} = 5$
Since 3 is the most frequent score, mode = 3
To find median, first, we re-arrange the scores in ascending order
2,3,3, 4, 5,8,10
Clearly the median is 4.
Range = 10 − 2 = 8 | 2. Find the mean, mode and median of 1,2,2,6,7,8,9,13
SOLUTION
mean, $\bar{x} = \frac{1+2+2+6+7+8+9+13}{8} = \frac{48}{8} = 6$
Since 2 is the most frequent score, mode = 2
1,2,2, 6,7,8,9,13
Median location = $\frac{n+1}{2} = \frac{8+1}{2} = 4.5th\ number$
To find the median, we find the average of the 4th and the 5th number.
Hence median = $\frac{6+7}{2} = 6.5$ |

EXERCISE 2B

Without using a calculator, determine the three measures of central tendency: mean, mode and median. Determine the range in each case as well.

| 1. 4, 5, 10, 7, 4

mean = 6
mode = 4
median = 5
range = 6 | 2. 10, -7, 7, 8, 2, 10

mean = 5
mode = 10
median = 7.5
range = 10 − (−7) = 17 |
| 3. 2, 5, 7, 11, 13, -11, 3, 2

mean = 4
mode = 2
median = 4
range = 24 | 4. 4, -4, 5, -5, 8, 12, 8

mean = 4
mode = 8
median = 5
range = 17 |

2C FINDING AVERAGES FOR A FREQUENCY TABLE

EXAMPLE 1

Without using your calculator find the mean, mode and median of the following frequency table.

Scores (x)	Frequency (f)	xf
0	10	0
1	10	10
2	6	12
3	2	6
	$\sum f = 28$	$\sum fx = 28$

SOLUTION

Mean, $\bar{x} = \frac{\sum xf}{\sum f} = \frac{28}{28} = 1$

The score 0 and 1 have the highest frequencies of 10,
∴ modes = 0 and 1 (bi-modal)

Median location = $\frac{n+1}{2} = \frac{28+1}{2}\ th\ score = 14.5th\ score$

To find the median score, keep adding up the frequencies until you go past 14.5

∴ *median* = 2

EXAMPLE 2

The dot frequency shows the daily pocket money received by a group of high school students.

Determine the mean, mode and median amount of pocket money. Which measure of central tendency would be most appropriate in this situation?

SOLUTION

We can first express the dot frequency as a frequency table and then calculate the averages.

Pocket money ($) x	4	5	7	8	10	13	14	15	
Frequency f	6	2	3	5	1	1	2	5	$\sum f = 25$
xf	24	10	21	40	10	13	28	75	$\sum fx = 221$

Mean, $\bar{x} = \frac{\sum xf}{\sum f} = \frac{221}{25} = 8.84$

The score 4 has the highest frequencies of 6, ∴ modes = $4

Median location = $\frac{n+1}{2} = \frac{25+1}{2}\ th\ score = 13th\ score$

To find the median score, keep adding up the frequencies until you go past 13

∴ *median* = $8

EXERCISE 2C

Without using your calculator find the mean, mode and median for the frequency tables and dot frequency diagrams below.

1.

Scores (x)	Frequency (f)	x f
1	6	6
2	4	8
3	7	21
4	8	32
	$\sum f = 25$	$\sum fx = 67$

Mean $= \frac{67}{25}$

Mode $= 4$

Median $= \frac{25+1}{2} = 13th\ score$
$= 3$

2.

Scores (x)	Frequency (f)	x f
5	1	5
10	2	20
15	5	75
20	7	140
	$\sum f = 15$	$\sum fx = 240$

Mean $= \frac{240}{15} = 16$

Mode $= 20$

Median $= \frac{15+1}{2} = 8th\ score$
$= 15$

3.

Scores (x)	Frequency (f)	x f
1	5	5
2	10	20
3	6	18
4	9	36
	$\sum f = 30$	$\sum fx = 79$

$mean = \frac{79}{30}$

$mode = 2$

Median $= \frac{30+1}{2} = 15.5th\ score$
$= 2.5$

4.

Scores (x)	Frequency (f)	x f
2	6	12
4	7	28
6	1	6
8	6	48
	$\sum f = 20$	$\sum fx = 94$

$mean = \frac{94}{20}$

$mode = 4$

Median $= \frac{20+1}{2} = 10.5th\ score$
$= 4$

5.

Scores (x)	Frequency (f)	x f
0	5	0
2	11	22
4	4	16
7	15	105
	$\sum f = 35$	$\sum fx = 143$

$mean = \frac{143}{35}$

$mode = 7$

Median $= \frac{35+1}{2} = 18th\ score$
$= 4$

6.

Scores (x)	f	x f
1	10	10
2	6	12
3	9	27
4	3	12
5	5	25
6	7	42
	$\sum f = 40$	$\sum fx = 128$

ROLL A DIE

$mean = \frac{128}{40}$ $mode = 1$

Median $= \frac{40+1}{2} = 20.5th\ score = 3$

SUMMARISING DATA: SOLUTIONS

7. The dot frequency diagram shows the number of books read by the pre-school playgroup. Calculate the mean, mode and median.

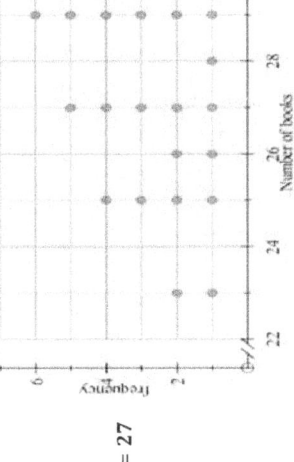

PRE-SCHOOL PLAYGROUP

$mean = \frac{595}{22} = 27.05$

$mode = 29$

$Median = \frac{22+1}{2} = 11.5th\ score = 27$

8. The stem and leaf on the right shows the marks of a group of Year 11 students in an Algebra assessment.

(a) state the lowest score,

   15

(b) state the highest score,

   48

(c) the mean, mode and median.

$mean = 33.33$

$mode = 36$

$Median = \frac{15+1}{2} = 8th\ score = 36$

```
1 | 5 8
2 | 5 8 9
3 | 0 2 6 6 6
4 | 0 0 2 5 8
```

9. The stem and leaf diagram on the right shows the distance run, in kilometres, by 22 members of a golf club to commemorate the 50th anniversary of the club.

(a) State the smallest distance run,

   8 km

(b) State the longest distance run,

   54 km

(c) The mean, mode and median.

$mean = 27.09$

$mode = 15$

$Median = \frac{22+1}{2} = 11.5th\ score = 21.5$

Distance Run (km)
```
0 | 8 9
1 | 1 3 4 5 5 5 6
2 | 0 0 3 8
3 | 0 3 8 9
4 | 2 9
5 | 1 3 4
```

MATHEMATICS APPLICATIONS UNIT 2

Without doing any calculations, compare the mean and the median for each of the following dot frequency diagrams. Choose one of the three answers displayed in the box.

> A : *mean = median*
> B : *mean > median*
> C : *mean < median*

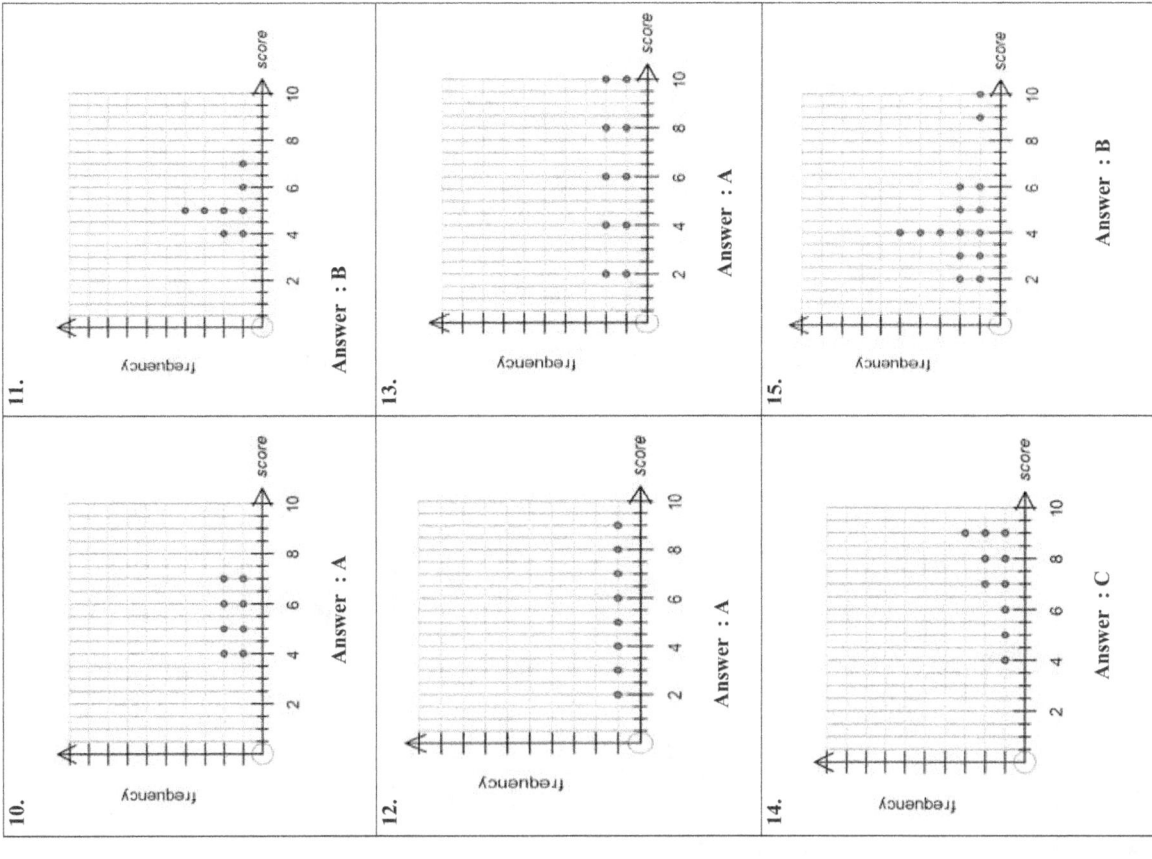

10. Answer : A

11. Answer : B

12. Answer : A

13. Answer : A

14. Answer : C

15. Answer : B

2D USING TECHNOLOGY TO FIND AVERAGES FOR RAW NUMBERS

So far we have tried as far as possible to limit the use of calculators. However, we can make use of technology to find the measures of central tendency without doing any calculations. The examples and steps are shown below for raw numbers. A calculator display has also been shown so as to give the reader an idea how the final results look like. The results have been displayed in two different boxes as we have to scroll down to see the results of the bottom box.

EXAMPLE

Use the statistical capability of your calculator to find the mean, mode and median of the following set of raw data.

11 13 15 14 21 28 29 35 14 14 28

SOLUTION

Use the following steps on your CAS to find the averages of raw numbers.

step 1 : Menu
step 2 : Statistics
step 3 : Enter the scores in list 1, do not order them.
step 4 : Calc
step 5 : One-variable
step 6 : OK

All the answers will be displayed as shown on the right.
At this stage of the chapter, we are only going to take the bold ones into account.

$\bar{x} = 20.181818$ is the mean

$\Sigma x = 222$ is the sum of all the numbers

$n = 11$ implies there are 11 scores in the data set.

$minX = 11$ means that the smallest number is 11

$Med = 15$ is the short form for median

$MaxX = 35$ shows the highest score

$Mode = 14$
$Mode\ N = 1$ means that there is a unique mode.
$Mode\ F = 3$ implies that the mode appears three times in the data set. That is, 14 is the mode and has a frequency of 3.

We could also find the mean by using the formula $\bar{x} = \frac{\Sigma x}{n} = \frac{222}{11} = 20.18$ as above.

Hence mean = 20.18,   mode = 14   and   median = 15.

```
x̄ = 20.181818
Σx = 222
Σx² = 5178
σₓ = 7.9637609
Sₓ = 8.3524629
n = 11
minX = 11
Q₁ = 14
Med = 15
```

```
Q₃ = 28
MaxX = 35
Mode = 14
Mode N = 1
Mode F = 3
```

EXERCISE 2D

Use the statistical capability of your calculator to complete the following table.

		mean	mode	median
1.	4,5,4,7,8,9,7,11,13,4	7.2	4	7
2.	10,12,13,14,21,15,17,19,18,14,14	15.18	14	14
3.	14,14,15,14,17,15,16,14,15,16	15	14	15
4.	11,13,11,15,11,20,25,28	16.75	11	14
5.	-20,-25,-36,-78,-89,-91,-105,-25	-58.625	-25	-57

6. For the dot frequency on the right, determine the mean, mode and median.

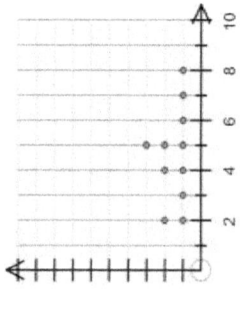

    mean = **4.38**

    mode = **2 and 4**

    Median = **4**

7. Determine the three measures of central tendency for the dot frequency diagram.

    mean = **4.64**

    mode = **5**

    Median = **5**

8. On a farm the number of workers picking oranges on eleven days last April were
3, 8, 7, 11, 9, 2, 11, 11, 9, 9, 5
Find the mean, mode and median.

    mean = **7.73**

    mode = **9**

    Median = **9**

2E USING TECHNOLOGY TO FIND AVERAGES FOR FREQUENCY TABLES

This section is very similar to the previous one. The only slight change is that we use more columns to display the data sets. A sample has been illustrated below to facilitate the teaching and learning of the capabilities of the statistical tool in our calculators.

EXAMPLE

Use the statistical capability of your calculator to find the mean, mode and median of the frequency table given below.

Score on die	1	2	3	4	5	6
Frequency	7	11	13	18	5	2

SOLUTION

Use the following steps on your CAS to find the averages.

step 1 : Menu
step 2 : Statistics
step 3 : Enter the score on die in list 1
step 4 : Enter frequency in list 2
step 5 : Calc
step 5 : One-variable   XList : list1
                         Freq : list2
step 6 : OK

Hence mean $\bar{x}$ = 3.16,
mode = 4 and
median = 3

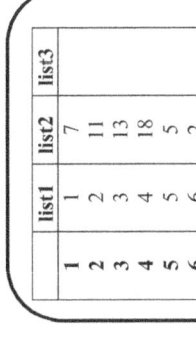
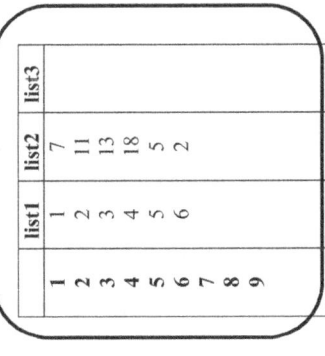
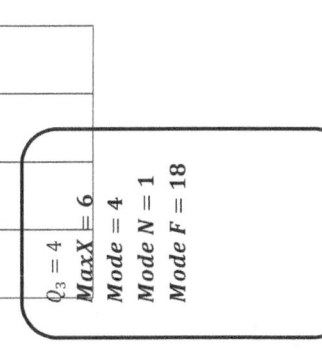

---

9. A study was made of the BMI of 12 athletes at the Olympic Games. The results, to the nearest whole number, were recorded as shown in the table.

BODY MASS INDEX		
28	16	19
17	32	22
25	31	30
24	29	26

For the group of 12 athletes, calculate

(i) mean   (ii) median
    24.92   **25.5**

A 13th athlete recorded a BMI Index of 40 and added to the existing 12 scores. What effect if any (do no calculate) would it have on the

(iii) mean   (iv) median
     increase   **increase**

10. Calculate the mean of each of the four dot frequency diagrams and hence rank them in ascending order.

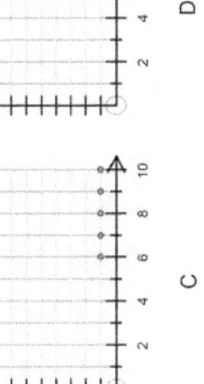
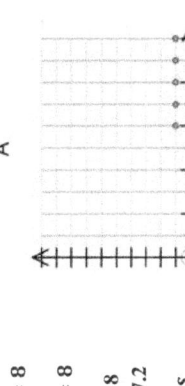
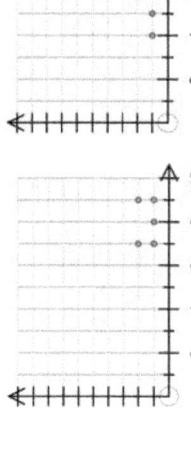

mean of A = 8
mean of B = 8
Mean of C = 8
Mean of D = 7.2

There are 6 answers

DABC,
DACB,
DBAC,
DBCA,
DCAB,
DCBA

SUMMARISING DATA: SOLUTIONS

EXERCISE 2E

Use the statistical capability of your calculator to find the mean, mode and median for each of the following.

1.

Scores (x)	Frequency (f)
0	10
1	15
2	18
3	11
4	14

mean = 2.06
mode = 2
median = 2

2.

Scores (x)	Frequency (f)
2	7
4	11
6	13
8	5
10	10

mean = 6
mode = 6
median = 6

3.

Scores (x)	Frequency (f)
8	7
9	12
10	20
12	6
15	10

mean = 10.65
mode = 10
median = 10

4.

Scores (x)	Frequency (f)
5	2
10	3
15	8
20	7
25	15

mean = 19.29
mode = 25
median = 20

5. Tickets for a show cost $5, $8, $10, $15 or $25. The number of tickets sold at each price is shown in the table.

Price ($)	5	8	10	15	25
Number of tickets sold	80	50	120	60	12

(a) Write down the modal price.

$10

(b) Use your calculator to find the median and mean prices.

***mean* = $9.94, *median* = $10**

6. A survey was carried out to find the number of children in each of 400 families living in an affluent suburb. The results are shown in the table below.

Number of children in family	0	1	2	3	4
Number of families	46	92	98	104	60

(a) Write down the modal number of children.

3

(b) Use your calculator to find the mean number of children per family.

2.1

7. The number of goals scored by a soccer team in each of 30 matches was as follows:

2 3 1 3 0 0 4 2 1 1 3 5 0 1 2
4 1 0 0 2 1 2 3 6 1 4 2 0 0 1

(a) Complete the table in the answer space.

Number of goals scored	0	1	2	3	4	5	6
Number of matches	7	**8**	**6**	**4**	**3**	1	1

(b) For this distribution, find the mean and state the mode.

***mean* = 1.83, *mode* = 1**

CHAPTER 2 : SUMMARISING DATA AND DESCRIBING DISTRIBUTIONS SOLUTIONS

2F   USING CAS TO FIND AVERAGES FOR GROUPED FREQUENCY TABLES

For a grouped frequency table, we use two columns same like we did for frequency table. However, the only difference is that we have to find the midpoints of each class and insert in list 1.

EXAMPLE

Use the statistical capability of your calculator to find the mean, modal class and the median class of the grouped frequency distribution.

Score	1-5	6-10	11-15	16-20	21-25	26-30
Frequency	8	12	19	3	8	10

SOLUTION

Use the following steps on your CAS to find the averages.

step 1 : Menu
step 2 : Statistics
step 3 : Enter midpoint of each class in list 1
step 4 : Enter frequency in list 2
step 5 : Calc
step 5 : One-variable   XList : list1
                         Freq : list2
step 6 : OK

$\bar{x} = 14.75$
$\sum x = 885$
$\sum x^2 = 17095$
$\sigma_x = 8.2069584$
$S_x = 8.27621679$
$n = 60$
$minX = 3$
$Q_1 = 8$
$Med = 13$
$Q_3 = 23$
$MaxX = 28$
$Mode = 13$
$Mode\ N = 1$
$Mode\ F = 19$

Hence mean $\bar{x} = 3.16$,
modal class = 11 - 15 and
median class = 11-15.

EXERCISE 2F

1. Use the statistical capability of your calculator to find the mean, modal class and the median class of the following grouped frequency tables.

Scores	Midpoint (x)	Frequency (f)
0-10	**5**	7
10-20	**15**	15
20-30	**25**	19
30-40	**35**	12
40-50	**45**	9

$mean = 25.16$
$modal\ class = 20 - 30$
Median class = 20-30

2.

Scores	Midpoint (x)	Frequency (f)
20-30	**25**	7
30-50	**40**	11
50-70	**60**	13
70-100	**85**	5
100-120	**110**	10

$mean = 63.48$
$modal\ class = 50 - 70$
Median class = 50 – 70

3.

Scores	Frequency (f)
10-19	7
20-29	12
30-39	25
40-49	6
50-59	10

$mean = 34.5$
$modal\ class = 30 - 39$
Median class = 30 – 39

4.

Scores	Frequency (f)
20-24	2
25-29	3
30-34	10
35-39	7
40-44	11

$mean = 35.3$
$modal\ class = 40 - 44$
Median class = 35 – 39

5. The heights of 40 children are shown in the table below.

Height (x cm)	$60 < x \leq 70$	$70 < x \leq 80$	$80 < x \leq 90$	$90 < x \leq 100$
Number of children	12	18	7	3

Calculate an estimate of the mean height of the children.
$$\bar{x} = 75.25$$

6. The length of time taken by all of the 200 pupils of a school to complete a task is given in the table below.

Time in minutes	$35 < x \leq 45$	$45 < x \leq 55$	$55 < x \leq 65$	$65 < x \leq 75$
Number of pupils	55	a	25	68

State the value of a and hence calculate an estimate of the mean time taken to complete the task.
$$a = 52, mean = 55.3$$

CHAPTER 2 : SUMMARISING DATA AND DESCRIBING DISTRIBUTIONS SOLUTIONS

7. The daily high temperature in degrees Fahrenheit registered at a Metrological station during a particular month is given below:

   61 70 62 64 70 74 70 62 67 65 68 71 66 59 78
   58 60 59 56 53 51 55 56 50 53 57 55 50 46

   Display the above information in the frequency table.

Interval	frequency
45-49	1
50-54	5
55-59	8
60-64	6
65-69	4
70-74	5
75-79	1

   (a) The true boundary of the class 50-54 is 49.5-54.5. State the true boundary of the median class.

   59.5 − 64.5

   (b) Calculate an estimate of the mean temperature.

   $\bar{x} = 61.33$

8. A test was set to group of students and marked out of 70. The minimum mark for a grade A is 58, and a minimum mark for a grade B is 40. Marks of 30 or more score a C grade. The results has been represented on a stem and leaf plot as under.

   Algebra Test
   ```
   0 | 6
   1 | 5
   2 | 8
   3 | 2 5 7 9 9
   4 | 0 3 3 5 5 6 6 8
   5 | 1 3 3 9
   6 | 4 9
   ```

   (a) How many students sat the test?

   24

   (b) State a suitable diagram that can used to illustrate the percentage of students in the three different grades. Determine the percentage of students in each grade.

   pie graph $A: \frac{3}{24} = 12.5\%$ $B: \frac{11}{24} = 45.83\%$ $C: \frac{5}{24} = 20.83\%$

   (c) Use your calculator to find the mean mark and state the median mark.

   mean $\bar{x} = 40.38$, median mark = 43

   (c) Is the mean or median a better indicator of how the students performed in the test?

   median

MATHEMATICS APPLICATIONS UNIT 2

2G APPLICATIONS

EXAMPLES

1. Four numbers have a mean 50. If three of the numbers are 20, 60 and 80, find the 4th number.
   Solution
   Total for the 4 numbers = 50 × 4 = 200
   Total of the three given numbers
   = 20 + 60 + 80 = 160
   The 4th number = 200 − 160 = 40.

2. In a test the 10 boys in a class score a mean mark of 60 and the 15 girls score a mean mark of 50. Calculate the mean mark for the whole group of 25.
   Solution
   Marks scored by boys = 10 × 60 = 600
   Marks scored by the girls = 15 × 50 = 750
   Total marks scored = 600 + 750 = 1350
   Mean mark = $\frac{1350}{25} = 54$

EXERCISE 2G

1. Five numbers have a mean 30. If four of the numbers are 10, 20, 30 and 25, find the 5th number.

   $(30 \times 5) - (10 + 20 + 30 + 25)$
   $= 65$

2. The mean of a set of 20 numbers is 65. When a 21st number is added, the mean increases to 70. Calculate the 21st number.

   $21 \times 70 = 1470$
   $20 \times 65 = 1300$
   21st Number $= 1470 - 1300 = 170$

3. The mean of a set of 9 numbers is 40. When a 10th number is added, the mean increases by 3. Calculate the 10th number.

   $9 \times 40 = 360$
   $10 \times 43 = 430$
   10th number. $= 430 - 360 = 70$

4. The mean of a set of 14 numbers is 63. When a 15th number is added, the mean decreases by 2. Calculate the 15th number.

   $14 \times 63 = 882$
   $15 \times 61 = 915$
   15th Number $= 915 - 882 = 33$

5. Paul scored 48%, 61% and 53% in his last three tests. He needs an average of 60% to pass his course. Find how much he needs to score in his fourth test to pass the course.

   $60 \times 4 = 240$
   $48 + 61 + 53 = 162$
   He needs $240 - 162 = 78\%$

6. Six numbers have a mean 40. If four of the numbers are 10, 20, 30 and 40, and the remaining two numbers are equal, find one of the remaining number.

   $6 \times 40 = 240$
   $10 + 20 + 30 + 40 = 100$
   $(240 - 100) \div 2 = 70$

7. The mean of a set of 10 numbers is 76. When an 11th number is added, the new mean is 80. Calculate the 11th number.

   $10 \times 76 = 760$
   $11 \times 80 = 880$
   11th number: $880 - 760 = 120$

8. In a test the 15 boys in a class score a mean mark of 62 and the 12 girls score a mean mark of 54. Calculate the mean mark for the whole group.

   $15 \times 62 = 930$
   $12 \times 54 = 648$
   $(930 + 648) \div 27 = 58.4$

CHAPTER 2 : SUMMARISING DATA AND DESCRIBING DISTRIBUTIONS SOLUTIONS

9. The mean weight of 10 boys is 64 kg and the mean weight of 5 girls is 59 kg. Find the mean weight for the whole group.

$10 \times 64 = 640$
$5 \times 59 = 295$
$(640 + 295) \div 15 = 62.3 \; kg$

10. Three classes A, B and C have 15, 21 and 12 students respectively. The three classes A, B and C have a mean mark of 52, 71 and 48 respectively. Calculate the mean mark for the three classes combined together.

$15 \times 52 = 780$
$21 \times 71 = 1491$
$12 \times 48 = 576$
$(780 + 1491 + 576) \div 48 = 59.31$

11. Three classes P, Q and R have 20, 15 and 25 students respectively. The three classes P, Q and R have a mean mark of 58, 74 and 52 respectively. Calculate the mean mark for the three classes combined together.

$20 \times 58 = 1160$
$15 \times 74 = 1110$
$25 \times 52 = 1300$
$(1160 + 1110 + 1300) \div 60 = 59.5$

12. A group of 10 boys ran an average of 22.4 km during a marathon. If the whole competition consisted of 25 participants and their mean distance ran was 19.6 km, determine the mean distance ran by the girls.

$10 \times 22.4 = 224$
$25 \times 19.6 = 490$
$(490 - 224) \div 15 = 17.73 \; km$

13. Given that the mean of 5, 10, 15, x and 25 is 15, determine the value of x.

$5 \times 15 = 75$
$75 - (5 + 10 + 15 + 25) = 20$
$\therefore x = 20$

14. The set of values 5, 8, 8, x, 13, 15 is arranged in ascending order. If the median is 9, state the value of x.

$x = 10$

15. For the set of numbers 10, 12, 14, x, 14, y, determine the value of x and y if the range is 10 and the median is 14.

$y = 20, \quad x = 14$

16. For the set of numbers $x, 10, 10, y, z, 14$, determine the value of x, y and z if
- the range is 9
- the median is 10
- The mean is 10.

$x = 5, y = 10$
$6 \times 10 = 60$
$60 - (5 + 10 + 10 + 10 + 14) = 11$
$\therefore z = 11$

17. For the set of numbers arranged in ascending order
$$a, b, 5, 7, 9, c, d, 12, 12, 17$$
- The range is 15
- The median is 10
- The mode is 12
- The mean is 9.

Work out the values of a, b, c and d.

$a = 2$
$c = 11$
$d = 12$
$10 \times 9 = 90$
$90 - (2 + 5 + 7 + 9 + 11 + 12 + 12 + 12 + 17) = 3$
$\therefore b = 3$

18. For the set of numbers arranged in ascending order
$$5, 6, a, 8, b, 11, 12, 15, c, d$$
- The range is 15
- The median is 10.5
- The mode is 20
- The mean is 11.4

Work out the values of a, b, c and d.

$d = 20$
$b = 10$
$c = 20$
$10 \times 11.4 = 114$
$114 - (5 + 6 + 8 + 10 + 11 + 12 + 15 + 20 + 20) = 7$
$\therefore a = 7$

CHAPTER 3

MEASURES OF DISPERSION OR SPREAD

3A CALCULATING STANDARD DEVIATION BY LONG HAND

The standard deviation calculation tells us how spread out the numbers is in a set of data. Mathematically, standard deviation is the average amount by which the scores in a data differ from the mean. A smaller value of the standard deviation implies that the data set has more consistency whereas the larger the standard deviation means that there is more spread among the data.

The symbol for standard deviation is σ (read as sigma) and can be calculated using the following formulae:

$$\sigma_x = \sqrt{\frac{\sum(x-\bar{x})^2}{n}} \quad \text{or} \quad \sqrt{\frac{\sum x^2}{n} - (\bar{x})^2}$$

To calculate standard deviation by hand follow these simple steps:
- Find the mean
- Subtract the mean from each score
- Square each of these differences
- Find the mean of these differences
- Find square root of this mean

EXAMPLE 1

Find the standard deviation of the numbers 5, 10, 15, 20 and 25.

SOLUTION

Mean $= \bar{x} = \frac{\sum x}{n} = \frac{75}{5} = 15$

The easiest way to find the standard deviation is to use a table as shown.

Scores (x)	mean ($\bar{x}$)	$x - \bar{x}$	$(x-\bar{x})^2$
5	15	-10	100
10	15	-5	25
15	15	0	0
20	15	5	25
25	15	10	100
			$\sum(x-\bar{x})^2 = 250$

Using the formula $\sigma_x = \sqrt{\frac{\sum(x-\bar{x})^2}{n}} = \sqrt{\frac{250}{5}} = 7.07 \ (2 \ d.p)$

We can also use the second formula $\sigma_x = \sqrt{\frac{\sum x^2}{n} - (\bar{x})^2}$ to calculate standard deviation in a much quicker way.

EXAMPLE 2

Calculate the standard deviation of 2, 3, 5, 7, 8.

SOLUTION

This time we are going to use the formula $\sigma_x = \sqrt{\frac{\sum x^2}{n} - (\bar{x})^2}$ to compute standard deviation.

Mean $= \bar{x} = \frac{\sum x}{n} = \frac{25}{5} = 5$

Scores (x)	x^2
2	4
3	9
5	25
7	49
8	64
	$\sum x^2 = 151$

$$\sigma_x = \sqrt{\frac{151}{5} - (5)^2} = 2.28$$

EXERCISE 3A

Calculate the standard deviation of each of the following using the long hand method. Use the given tables to show your workings.

1. 4, 6, 8, 10, 12

Scores (x)	mean ($\bar{x}$)	$x - \bar{x}$	$(x-\bar{x})^2$
4	8	-4	16
6	8	-2	4
8	8	0	0
10	8	2	4
12	8	4	16
			$\sum(x-\bar{x})^2 = 40$

$$\sigma_x = \sqrt{\frac{\sum(x-\bar{x})^2}{n}} = \sqrt{\frac{40}{5}} = 2.83$$

2. 10, 10, 10, 10, 10

Scores (x)	mean ($\bar{x}$)	$x - \bar{x}$	$(x-\bar{x})^2$
10	10	0	0
10	10	0	0
10	10	0	0
10	10	0	0
10	10	0	0
			$\sum(x-\bar{x})^2 = 0$

$$\sigma_x = \sqrt{\frac{\sum(x-\bar{x})^2}{n}} = \sqrt{\frac{0}{5}} = 0$$

MATHEMATICS APPLICATIONS UNIT 2

Without making any calculations determine which graph in each question has the greater mean or greater standard deviation.

	DOT FREQUENCY DIAGRAM		Greater mean	Greater standard deviation
	GRAPH A	GRAPH B		
8.			SAME MEAN	GRAPH A
9.			GRAPH B	SAME
10.			GRAPH B	GRAPH A
11.			SAME MEAN	GRAPH B
12.			GRAPH B	SAME

CHAPTER 3 : MEASURES OF DISPERSION OR SPREAD : SOLUTIONS

3. 10, 12, 15, 18, 20, 21

Scores (x)	x^2
10	100
12	144
15	225
18	324
20	400
21	441
	$\sum x^2 = 1634$

$$\sigma_x = \sqrt{\frac{\sum x^2}{n} - (\bar{x})^2}$$

$$\sigma_x = \sqrt{\frac{1634}{6} - (16)^2} = 4.04$$

5. 6, 6, 6, 6

Scores (x)	x^2
6	36
6	36
6	36
6	36
	$\sum x^2 = 144$

$$\sigma_x = \sqrt{\frac{144}{4} - (6)^2} = 0$$

7. 2, 10, 17, 23, 28

Scores (x)	mean ($\bar{x}$)	$x - \bar{x}$	$(x - \bar{x})^2$
2	16	-14	196
10	16	-6	36
17	16	1	1
23	16	7	49
28	16	12	144
			$\sum(x-\bar{x})^2 = 426$

4. 40, 42, 45, 48, 51, 62

Scores (x)	x^2
40	1600
42	1764
45	2025
48	2304
51	2601
62	3844
	$\sum x^2 = 14138$

$$\sigma_x = \sqrt{\frac{14138}{6} - (48)^2} = 7.23$$

6. 10, 20, 30, 60

Scores (x)	x^2
10	100
20	400
30	900
60	3600
	$\sum x^2 = 5000$

$$\sigma_x = \sqrt{\frac{5000}{4} - (30)^2} = 18.71$$

$$\sigma_x = \sqrt{\frac{\sum(x-\bar{x})^2}{n}} = \sqrt{\frac{426}{5}} = 9.23$$

CHAPTER 3 : MEASURES OF DISPERSION OR SPREAD : SOLUTIONS

3B USING TECHNOLOGY TO CALCULATE STANDARD DEVIATION OF RAW NUMBERS

Now we are going to make use of the statistical capabilities of our calculators to determine the standard deviation in a much quicker and less painstaking way.

EXAMPLE

Use the statistical capability of your calculator to find the standard deviation of the following set of raw data. Also state the value of the variance.

$$11 \quad 13 \quad 15 \quad 14 \quad 21 \quad 28 \quad 29 \quad 35 \quad 14 \quad 14 \quad 28$$

SOLUTION

Use the following steps on your CAS to find the averages of raw numbers.

step 1 : Menu
step 2 : Statistics
step 3 : Enter the scores in list 1
step 4 : Calc
step 5 : One-variable
step 6 : OK

$\bar{x} = 20.181818$
$\Sigma x = 222$
$\Sigma x^2 = 5178$
$\boldsymbol{\sigma_x = 7.9637609}$
$S_x = 8.3524629$
$n = 11$
$minX = 11$
$Q_1 = 14$
$Med = 15$

All the answers will be displayed as shown on the right. Hence standard deviation (shown in bold) = 7.96

Variance is the square of the value of the standard deviation.

$\therefore$ Variance $= \sigma^2 = 7.96^2 = 63.4$

EXERCISE 3B

1. For each of the following, find the mean ($\bar{x}$) and the standard deviation (σ_x) and the variance $\sigma^2(x)$ using your calculators. Give your answers to 1 decimal place.

		mean	Standard deviation	Variance
1.	2,3,4,5,8,8,10,15	6.9	4.0	16.1
2.	1,5,4,7,8,11,15,20	8.9	5.8	33.9
3.	4,10,15,20,25,50	20.7	14.7	216.1
4.	6,8,7,8,9,10,11,13	9	2.1	4.5
5.	10,12,14,15,16,17,18	14.6	2.6	6.8
6.	-4,-5,-4,-7,-8,-9,-7,-11	-6.9	2.3	5.4
7.	11,13,11,15,11,20,25,28	16.7	6.3	40.2

MATHEMATICS APPLICATIONS UNIT 2

For each of the following determine the mean and the standard deviation using your calculator.

8.

$\bar{x} = 4.27$
$\sigma_x = 2.43$

9.

$\bar{x} = 4.8$
$\sigma_x = 2.71$

10. During the Olympics diving competition, three athletes recorded the following number of points in their 7 jumps during their final round.

Jumps	J1	J2	J3	J4	J5	J6	J7
Alex	8.5	8.5	9.0	8.5	8.5	8.0	7.5
Zhiao	7.0	7.5	8.0	8.5	7.5	7.5	8.5
Mitcham	10	9.5	10	9	10	9.5	9.5

(a) Rank the three athletes by calculating their mean scores over their 7 jumps.

Alex : **8.36**
Zhiao : **7.79**
Mitcham : **9.64**
Rank : ***Mitcham, Alex, Zhiao***

(b) Which of the three divers had a more consistent result? Show workings to support your answer.

Alex : $\sigma_x = 0.44$
Zhiao : $\sigma_x = 0.52$
Mitcham : $\sigma_x = 0.35$

Clearly Mitcham has a more consistent result as his standard deviation is lower.

CHAPTER 3: MEASURES OF DISPERSION OR SPREAD : SOLUTIONS

3C USING TECHNOLOGY TO CALCULATE STANDARD DEVIATION FOR FREQUENCY TABLES

In this section, we use two columns instead of one to determine the standard deviation. The steps that the reader needs to follow on their calculator together with the calculator display have been presented in a well set example below.

EXAMPLE

Use the statistical capability of your calculator to find the standard deviation of the frequency table given below.

Score on die	1	2	3	4	5	6
Frequency	7	11	13	18	5	2

SOLUTION

Use the following steps on your CAS to find the averages.

step 1 : Menu
step 2 : Statistics
step 3 : Enter the score on die in list 1
step 4 : Enter frequency in list 2
step 5 : Calc
step 5 : One-variable   XList : list1
                         Freq : list2
step 6 : OK

Hence standard deviation = 1.29

	list1	list2	list3	
1	1	7		
2	2	11		
3	3	13		
4	4	18		
5	5	5		
6	6	2		
7				
8				
9				

$\bar{x} = 3.1607143$
$\Sigma x = 177$
$\Sigma x^2 = 653$
$\sigma_x = \mathbf{1.2925167}$
$S_x = 1.3042139$
$n = 56$
$minX = 1$
$Q_1 = 2$
$Med = 3$

EXERCISE 3C

Use the statistical capability of your calculator to calculate the mean and standard deviation of the frequency tables given below.

1.

Scores (x)	Frequency (f)
0	8
1	11
2	12
3	15
4	10

$\bar{x} = 2.14 \qquad \sigma_x = 1.32$

2.

Scores (x)	Frequency (f)
2	5
4	11
6	10
8	9
10	6

$\bar{x} = 6 \qquad \sigma_x = 2.50$

3.

Scores (x)	Frequency (f)
8	6
9	11
10	13
12	4
15	15

$\bar{x} = 11.22 \qquad \sigma_x = 2.68$

4.

Scores (x)	Frequency (f)
5	2
10	4
15	3
20	7
25	18

$\bar{x} = 20.15 \qquad \sigma_x = 6.36$

5.

Scores	1	3	4	6	8
frequency	1	2	5	4	12

$\bar{x} = 6.125 \qquad \sigma_x = 2.15$

CHAPTER 3 : MEASURES OF DISPERSION OR SPREAD : SOLUTIONS

Use the statistical capability of your calculator to calculate the mean, standard deviation of the following dot frequency diagrams.

6.

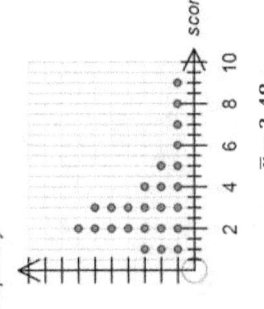

$\bar{x} = 3.48$
$\sigma_x = 2.10$

7.

$\bar{x} = 5.15$
$\sigma_x = 2.78$

5. Tickets for a show at the Writing Cinemas cost $10, $12, $14, $20 or $32. The number of tickets sold during the premiere of a popular horror movie at each price is shown in the table.

Price ($)	10	12	14	20	32
Number of tickets sold	110	50	30	180	12

Determine the mean and the standard deviation of the ticket prices to the nearest integer.

$\bar{x} = 15.98$
$\sigma_x = 5.32$

6. A survey was carried out to find the number of pets in each of 120 families living in a village. The results are shown in the table below.

Number of pets owned	0	1	2	3	4	5	6
Number of families	10	25	15	10	40	13	7

Determine the mean and the standard deviation of the number of pets owned by the villagers.

$\bar{x} = 2.93$
$\sigma_x = 1.73$

7. The table below shows the number of run scored by the 15 cricketers during a friendly match.

Number of runs	0	10	20	25	50	75	100
Number of players	2	5	0	4	2	1	1

Determine the mean and the standard deviation of the number of runs scored by the cricketers.

$\bar{x} = 28.3$
$\sigma_x = 27.8$

8. Petra rolled an octahedral die 50 times and recorded her scores below.

```
2 5 1 3 4 8 5 5 7 2
1 3 3 2 3 5 3 3 6 2
4 8 3 3 3 4 5 4 7 8
1 2 3 4 5 6 7 8 3 4
7 3 4 3 3 5 3 3 3 3
```

(a) Complete the table in the answer space.

Number on top of die	1	2	3	4	5	6	7	8
Frequency	**3**	**5**	**18**	**7**	**7**	**2**	**4**	**4**

(b) For this distribution, find the mean and standard deviation.

$\bar{x} = 4.04$
$\sigma_x = 1.90$

(c) Do you consider this die to be fair or biased? Justify your answer.

biased in favour of 3

9. A special six-sided die is rolled 80 times. The results are tabulated below:

Number on top of die	1	2	3	4	5	6
Frequency	7	10	3	25	15	20

(a) Determine:

(i) the mean of these data.

$\bar{x} = 4.1375$

(ii) the standard deviation of these data.

$\sigma_x = 1.58$

(b) What conclusion can you draw about the die? Is it fair or biased?

biased against 1, 2 and 3

CHAPTER 3 : MEASURES OF DISPERSION OR SPREAD : SOLUTIONS

3D USING TECHNOLOGY TO CALCULATE STANDARD DEVIATION FOR GROUPED FREQUENCY TABLES.

By now the reader must be very familiar as to how the statistical capabilities of the calculator help to compute the value of the standard deviation. For a grouped frequency table, we still use two columns, the first column being the midpoints of each class interval and the second column used for frequency as shown.

EXAMPLE

Use the statistical capability of your calculator to find the mean and standard deviation of the grouped frequency distribution.

Score	10-19	20-29	30-39	40-49	50-59	60-69
Frequency	5	11	23	4	1	7

SOLUTION

Use the following steps on your CAS to find the averages.

step 1 : Menu
step 2 : Statistics
step 3 : Enter midpoint of each class in list 1
step 4 : Enter frequency in list 2
step 5 : Calc
step 5 : One-variable   XList : list1
                         Freq : list2
step 6 : OK
Hence   mean $\bar{x}$ = 35.68,
        standard deviation = 14.09

One Variable

	list1	list2	list3
1	14.5	5	
2	24.5	11	
3	34.5	23	
4	44.5	4	
5	54.5	1	
6	64.5	7	
7			
8			

One variable

$\bar{x} = 35.676471$
$\Sigma x = 1819.5$
$\Sigma x^2 = 75042.75$
$\sigma_x = 14.093116$
$S_x = 14.233349$
$n = 51$
$minX = 14.5$
$Q_1 = 24.5$
$Med = 34.5$

EXERCISE 3D

Use the statistical capabilities of your calculator to the mean ($\bar{x}$) and the standard deviation (σ_x) of the following grouped frequency tables.

1.

Scores (x)	Midpoint of class	Frequency (f)
0-10	5	2
10-20	15	11
20-30	25	13
30-40	35	10
40-50	45	8

$\bar{x} = 27.5$
$\sigma_x = 11.51$

2.

Scores (x)	Midpoint of class	Frequency (f)
20-30	25	6
30-50	40	11
50-70	60	15
70-100	85	4
100-120	110	10

$\bar{x} = 63.70$
$\sigma_x = 28.98$

3.

Scores (x)	midpoint	Frequency (f)
10-19	14.5	7
20-29	24.5	12
30-39	34.5	25
40-49	44.5	6
50-59	54.5	10

$\bar{x} = 34.5$
$\sigma_x = 11.97$

Note that a two variable statistics will give two sets of averages. Thus, by using two two variable statistics we would obtain two means, two modes, two standard deviations etc... one for each column.

CHAPTER 3 : MEASURES OF DISPERSION OR SPREAD : SOLUTIONS

4.

Scores (x)	midpoint	Frequency (f)
20-30	25	5
30-40	35	4
40-60	50	10
60-70	65	8
70-100	85	13

$\bar{x} = 59.75$

$\sigma_x = 21.18$

5. The heights of 50 trees in a nursery are shown in the table below.

Height (x cm)	$60 < x \leq 70$	$70 < x \leq 80$	$80 < x \leq 90$	$90 < x \leq 100$
Number of tress	8	22	13	7

Determine the mean and the standard deviation of the height of trees correct to 1 decimal place.

$\bar{x} = 78.8$

$\sigma_x = 9.1$

6. The time taken by a group of 100 participants to swim a 200m race is given in the table below.

Time in seconds	$35 < x \leq 45$	$45 < x \leq 55$	$55 < x \leq 65$	$65 < x \leq 75$
Number of participants	12	18	20	50

Determine the mean and the standard deviation of the time taken by those participants correct to one decimal place.

$\bar{x} = 60.8$

$\sigma_x = 10.7$

7. The table below shows the age distribution of the teaching staff in a certain school in the metropolitan area. Determine the mean and the standard deviation of the ages of the staff.

Age (x years)	$20 < x \leq 30$	$30 < x \leq 40$	$40 < x \leq 50$	$50 < x \leq 60$
Number of staff	8	12	15	5

$\bar{x} = 39.25$

$\sigma_x = 9.46$

8. The length of time spent by shoppers at a shop during Christmas time were recorded and tabulated below. Determine the mean and the standard deviation of the time taken by those shoppers.

Time (x hours)	$0.5 < x \leq 1.5$	$1.5 < x \leq 2.5$	$2.5 < x \leq 3.5$	$3.5 < x \leq 4.5$
Number of people	200	150	50	20

$\bar{x} = 1.74$

$\sigma_x = 0.85$

Determine an estimate of the mean ($\bar{x}$) and standard deviation (σ_x) for the following histogram.

9.

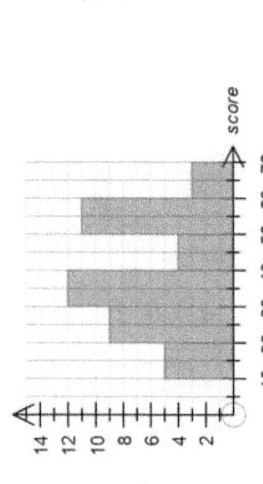

$\bar{x} = 38.6$

$\sigma_x = 14.9$

10. The table below shows the prices and number of properties sold in two different states.

		State A	State B
Prices ($000)	$250 \leq P < 300$	10	1
	$300 \leq P < 350$	8	18
	$350 \leq P < 400$	15	3
	$400 \leq P < 450$	5	10
	$450 \leq P < 500$	16	15
	$500 \leq P < 550$	2	5

(a) Calculate the mean and standard deviation of the property prices in both states.

State A $\bar{x} = 388.4$ $\sigma_x = 76.5$
State B $\bar{x} = 408.7$ $\sigma_x = 73.9$

(b) Write down the modal class for State A.

$450 \leq P < 500$

(c) Determine the median class for State B.

$400 \leq P < 450$

(d) In which of the two states were the prices less variable? Explain.

State B as it has a lower standard deviation.

(e) State A sold a property for $980 000 but was not recorded in the table. What effect would this have to your answers in (a)?

Both the mean and standard deviation will increase.

CHAPTER 3 : MEASURES OF DISPERSION OR SPREAD : SOLUTIONS

11. Table 1 below shows a summary statistics for maximum daily temperatures in March 2013 and March 2014 for a New Zealand town. The maximum daily temperatures (°C) in March 2013 for the town are summarised in Table 2.

Table 1 : Maximum daily temperatures (°C), March 2013-2014

	March 2013	March 2014
Mean	**26.5**	22.6
Standard deviation	**6.2**	5.1

Table 2 : Maximum daily temperature (°C), March 2013

Temperature T (°C)	Frequency
$15 \leq T < 19$	4
$19 \leq T < 23$	9
$23 \leq T < 27$	1
$27 \leq T < 31$	5
$31 \leq T < 35$	12

(a) Use the data in Table 2 to

(i) Calculate the mean and standard deviation of the temperatures for March 2013 and enter the results in Table 1.

(ii) State the modal class.

$$31 \leq T < 35$$

(iii) In which class interval would the median lie?

$$27 \leq T < 31$$

(b) In which of the two years were the March temperatures in the town less variable? Explain your choice.

March 2014 as it has a lower standard deviation.

(c) Melissa is a new statistician recruit. She had a quick glimpse of the data and concluded that the year 2014 tended to be cooler than the year 2013. Is she correct? Justify your answer.

No. Information for one month is not enough to justify which year is cooler.

3E   STANDARD DEVIATION FROM MEAN

Very often we come across question such as "how many scores lie within 2 standard deviations from the mean" or "how many scores are 0.5 standard deviation above the mean" and so on. In this part of the chapter, an attempt has been made to answer all those queries.

EXAMPLE

An ordinary die was rolled 50 times and the results recorded in the table below.

Score on die (x)	1	2	3	4	5	6
frequency	3	10	7	15	11	4

(a) Determine the mean and the standard deviation for the above frequency table.
(b) How many scores are 0.5 standard deviation above the mean?
(c) How many scores are 1 standard deviation below the mean?
(d) How many scores are 2 standard deviations within the mean?

Solution

(a) $\bar{x} = 3.66 \qquad \sigma_x = 1.38$

(b) Multiply the standard deviation by 0.5 and add to the mean as shown

$$\bar{x} + 0.5\,\sigma_x = 3.66 + 0.5(1.38) = 4.35$$

Find from the table the number of scores greater than 4.35 (scores 5 or 6) = 11 + 4 = 15.

(c) Subtract one standard deviation from the mean as shown

$$\bar{x} - \sigma_x = 3.66 - 1.38 = 2.28$$

Now find from the table the number of scores less than 2.28 (scores 1 and 2) = 3 + 10 = 13.

(d) Add and subtract 2 standard deviations from the mean as shown

$$\bar{x} - 2\,\sigma_x = 3.66 - 2(1.38) = 0.9$$
$$\bar{x} + 2\,\sigma_x = 3.66 + 2(1.38) = 6.42$$

Now find from the table the number of scores lying between 0.9 and 6.42, which are obviously all 50.

SUMMARY

Hence to find out how many scores lie above, below or within the mean, the following table will definitely help.

Above the mean	ADD
Below the mean	SUBTRACT
Within the mean	ADD & SUBTRACT

CHAPTER 3 : MEASURES OF DISPERSION OR SPREAD : SOLUTIONS

EXERCISE 3E

1.

Score on die	1	2	3	4	5	6
frequency	2	8	10	7	11	2

(a) Determine the mean and the standard deviation for the above frequency table.

$$\bar{x} = 3.6 \quad \sigma_x = 1.3$$

(b) How many scores are 0.5 standard deviation above the mean?

$$\bar{x} + 0.5\,\sigma_x = 3.6 + 0.5(1.3) = 4.25$$

Answer : 11 + 2 = 13

2.

Score	1	2	3	4	5
frequency	3	11	9	4	3

(a) Determine the mean and the standard deviation for the above frequency table.

$$\bar{x} = 2.8 \quad \sigma_x = 1.1$$

(b) How many scores is one standard deviation below the mean?

$$\bar{x} - 0.5\,\sigma_x = 2.8 - 0.5(1.1) = 2.25$$

Answer : 11 + 3 = 14

3. Mr Andrews had a maths test out of 8 marks and the class results have been tabulated below.

Test Score	0	1	2	3	4	5	6	7	8
frequency	2	3	7	1	2	5	6	1	2

(a) How many students took the test?

29

(b) Use your calculator to determine the mean and the standard deviation.

$$\bar{x} = 3.9 \quad \sigma_x = 2.3$$

(c) How many scores are 2 standard deviations **within** the mean?

$$\bar{x} - 0.5\,\sigma_x = 3.9 - 0.5(2.3) = 2.75$$
$$\bar{x} + 0.5\,\sigma_x = 3.9 + 0.5(2.3) = 5.05$$

Answer : 1 + 2 + 5 = 8

4. Nine Applications Mathematics students' performance in a recently held examination is given below.

Mercury	54	Eli	55	Valerie	7
Venus	52	Amy	39	John	13
Alex	13	Norman	16	Peppa	21

(a) Calculate the mean and standard deviation of the students 'marks.

$$\bar{x} = 30 \quad \sigma_x = 18.7$$

(b) An outlier for this data is defined to be a score which is at least two standard deviations above or below the mean. Which, if any, of the above student can be classified as an outlier? Show all working.

$$\bar{x} - 2\,\sigma_x = 30 - 2(18.7) = -7.4$$
$$\bar{x} + 2\,\sigma_x = 30 + 2(18.7) = 67.4$$

There are no outliers

5. In a study of family size at New Idea High School, a teacher gathers information from students in a particular year group.

No. of children per family	1	2	3	4	5	8
No. of families	7	45	35	8	2	1

(a) Find the total number of children questioned in this survey.

98

(b) Find the mean number of children per family, giving your answer to 2 decimal places.

$$\bar{x} = 2.57$$

(c) Determine the standard deviation of the number of children per family.

$$\sigma_x = 0.99$$

(d) An outlier for this data is defined to be a score which is at least 1.5 standard deviations above the mean. How many families can be classified as outliers?

$$\bar{x} + 1.5\,\sigma_x = 2.57 + 1.5(0.99) = 4.055$$

Three families can be classified as outliers

(e) Describe what happens to the mean and standard deviation if the student from the 8-child family leaves the group.

Both the mean and standard deviation will decrease.

CHAPTER 4

BOX PLOTS AND HISTOGRAMS

4A BOX AND WHISKER PLOTS

A box plot also referred to as a box and whisker plot is a graph that summarises a set of data along a number line. It uses values called quartiles, which divide the data into four equally sized groups.

To make a box plot from a set of data, we need what is called the five-number summary consisting of

- The smallest value
- The highest value
- The lower quartile
- The median
- The upper quartile

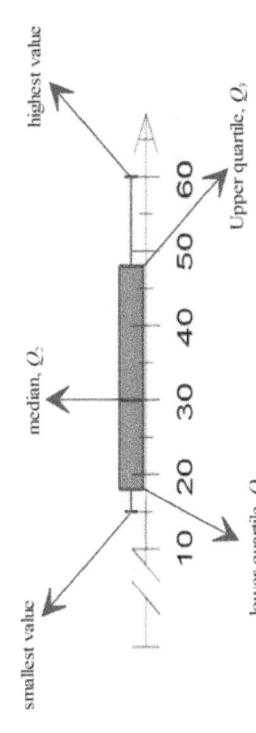

Each section in a box plot represents 25% of the whole data set as shown below.

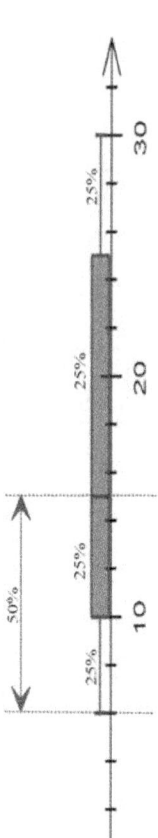

For example, if the above box plot represents the results of 40 students in a Mathematics Applications Assessment then we can draw the following conclusions:

- 25% (means 10 students) scored 10 marks or lower.
- 50% scored less than 15 marks.
- The top 25% students scored between 25 and 30 marks.
- 100% scored in the range 6 – 30 marks.
- Inter quartile range which is the width of the box and it is the central 50% of the scores can be calculated as
  $IQR = Q_3 - Q_1 = 25 - 10 = 15$.

EXAMPLE 1

Draw a box plot for the following set of scores
22, 23, 25, 27, 29, 35, 38, 40, 42, 45, 50

SOLUTION

First, be sure that your data is arranged from least to greatest.
Then we find the median or middle value that splits the data set into two equal groups. The 6th value of 35 divides this list into two equal groups of 5 numbers each as shown below.

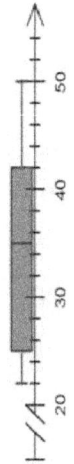

Next we find the median of the lower half of the data set known as the lower quartile (Q_1). Similarly, we find the median of the upper half of the data set. The latter is termed as upper quartile (Q_3).

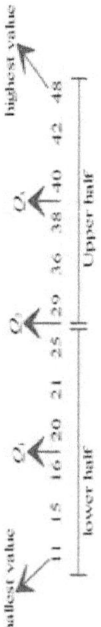

Next we find the smallest and the highest value of the data set. These 5 values as shown above and previously called as the five-number summary is everything we need to construct a box plot. To construct the actual box plot, we will need to draw an ordinary number line that extends far enough in both directions to include all the numbers in the data set.
The lower quartile, median and upper quartiles form the box. We join the lowest value of 22 to the box by means of a line to obtain our left whisker. Similarly, joining 50 to the box produces the right whisker as shown. Hence the diagram below is our box plot showing the five-number summary.

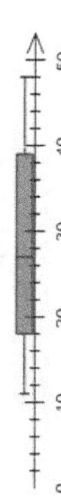

EXAMPLE 2

Draw a box and whisker plot for the set of 12 scores given below.
15, 11, 16, 21, 36, 20, 25, 38, 42, 48, 40, 29

SOLUTION

First, we re-arrange the numbers in ascending order as shown. Since there are 12 scores there will two groups of six numbers each as shown. The median lies between 25 and 29. So we find the average of these two numbers. Hence $Q_2 = \frac{25+29}{2} = 27$. Similarly $Q_1 = 18$ and $Q_3 = 39$.

With 11 being the least value and 48 the largest, we can construct our box plot as shown below.

CHAPTER 4 : BOX PLOTS AND DESCRIBING DISTRIBUTIONS : SOLUTIONS

EXERCISE 4A

Complete the table below stating the smallest value (S), the largest value (L), the lower quartile (Q_1), the median (Q_2) and the upper quartile (Q_3) and interquartile range (IQR) for each box plot.

	S	L	Q_1	Q_2	Q_3	IQR
1.	10	40	16	23	36	20
2.	8	45	16	29	39	23
3.	16	52	20	36	40	20
4.	7	20	10	13	15	5

5. Three classes had a common investigation (out of 20 marks) on box plots and the results shown below.

(a) In one of the classes one student scored 100%. Which class was he/she in?

Class A

(b) Which class had the lowest range?

Class A = 13, Class B = 16 and Class C = 15

∴ *Class A has the lowest range.*

MATHEMATICS APPLICATIONS UNIT 2

6. The box plots show the performance, measured in minutes, of two clubs P and Q at a swimming event held during the summer school holidays.

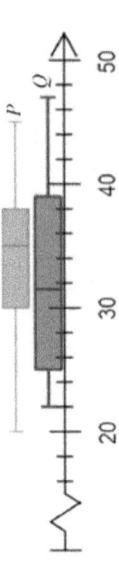

Tick true or false for each of the following:

Statement	True	False
Club Q had the fastest swimmer		✓
Club P has a larger median	✓	
Both clubs have the same range of swimming times	✓	
Interquartile range for Club P is 10		✓
50% of the swimmers in Club P took more than 35 minutes to finish their race.	✓	

7. Draw box plots for each of the following data sets.

(a) 12, 15, 15, 18, 20, 22, 26, 28, 28, 30, 34

(b) 11, 12, 15, 16, 18, 18, 22, 26, 30

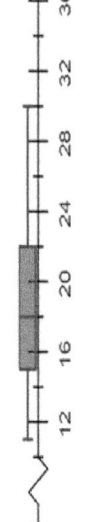

(c) 22, 20, 23, 28, 35, 22, 26, 38, 45, 46, 40

(d) 20, 16, 22, 28, 46, 40, 30, 20, 18, 36, 35, 30

CHAPTER 4 : BOX PLOTS AND DESCRIBING DISTRIBUTIONS : SOLUTIONS

4B IDENTIFYING OUTLIERS

An outlier can be defined as an extremely high or extremely low value in our data set. Outliers can be identified by using one of the following formulae:

Any number greater than $Q_3 + 1.5 \times IQR$ or $Q_3 + 1.5(Q_3 - Q_1)$ will be an outlier.
Or if a number is less than $Q_1 - 1.5 \times IQR$ or $Q_1 - 1.5(Q_3 - Q_1)$ it will be considered an outlier too.

EXAMPLE

Peter recorded the number of phone calls he received during the first 12 days at work.

10, 11, 12, 11, 14, 11, 15, 14, 13, 22, 14, 16

Determine if this data contains any outliers.

SOLUTION

Re-arranging the numbers in ascending order, we have

$$Q_1 \quad Q_2 \quad Q_3$$

10 11 | 11 12 13 | 14 14 14 | 15 16 22

$\therefore Q_1 = 11, Q_2 = 13.5 \text{ and } Q_3 = 14.5$.

To find out if any of these 12 scores are extremely high in value, we compute
$Q_3 + 1.5 \times IQR = Q_3 + 1.5(Q_3 - Q_1)$
$= 14.5 + 1.5 \times (14.5 - 11) = 19.75$

Any number greater than 19.75 will be an outlier.
From the data set only 22 is greater than 19.75, therefore 22 is an outlier.

Similarly, to find out if any of the 12 scores are extremely low in value, we compute
$Q_1 - 1.5 \times IQR = Q_1 - 1.5(Q_3 - Q_1)$
$= 11 - 1.5 \times (14.5 - 11) = 5.75$

So any number less than 5.75 will be called an outlier.
From the data set 10 being the smallest data, there is nothing less than 5.75. Hence there is no lower outlier.

A usual box plot for the above data will be as follows. The highest value will be 16 instead of 22. The outlier 22 will be marked with a dot on the diagram.

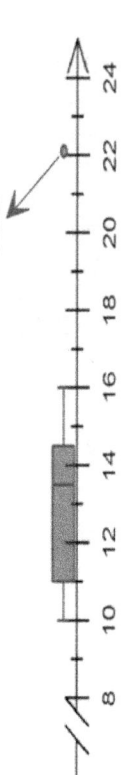

EXERCISE 4B

For the following sets of data, determine if there are any outliers.

1. 111, 21, 29, 19, 121, 25, 29, 27, 17

    $Q_1 = 20, Q_2 = 27 \text{ and } Q_3 = 70$

    $Q_3 + 1.5 \times IQR$
    $= 70 + 1.5 \times (70 - 20) = 145$

    $Q_1 - 1.5 \times IQR$
    $= 20 - 1.5(70 - 20) = -55$

    There are no outliers.

2. 41, 31, 49, 199, 32, 35, 39, 37, 43

    $Q_1 = 33.5, Q_2 = 39 \text{ and } Q_3 = 46$

    $Q_3 + 1.5 \times IQR$
    $= 46 + 1.5 \times (46 - 33.5) = 64.75$

    $Q_1 - 1.5 \times IQR$
    $= 33.5 - 1.5(46 - 33.5) = 14.75$

    199 is an outlier.

3. 222, 121, 212, 242, 225, 209, 277, 237

    $Q_1 = 210.5, Q_2 = 223.5 \text{ and } Q_3 = 239.5$

    $Q_3 + 1.5 \times IQR$
    $= 239.5 + 1.5 \times (239.5 - 210.5) = 283$

    $Q_1 - 1.5 \times IQR$
    $= 210.5 - 1.5(239.5 - 210.5) = 167$

    121 is an outlier.

4. 19, 1, 2, 3, 5, 4, 7, 3, 2, 1, 20

    $Q_1 = 2, Q_2 = 3 \text{ and } Q_3 = 7$

    $Q_3 + 1.5 \times IQR$
    $= 7 + 1.5 \times (7 - 2) = 14.5$

    $Q_1 - 1.5 \times IQR$
    $= 2 - 1.5(7 - 2) = -5.5$

    19 and 20 are outliers.

5. 44, 41, 49, 19, 42, 45, 49, 47, 46

    $Q_1 = 41.5, Q_2 = 45 \text{ and } Q_3 = 48$

    $Q_3 + 1.5 \times IQR$
    $= 48 + 1.5 \times (48 - 41.5) = 57.75$

    $Q_1 - 1.5 \times IQR$
    $= 41.5 - 1.5(48 - 41.5) = 31.75$

    19 is an outlier.

6. 56, 76, 12, 82, 53, 70, 53, 53

    $Q_1 = 53, Q_2 = 54.5 \text{ and } Q_3 = 73$

    $Q_3 + 1.5 \times IQR$
    $= 73 + 1.5 \times (73 - 53) = 103$

    $Q_1 - 1.5 \times IQR$
    $= 53 - 1.5(73 - 53) = 23$

    12 is an outlier.

CHAPTER 4 : BOX PLOTS AND DESCRIBING DISTRIBUTIONS : SOLUTIONS

4C SKEWNESS

Skewness can be defined as a measure of asymmetry in a statistical distribution, in which the curve appears distorted or skewed either to the left or to the right. In this part of the chapter, an attempt has been made to analyse three different types of scenarios: symmetric distributions, positively skewed distributions and negatively skewed distributions.

To establish skewness, we are going to use three different methods
  ❖ Measures of central tendency
  ❖ The quartiles
  ❖ Graphical representation

SYMMETRIC DISTRIBUTIONS			For a symmetric distribution ➢ Mode = Median = Mean ➢ $Q_3 - Q_2 = Q_2 - Q_1$ ➢ The whiskers are of equal length
POSITIVELY SKEWED			For a positively skewed distribution, most of the data are bunched to the left. ➢ Mode < Median < Mean ➢ $Q_3 - Q_2 > Q_2 - Q_1$ ➢ The right whisker is longer than the left one.
NEGATIVELY SKEWED			For a negatively skewed distribution, most of the data are bunched to the right. ➢ Mode > Median > Mean ➢ $Q_3 - Q_2 < Q_2 - Q_1$ ➢ The right whisker is shorter than the left one.

EXERCISE 4C

Without doing any statistical computations, tick whether the following sets of data will be symmetrical, positively skewed or negatively skewed.

	symmetric	Positively skewed	Negatively skewed
1. 2, 5, 5, 6, 7, 8, 8, 10, 16, 20		✓	
2. 4, 4, 6, 8, 8, 10, 10, 12, 12	✓		
3. 2, 5, 8, 15, 18, 21, 23, 25	✓		
4. 10, 12, 12, 15, 15, 18, 18, 20			✓

MATHEMATICS APPLICATIONS UNIT 2

4D COMPARING BOX PLOTS

To compare box plots comment on the following:

  ❖ Median
  ❖ Spread : include range or interquartile range
  ❖ Skewness : symmetrical, positively skewed or negatively skewed

EXAMPLE

Compare the annual rainfall of two states A and B. The data for the past ten years has been plotted as box plots.

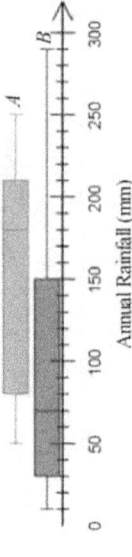

Annual Rainfall (mm)

SOLUTION

Median	spread	skewness
• Median for state A is greater than median for state B	• State B has a greater range (280mm) compared to State A (200mm) • IQR for State A (130 mm) is greater than IQR for State B (120 mm)	• State B's box plot is positively skewed whereas state A's box plot is negatively skewed.

EXERCISE 4D

1. Compare the distributions of the two box plots.

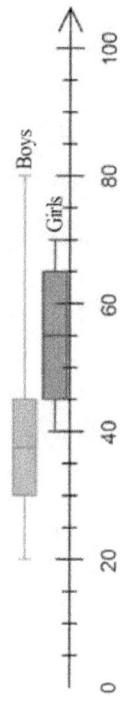

Median	spread	skewness
• Median for girls is greater than median for boys.	• Boys has a greater range (60) compared to girls (30) • IQR for girls (20) is greater than IQR for boys (15)	• While the girls box plot is symmetrical, the boy's box plot is positively skewed.

4E DESCRIBING OTHER DISTRIBUTIONS

In this part of the chapter, the reader will be given histograms, dot plots, bar graphs or stem and leaf plots and their objective would be to describe those using appropriate statistical terms as explained through an example below.

To describe a given distribution of scores, comment on each of the following to have a complete overview of the distributions.

➢ **Modality** : Uni or multi modal
➢ **Location**: calculate the value of the mean and median using your calculator.
➢ **Spread**: state the range and calculate the standard deviation, or interquartile range.
➢ **Shape** : symmetric versus positively or negatively skewed. Mention about any other feature of the graph such as gaps, clusters, outliers etc..

EXAMPLE

The histogram below shows the amount of pocket money received by a group of students in a particular class. Give a brief description of the distribution of the scores.

	list1	list2	list3
1	7.5	8	
2	12.5	5	
3	17.5	6	
4	22.5	0	
5	27.5	3	
6	32.5	0	
7	37.5	1	
8			
9			

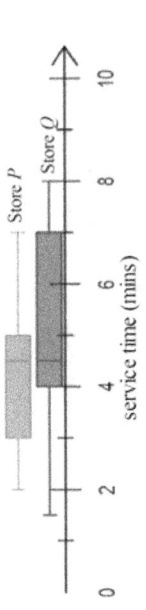

SOLUTION

Use your calculator and the statistical capability to compute most of the measures given below.

$\bar{x} = 15.108696$
$\sum x = 347.5$
$\sum x^2 = 6743.75$
$\sigma_x = 8.0581535$
$S_x = 8.2392582$
$n = 23$
$minX = 7.5$
$Q_1 = 7.5$
$Med = 12.5$

Modality
✓ Clearly the distribution is uni modal

Location
✓ The mean pocket money is $ 15.10
✓ the median class is $10 - 15

Spread
✓ The scores have a range of 35 (5–40)
✓ The standard deviation is 8.06

Shape
✓ There is gaps at 20-25 and 30-35
✓ Clusters between 5-20
✓ Positively skewed

---

2. The box plots below shows the running time, in minutes, of two groups A and B in a marathon. Compare the distributions of the two box plots and recommend which group is better.

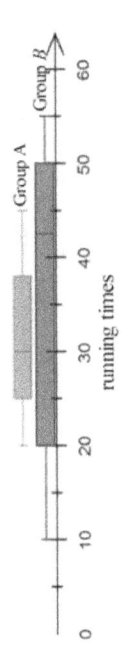

Median	spread	skewness
• Median for Group B is greater than median for Group A.	• Group B has a greater range (45) compared to Group A (25) • IQR for Group B (30) is greater than IQR for Group A (12.5).	• Group B's box plot is negatively skewed whereas Group A's box plot is positively skewed.

Although Group B has the fastest running time, 75% of Group A runners finish under 37.5 minute, showing more consistency.

3. The box plot below shows the service waiting time in two stores in the city. Compare the distributions of the two box plots and recommend which store is better.

Median	spread	skewness
• Median for both stores is the same.	• Store Q has a greater range (6.5 mins) compared to store P (5 mins) • IQR for Store P is (2) is less than IQR for Store Q (3).	• Store P's box plot is positively skewed whereas Store Q's box plot is negatively skewed.

Store Q seems to have the fastest service time. However, having a larger range means that some customers have to wait longer to be served. In addition, 75% of store P customers are served under 5 minutes compared to 7 minutes for store Q. Store P seems to be more consistent.

CHAPTER 4 : BOX PLOTS AND DESCRIBING DISTRIBUTIONS : SOLUTIONS

EXERCISE 4E

1. The histogram below shows the distances, in kilometres, walked monthly by a group of athletes in a particular sporting club in Bunbury. Describe the distribution of the distance walked.

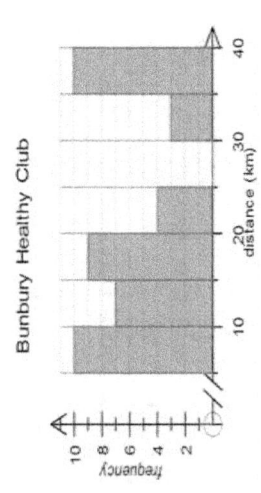
Bunbury Healthy Club

SOLUTION

Modality
✓ Clearly the distribution is bi modal having modal classes 5-10 and 35-40.

Location
✓ The mean distance walked 20.52 km ✓ the median class is 15-20 km

Spread
✓ The scores have a range of \$35 (5–40) ✓ The standard deviation is 11.36

Shape
✓ There is a gap at 25-30. ✓ Clusters between 5-25 ✓ Positively skewed

2. For the dot frequency diagram on the right, describe the distribution of the numerical data.

SOLUTION

Modality
✓ Clearly the distribution is uni modal. Mode is 2.

Location
✓ The mean is 3.61 ✓ the median is 3

Spread
✓ The scores have a range of 8. ✓ The standard deviation is 2.36

Shape
✓ Clusters between 1-3. ✓ Positively skewed

3. The histogram below shows the speed, in km/h, of vehicles in a busy highway (speed limit 70 km/h) at 9 a.m. Give a brief description of the distribution of the speed of the vehicles.

Final Destination Highway

SOLUTION

Modality
✓ Clearly the distribution is uni-modal having modal classes 70-80.

Location
✓ The mean speed is 72.2 km/h ✓ the median class is 70-80 km/h

Spread
✓ The scores have a range of 40–100 ✓ The standard deviation is 13.36

Shape
✓ There is a gap at 50-60. ✓ Clusters between 60-90 ✓ negatively skewed

4. For the given stem plot, describe the distribution of the numerical data.

Maths Test Score

```
4 | 0 2 5 8 9
3 | 2 2 5 8 8 9
2 | 1 5
1 | 1
0 | 9
```

SOLUTION

Modality
✓ Clearly the distribution is bi modal. Mode is 32 and 38.

Location
✓ The mean is 33.6 ✓ the median is 38

Spread
✓ The scores have a range of 40. ✓ The standard deviation is 11.86

Shape
✓ Clusters between 32-49. ✓ negatively skewed

CHAPTER 5

STATISTICAL INVESTIGATION PROCESS

As a statistical project or investigation, one of the tasks may be listed as follows:

- ❖ Are left handers tennis players better than right handers?
- ❖ Are boys better than girls in Mathematics?
- ❖ Children's favourite pet.
- ❖ Fitness level of soccer players of varying ages and gender.
- ❖ Genre of movies preferred by youngsters.

To be able to come up with a great statistical report, the reader must try and put into practice majority of the statistical terms and concepts learnt in the previous four chapters of this book. An attempt has been made to enumerate the different steps that will help during the investigation process.

STEP 1 – IDEAS TO EXPLORE

Carefully decide upon an idea that would be of your interest or your group investigating. Choose a topic that you are fond and familiar with. This will help and lend itself to data collection and statistical analysis. For example

- People's preferred novels: action, drama, horror etc...
- Types of vehicles at a busy intersection at different times.
- Popularity of Mathematics in Years 10 - 12.
- Characteristics of advertisement shown on television.

STEP 2 – SAMPLE SELECTION

Choose a sampling technique appropriate to your situation. Stratified sampling, random sampling, draw names out of a hat … and always choose a suitable sample size. The sample must be representative of the whole population. For example, if we are find out about the popularity of Mathematics in Years 10-12, stratified sampling might be the best where each year level would have some say in the sample.

STEP 3 - STATISTICAL INSTRUMENTS

Design the tools and instruments that are going to be used to gather the required data. For example,
- ➢ Questionnaires
- ➢ interviews
- ➢ Surveys
- ➢ Experiments and so on.

STEP 4 – DATA COLLECTION AND ORGANISATION

Collect and organise your data the simplest possible way so that it can be understood easily. For example make use of
- ➢ Tables
- ➢ Lists
- ➢ charts
- ➢ diagrams etc.

STEP 5 – STATISTICAL REPRESENTATION

Use statistical measures and graphs where appropriate. Now it's time for you to use all the statistics you have learnt so far. Try and include:

- ➢ Bar charts, pie graphs, dot frequency diagrams.
- ➢ Make use of technology to calculate all the measures of central tendency (mean, mode and median) wherever they are applicable.
- ➢ Frequency distributions and grouped frequency distributions, histograms, stem and leaf plots, box and whisker plots.
- ➢ Scatter graphs, tree diagrams, two-way tables.

STEP 6 – STATISTICAL REPORT

With the aid of the statistical representation mentioned above
- ➢ make reasonable conclusions and observations about the data you have gathered
- ➢ Evaluate the statistical instruments you used to gather the data.

Ask yourself and seek help from the members of your group
- ➢ Are your data reliable?
- ➢ Are the data collected valid or irrelevant?
- ➢ What could be done to improve the exercise?

STEP 7 – THE FINAL PRODUCT

Last you need to collate your project into a logical, neat and concise statistical report. To give your work some more weight, it is advisable that you include a title page and a good quality final copy of all your data, statistics and conclusions.

EXERCISE 5A

Using the 7 steps described above, choose one of the following topics to investigate.

1. Are rich people happier?
2. Are boys better than girls in Physics?
3. Children's favourite toy.
4. Fitness level of footy players of varying ages.
5. Genre of movies preferred by mature people.
6. Do children coming from split families tend to have lower academic progress?
7. Should IPAD be banned from classes?

CHAPTER 6

SOLVING EQUATIONS

6A SOLVING EQUATIONS USING TECHNOLOGY (CAS)

In this section, we are going to make use of the solve facility in our calculator to solve equations. In the subsequent stage of the chapter, it is recommended to try each question by a step by step approach and try to isolate the unknown.

To solve equations in CAS make use of the following steps:

$$MAIN \rightarrow ACTION \rightarrow ADVANCED \rightarrow SOLVE$$

Most calculators have similar solve facilities available which the reader can try and adapt accordingly.

EXAMPLES

Use the solve facility on your calculators to solve the following equations.

(a) $5x + 12 = 2(x - 3) + 9$

$$Solve\ (5x + 12 = 2(x - 3) + 9, x)$$
$$\{x = -3\}$$

(b) $2(3x + 7) - 2(x - 1) = 5(x + 3)$

$$Solve\ (2(3x + 7) - 2(x - 1) = 5(x + 3), x)$$
$$\{x = 1\}$$

(c) Given that $v^2 = u^2 + 2as$, find the value of a when $v = 20$, $u = 8$ and $s = 12$.

$$MENU \rightarrow NUMSOLVE \rightarrow INSERT\ EQUATION \rightarrow INPUT\ GIVEN\ VALUES \rightarrow SOLVE$$

Remember to click the bubble at a as we are solving for a.

Equation:
$v^2 = u^2 + 2 \times a \times s$

○ $v = 20$
○ $u = 8$
● $a =$
○ $s = 12$

Lower = -9E+999
Upper = 9E+999

$a = 14$

EXERCISE 6A

Use the solve facility on your calculator to solve the following equations.

1. $5x + 3 = -17$ $\quad x = -4$	2. $2x + 5 = x + 11$ $\quad x = 6$
3. $5(x - 4) = 3(x + 6)$ $\quad x = 19$	4. $3x + 4 = x + 16$ $\quad x = 6$
5. $\frac{2x+3}{5} = \frac{x-1}{3}$ $\quad x = -14$	6. $\frac{5}{x} = 10$ $\quad x = 0.5$
7. $\frac{x}{2} + \frac{x+1}{5} = \frac{2}{3}$ $\quad x = \frac{2}{3}$	8. $\frac{x}{3} + \frac{x}{2} = 10$ $\quad x = 12$

Use Num Solve to find the missing pronumerals in each case.

9. Given $v = u + at$, find a given that $v = 55$, $u = 25$ and $t = 3$. $\quad a = 10$	10. If $E = \frac{1}{2}mv^2$, find the value of m when $E = 125$ and $v = 5$. $\quad m = 10$
11. In the formula $y = mx + c$, find m given that $y = 18$, $x = 10$ and $c = 13$. $\quad m = 0.5$	12. In the formula $E = mc^2$, find m given that $E = 2000$ and $c = 5$. $\quad m = 80$
13. Given that $v^2 = u^2 + 2as$, find the value of s when $v = 30$, $u = 12$ and $a = 10$. $\quad s = 37.8$	14. Given that $A = \frac{a+b}{2} \times h$, find the value of h when $A = 88$, $a = 11$ and $b = 5$. $\quad h = 11$
15. $V = \pi r^2 h$, find the value of h given that $V = 200$ and $r = 5$. $\quad h = 2.55$	16. $V = 2\pi r^2 + 2\pi rh$, find the value of h given that $V = 500$ and $r = 4$. $\quad h = 15.89$

6B SIMPLE LINEAR EQUATIONS

A linear equation is a polynomial of degree 1. In order to solve for the unknown variable, we have to isolate the variable as shown in the examples below.

EXERCISE 6B
Solve the following linear equations.

1. $3x + 1 = 16$ **SOLUTION** Subtract 1 on both sides $3x + 1 - 1 = 16 - 1$ $3x = 15$ $x = 5$	2. $10 - 2x = 30$ **SOLUTION** Subtract 10 on both sides $10 - 2x - 10 = 30 - 10$ $-2x = 20$ $x = -10$	13. $2(x+3) - 8 = 10$ $2x + 6 = 18$ $2x = 12$ $x = 6$	14. $3(1 - 5x) = 15$ $1 - 5x = 5$ $-5x = 4$ $x = -\dfrac{4}{5}$
3. $4x - 3 = 17$ $4x = 20$ $x = 5$	4. $2x - 7 = 11$ $2x = 18$ $x = 9$	15. $2x + 3x = 50$ $5x = 50$ $x = 10$	16. $3y + 4y - 2 = 19$ $7y - 2 = 19$ $7y = 21$ $\therefore y = 3$
5. $3(x + 4) = 9$ Divide by 3 on both sides, $x + 4 = 3$ $x = -1$	6. $5(2x - 1) = 15$ Divide by 5 on both sides, $2x - 1 = 3$ $2x = 4$ $x = 2$	17. $2(3x + 1) - 3(1 - 5x) = -43$ $6x + 2 - 3 + 15x = -43$ $21x - 1 = -43$ $21x = -42$ $x = -2$	18. $4(x + 2) - 7x = -16$ $4x + 8 - 7x = -16$ $-3x = -24$ $x = 8$
7. $\dfrac{x}{5} = 6$ Cross multiply, $x = 30$	8. $\dfrac{y}{2} = -8$ Cross multiply, $y = -16$	19. $\dfrac{2x - 1}{5} - 3 = 7$ $\dfrac{2x - 1}{5} = 10$ $2x - 1 = 50$ $2x = 51$ $x = 25.5$	20. $\dfrac{10 - 2x}{5} = 4$ $10 - 2x = 20$ $-2x = 10$ $x = -5$
9. $\dfrac{z}{4} = -6$ Cross multiply, $z = -24$	10. $\dfrac{20}{x} = 5$ $5x = 20$ $x = 4$	21. $\dfrac{7x + 10}{5} - 2 = 7$ $\dfrac{7x + 10}{5} = 9$ $7x + 10 = 45$ $7x = 35$ $x = 5$	22. $\dfrac{12 - 3x}{6} = 5$ $12 - 3x = 30$ $-3x = 18$ $x = -6$
11. $\dfrac{x - 6}{3} = 2$ $x - 6 = 6$ $x = 12$	12. $\dfrac{x + 4}{2} = 3$ $x + 4 = 6$ $x = 2$		

CHAPTER 6 : SOLVING EQUATIONS: SOLUTIONS

6C EQUATIONS HAVING UNKNOWN ON BOTH SIDES

EXERCISE 6C
Solve the following linear equations.

1. $3x + 7 = x + 13$ **SOLUTION** Subtract x on both sides $3x - x + 7 = x - x + 13$ $2x + 7 = 13$ Subtract 7 on both sides $2x + 7 - 7 = 13 - 7$ $2x = 6$ $\therefore x = 3$	**2.** $5x - 3 = 2x + 12$ **SOLUTION** Subtract 2x on both sides $5x - 2x - 3 = 2x - 2x + 12$ $3x - 3 = 12$ Add 3 on both sides $3x - 3 + 3 = 12 + 3$ $3x = 15$ $\therefore x = 5$
3. $2x + 11 = 3x + 5$ $11 - 5 = 3x - 2x$ $x = 6$	**4.** $6x - 1 = 4x + 11$ $6x - 4x = 11 + 1$ $2x = 12$ $x = 6$
5. $2(3x - 4) = 5x + 6$ $6x - 8 = 5x + 6$ $6x - 5x = 6 + 8$ $x = 14$	**6.** $3(2x + 5) = 4x - 10$ $6x + 15 = 4x - 10$ $6x - 4x = -10 - 15$ $2x = -25$ $x = -12.5$
7. $7(2x + 5) = 3(x + 4)$ $14x + 35 = 3x + 12$ $14x - 3x = 12 - 35$ $11x = -23$ $x = -\dfrac{23}{11}$	**8.** $5(x + 3) = 4(x + 1)$ $5x + 15 = 4x + 4$ $5x - 4x = 4 - 15$ $x = -11$

6D SOLVING EQUATIONS USING CROSS MULTIPLICATION

EXERCISE 6D
Solve the following linear equations.

1. $\dfrac{3x + 1}{2} = \dfrac{2x - 6}{3}$ **SOLUTION** Cross multiplying, we have $3(3x + 1) = 2(2x - 6)$ Expand both sides $9x + 3 = 4x - 12$ Subtract 4x on both sides $9x - 4x + 3 = 4x - 4x - 12$ $5x + 3 = -12$ Subtract 3 on both sides $5x + 3 - 3 = -12 - 3$ $5x = -15$ $\therefore x = -3$	**2.** $\dfrac{2x + 1}{7} = \dfrac{x - 1}{4}$ Cross multiplying, we have $4(2x + 1) = 7(x - 1)$ $8x + 4 = 7x - 7$ $8x - 7x = -7 - 4$ $x = -11$
3. $\dfrac{10 - 3x}{5} = \dfrac{2x + 1}{2}$ Cross multiplying, we have $2(10 - 3x) = 5(2x + 1)$ $20 - 6x = 10x + 5$ $20 - 5 = 10x + 6x$ $16x = 15$ $x = \dfrac{15}{16}$	**4.** $\dfrac{x + 4}{3} = \dfrac{5 - 2x}{4}$ Cross multiplying, we have $4(x + 4) = 3(5 - 2x)$ $4x + 16 = 15 - 6x$ $4x + 6x = 15 - 16$ $10x = -1$ $x = \dfrac{-1}{10}$
5. $\dfrac{4}{2x - 1} = \dfrac{5}{x + 3}$ Cross multiplying, we have $4(x + 3) = 5(2x - 1)$ $4x + 12 = 10x - 5$ $12 + 5 = 10x - 4x$ $6x = 17$ $x = \dfrac{17}{6}$	**6.** $\dfrac{8}{2x + 3} = \dfrac{2}{x + 1}$ Cross multiplying, we have $8(x + 1) = 2(2x + 3)$ $8x + 8 = 4x + 6$ $8x - 4x = 6 - 8$ $4x = -2$ $x = -0.5$

6E USING LCM TO SOLVE EQUATIONS

Many equations cannot be solved by the method of cross multiplication. The easiest way would be to find the LCM of the denominators and multiply each term by the LCM.

EXERCISE 6E

1. Solve $\quad \dfrac{x}{2} + \dfrac{x+1}{5} = \dfrac{2}{3}$

   SOLUTION
   The LCM of 2,3 and 5 is 30.
   Multiplying each term by 30, we have
   $$\dfrac{x}{2} \times 30 + \dfrac{x+1}{5} \times 30 = \dfrac{2}{3} \times 30$$
   $$15x + 6(x+1) = 20$$
   $$15x + 6x + 6 = 20$$
   $$21x = 14$$
   $$\therefore x = \dfrac{14}{21} = \dfrac{2}{3}$$

2. Solve $\quad \dfrac{4x-3}{6} = \dfrac{x-1}{3} + 2$

   SOLUTION
   LCM of 3 and 6 is 6.
   So we multiply each term by 6.
   $$\dfrac{4x-3}{6} \times 6 = \dfrac{x-1}{3} \times 6 + 2 \times 6$$
   $$4x - 3 = 2(x-1) + 12$$
   $$4x - 3 = 2x - 2 + 12$$
   $$4x - 3 = 2x + 10$$
   $$2x = 13$$
   $$x = 6.5$$

3. Solve $\dfrac{x}{2} + \dfrac{2x+1}{3} = \dfrac{1}{4}$

   The LCM of 2,3 and 4 is 12.
   Multiplying each term by 12, we have
   $$\dfrac{x}{2}(12) + \dfrac{2x+1}{3}(12) = \dfrac{1}{4}(12)$$
   $$6x + 4(2x+1) = 3$$
   $$6x + 8x + 4 = 3$$
   $$14x = -1$$
   $$\therefore x = \dfrac{-1}{14}$$

4. Solve $\dfrac{3x+2}{5} - 1 = \dfrac{4x-1}{4}$

   The LCM of 4 and 5 is 20.
   Multiplying each term by 20, we have
   $$\dfrac{3x+2}{5}(20) - 1(20) = \dfrac{4x-1}{4}(20)$$
   $$4(3x+2) - 20 = 5(4x-1)$$
   $$12x + 8 - 20 = 20x - 5$$
   $$8 - 20 + 5 = 20x - 12x$$
   $$\therefore 8x = -7$$
   $$x = \dfrac{-7}{8}$$

5. Solve $\dfrac{3x-4}{2} + \dfrac{1}{3} = \dfrac{x}{4}$

   $$\dfrac{3x-4}{2}(12) + \dfrac{1}{3}(12) = \dfrac{x}{4}(12)$$
   $$6(3x-4) + 4 = 3x$$
   $$18x - 24 + 4 = 3x$$
   $$18x - 3x = 24 - 4$$
   $$\therefore 15x = 20$$
   $$x = \dfrac{4}{3}$$

6. Solve $\dfrac{x}{2} + \dfrac{x}{3} = 5$

   $$\dfrac{x}{2}(6) + \dfrac{x}{3}(6) = 5(6)$$
   $$3x + 2x = 30$$
   $$5x = 30$$
   $$x = 6$$

6F APPLICATIONS

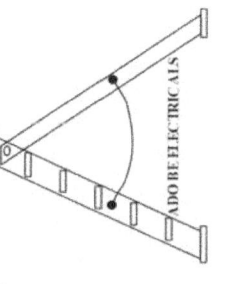

1. John is a full time electrician and the cost (C $) for hiring him is made up of a callout fee of $80 and a fixed rate of $42 per hour. The equation for hiring John for h hours is given by $C = 0.95(80 + 42h)$.

   (a) John offers a discount to all his clients. State how much discount he allows to his clients.

   5%

   (b) Calculate the cost of hiring John for 6 hours.

   $$C = 0.95(80 + 42 \times 6) = \$315.40$$

   (c) Mrs Will paid $335.35 for some repairs undergone. For how many hours did she hire the electrician?

   Solve $(335.35 = 0.95(80 + 42h), h$
   $$h = 6.5$$

2. Julia is employed by JK Real Estates. She earns $300 per week plus 1.25% commission on selling a property.
   Her monthly revenue ($R) for selling a property worth $P can be modelled by the equation
   $$R = 3000 + 0.0125P$$

   (a) In a particular week she sold a property for $280 000. Determine her revenue for that week.
   $$R = 3000 + 0.0125 \times 280\,000 = \$6500$$

   (b) During another week, she earned $6050 when she managed to sell one property only. Calculate the price the property was sold for.
   Solve $(3000 + 0.0125P = 6050, P)$
   $$P = \$244\,000$$

CHAPTER 6 : SOLVING EQUATIONS: SOLUTIONS

3. When an agent of a particular software company sells IT products to the general public, his monthly salary is given by
   $S = 2500 + 3N$, where N is the amount he sells in thousands of dollars.
   Find the agent's salary if he sells
   (a) $125 000 worth of goods during the month of July,

   $S = 2500 + 3 \times 125 = \$2875$

   (b) $90 000 worth of IT products in August.

   $S = 2500 + 3 \times 90 = \$2770$

   (c) During the month of December his salary was $2980. Find out the value of the products he sold.

   $Solve\ (2980 = 2500 + 3N, N)$
   $N = \$160\,000$

4. Jimmy is a broker and is self-employed. His income depends on the value of the loan he is able to get approved for his clients. His income can be summarised by the formula
   $I = 400 + 1.5L$, where L is the value of the loan in hundreds of dollars.
   (a) Calculate Jimmy's income if he settles a loan of $250 000.

   $I = 400 + 1.5(2500) = \$4150$

   (b) For one of his transaction, Jimmy was paid $5200. Calculate the value of the loan he got approved for his client.

   $Solve\ (5200 = 400 + 1.5L, L)$
   $L = 3200$
   $Value\ of\ loan = \$320\,000$

MATHEMATICS APPLICATIONS UNIT 2

5. The centripetal force F acting on a particle travelling at a speed of v m/s, having mass m kg and radius r metres is given by $F = \frac{mv^2}{r}$.

   (a) Determine the centripetal force of a car of 800 kg if it travels a curve of radius 400m at 12 m/s.

   $F = \frac{800 \times 12^2}{400} = 288\ N$

   (b) If the centripetal force of an object of mass 60 kg travelling in a circular path of radius 30m is 200N, determine the particle's velocity.

   $Solve\ (200 = \frac{60 \times v^2}{30}, v)$
   $v = 10\ m/s$

6. Appleby Senior High usually hires a school bus to pick and drop its students. The cost ($C) associated with hiring the bus for d days and travelling k kilometres is given by
   $C = 50 + 25d + 10k$

   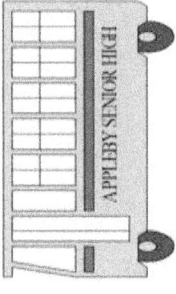

   (a) Explain the significance of 50 in the above equation.

   a fixed cost

   (b) Calculate the cost of hiring the bus for 10 days and travelling a total distance of 165 km.

   $C = 50 + 25(10) + 10(165) = \$1950$

   (c) In a particular month, the school hired the bus for 21 days for a total cost of $3005. Determine the number of kilometres the bus travelled.

   $Solve\ (3005 = 50 + 25(21) + 10k, k)$
   $k = 243\ km$

   (d) The school Principal was concerned that the cost of hiring the school bus was way past the available budget. She appointed all the top mathematics teachers and assigned them a task to work out a minimal spanning tree whereby all the students using the school bus could be picked up and dropped by travelling the least possible distance. If the school's budget is $2500 for 21 days, determine the maximum distance that the bus could travel without exceeding the budget.

   $Solve\ (2500 = 50 + 25(21) + 10k, k)$
   $k = 192.5\ km$

CHAPTER 7

USING EQUATIONS TO SOLVE PROBLEMS

7A   ALGEBRAIC EQUATIONS AND EXPRESSIONS

1. Complete the following table by converting the mathematical statements into **algebraic expressions**. Use x to represent the number.

Mathematical statement	Algebraic Expression
(a) Add six to the number	$x + 6$
(b) Multiply the number by three then subtract five	$3x - 5$
(c) Double the number then add three	$2x + 3$
(d) Twice a number divided by five	$\dfrac{2x}{5}$
(e) Two less than five times a number	$5x - 2$
(f) Six more than five times a number	$5x + 6$
(g) Three times a number minus eleven	$3x + 11$
(h) Two less than three times a number	$3x - 2$
(i) Five times the sum of x and 3	$5(x + 3)$
(j) Three less than twice the sum of x and five	$2(x + 5) - 3$

2. Complete the following table by converting the mathematical statements into **algebraic equations**. Use x to represent the number.

Mathematical statement	Algebraic Equation
(a) Think of a number, add two and the result is 10	$x + 2 = 10$
(b) Think of a number, subtract five and the result is 11	$x - 5 = 11$
(c) When a number is decreased by six and the result multiplied by four, the final answer is 32	$4(x - 6) = 32$
(d) When a number is increased by two and the result multiplied by three, the final answer is 15	$3(x + 2) = 15$
(e) Think of a number, multiply it by three, subtract four from the result and the final answer is 8	$3x - 4 = 8$
(f) When a certain number is subtracted from 25 and the result multiplied by six, the final result is 60.	$6(25 - x) = 60$
(g) Think of a number, add one then divide the result by two and the final result is 4	$\dfrac{x + 1}{2} = 4$
(h) When two consecutive numbers are added together the answer is 25.	$x + x + 1 = 25$
(i) The sum of three consecutive even numbers is 36.	$2x + 2x + 2 + 2x + 4 = 36$

3. For these questions, rewrite the statements as equations involving x and hence solve them.

(a) When a number is decreased by four and the result multiplied by three, the final answer is fifteen. $$3(x - 4) = 15$$ $$3x - 12 = 15$$ $$3x = 27$$ $$\therefore x = 9$$ OR USING TECHNOLOGY $Solve\ (3(x - 4) = 15, x)$ $\{x = 9\}$	(b) When a number is increased by nine and the result multiplied by five, the final answer is 50. $Solve\ (5(x + 9) = 50, x)$ $\{x = 1\}$
(c) When two consecutive numbers are added together the answer is 101. $Solve\ (x + x + 1 = 101, x)$ $\{x = 50\}$ *the numbers are 50 and 51*	(d) Think of a number, multiply it by two, subtract seven from the result and the final answer is 11. $Solve\ (2x - 7 = 11, x)$ $\{x = 9\}$
(e) The sum of three consecutive even numbers is 30. $Solve\ (2x + 2x + 2 + 2x + 4 = 30, x)$ $\{x = 4\}$ *the numbers are 8, 10 and 12*	(f) When a certain number is subtracted from ten and the result multiplied by three, the final result is -30. $Solve\ (3(10 - x) = -30, x)$ $\{x = 20\}$
(g) The sum of two consecutive odd numbers is 44. $Solve\ (2x + 1 + 2x + 3 = 44, x)$ $\{x = 10\}$ *the numbers are 21 and 23*	(h) Think of a number, add eleven then divide the result by two and the final result is 10. $Solve\ \left(\dfrac{x + 11}{2} = 10, x\right)$ $\{x = 9\}$
(i) Think of a number, add one then divide the result by four and my final answer is 7. $Solve\ \left(\dfrac{x + 1}{4} = 7, x\right)$ $\{x = 27\}$	

CHAPTER 7 USING EQUATIONS TO SOLVE PROBLEMS : SOLUTIONS

7B MATH PYRAMIDS

How does the pyramid work?

Add the 2 numbers next to each other in the bottom row and write the answer in the circle above those 2 numbers as shown on the left.

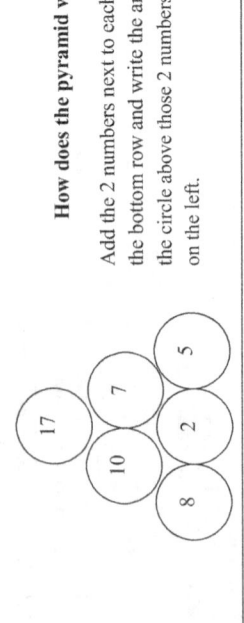

EXERCISE 7B
Complete the following pyramids.

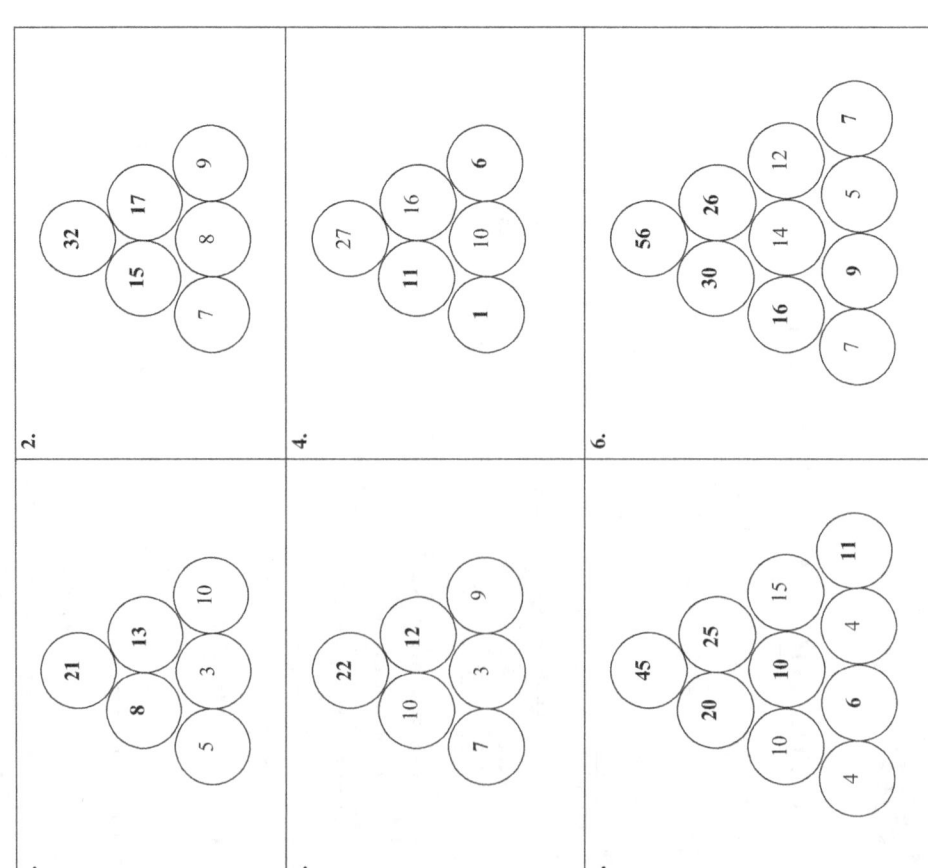

MATHEMATICS APPLICATIONS UNIT 2

In the following pyramids work out the value of x. Question 7 has been done as an example.

7.
$x + 5 + 7 = 20$
$x + 12 = 20$
$\therefore x = 8$

8.
$x + 9 + 22 = 35$
$x + 31 = 35$
$\therefore x = 4$

9.
$2(x + 10) = 50$
$x + 10 = 25$
$\therefore x = 15$

10.
$4x = 100$
$\therefore x = 25$

11.
$x + 8 + 19 = 39$
$x + 27 = 39$
$\therefore x = 12$

12.
$x + 12 + x + 8 = 40$
$2x + 20 = 40$
$\therefore x = 10$

13.
$2(x + 20) = 72$
$x + 20 = 36$
$\therefore x = 16$

14.
$8x = 24$
$\therefore x = 3$

CHAPTER 7 USING EQUATIONS TO SOLVE PROBLEMS : SOLUTIONS

15. Study the two number patterns below carefully and then complete the sentences.

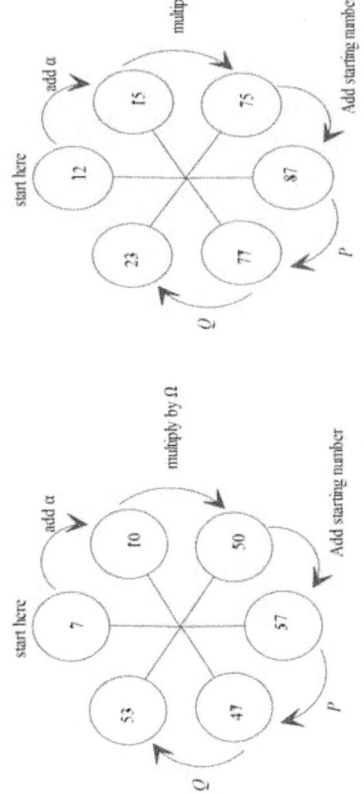

(a) The value of $\alpha = 3$

(b) The value of $\Omega = 5$

(c) What instruction must P represent? **Subtract 10**

(d) What does Q stand for? **Subtract from 100**

Now complete the following number patterns diagrams.

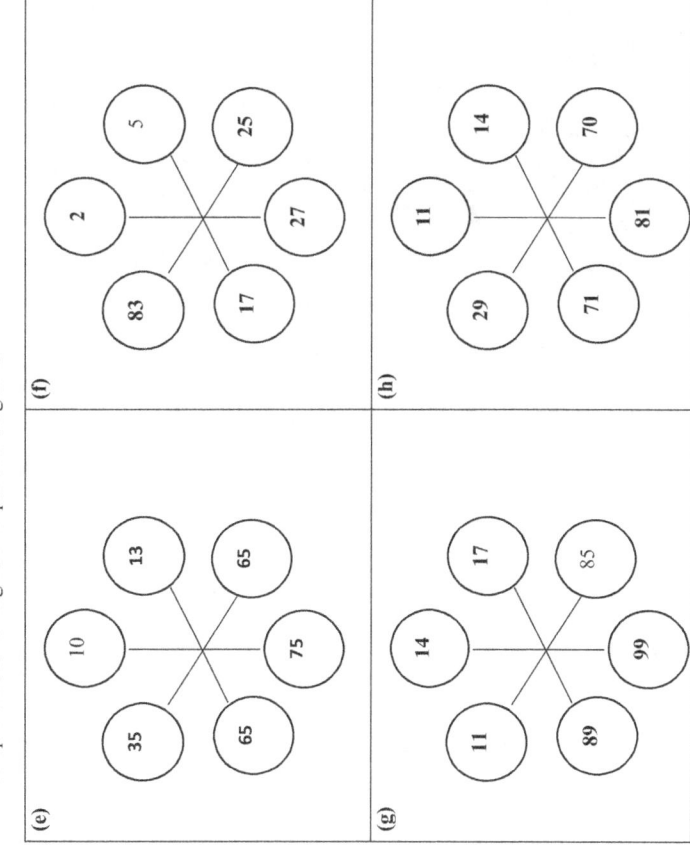

MATHEMATICS APPLICATIONS UNIT 2

7C   APPLICATIONS

EXAMPLE 1

Peter is 6 years younger than his sister Jane. In ten years' time, the sum of their ages will be 40 years. Find their present ages.

SOLUTION

First, we express the information in a table as shown.

	Peter	Jane
now	x	x + 6
In 10 years' time	x + 10	x + 6 + 10 = x + 16

Since the sum of their ages will be 40 years after 10 years,

$$x + 10 + x + 16 = 40$$
$$2x + 26 = 40$$
$$2x = 14$$
$$x = 7$$

Hence Peter is 7 years old and his Jane is 13 years of age.

EXAMPLE 2

Tickets for a show were priced at $4, $7 and $12. The number of $4 tickets sold was three times the number of $7 tickets. The number of $12 tickets sold was 20 more than the number of $7 tickets.

The number of $7 tickets sold was x.

(a) Find an expression, in terms of x, for the total sum of money received from the sale of the tickets.

Price	$4	$7	$12
Number of tickets	3x	x	x + 20

$$\begin{aligned}\text{Total sum of money received} &= 4 \times 3x + 7 \times x + 12(x + 20) \\ &= 12x + 7x + 12x + 240 \\ &= 31x + 240\end{aligned}$$

(b) Given that $1790 was received from the sale of the tickets, form an equation in x. Solve this equation and hence find the total number of tickets that were sold.

Total number of tickets sold
$= 3(50) + 50 + (50 + 20) = 270\ tickets.$

$$\boxed{\begin{array}{c}solve\,(31x + 240 = 1790, x) \\ \{x = 50\}\end{array}}$$

CHAPTER 7 USING EQUATIONS TO SOLVE PROBLEMS : SOLUTIONS

EXERCISE 7C

1. David is 3 years younger than his sister Rita. Given that the sum of their ages is 31 years, determine their respective ages.

Let David' sage $= x$
Rita's age $= x + 3$
$Solve\ (x + x + 3 = 31, x)$
$\therefore x = 14$
David is 14 and Rita is 17.

2. Rahul is x years old. John, his father, is twice as old as him. In two years' time the sum of their ages will be 79 years. Form an equation in terms of x and hence determine their present ages.

	Rahul	John
Now	x	$2x$
In two years	$x + 2$	$2x + 2$

$Solve\ (x + 2 + 2x + 2 = 79, x)$
$\therefore x = 25$
Rahul is 25 and John is 50 at present.

3. The length of a rectangle is 5 cm more than its width. If the perimeter of rectangle is 38 cm, find the dimensions of the rectangle.

$Solve\ (2x + 2(x + 5) = 38, x)$
$\therefore x = 7$
width is 7 cm and length is 12 cm.

4. A 240 m long wire is used to fence a rectangular plot whose length is twice its width. Find the length and width of the plot.

$Solve\ (x + x + 2x + 2x = 240, x)$
$\therefore x = 40\ m$
width is 40m and length is 80m.

5. The diagram shows an equilateral triangle of side x cm and a square whose edge is 2cm greater than that of the triangle. Given that the perimeter of square is 20 cm more than the perimeter of the triangle, determine the value of x.

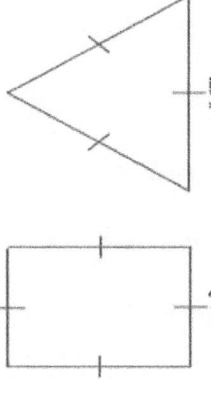

Perimeter of triangle $= 3x$
Perimeter of square $= 4(x + 2)$
$Solve\ (3x = 4(x + 2), x)$
$\therefore x = 12\ cm$

6. A cell phone cap plan costs $30 per month for unlimited calls plus $0.12 per text message.
(a) Write a linear model that represents the monthly cost of this cell phone plan if the user sends x text messages.

$C = 30 + 0.12x$

(b) A particular user paid $102 during the month of December. How many texts did he send?

$Solve\ (30 + 0.12x = 102, x)$
$\therefore x = 600$
He sent 600 messages.

7. A computer salesperson earns a basic salary of $30 000 per year plus a commission of $190 for every computer he sells.
(a) Write an equation that shows the total amount of income the salesperson earns, if he sells x computers in a year.

$I = 30000 + 190x$

(b) How many computers would the salesperson need to sell to earn a total income of $87000?

$Solve\ (30000 + 190x = 87000, x)$
$x = 300$
He sells 300 computers.

8. At an outdoor cinema, children's tickets cost $4 each and adult tickets cost $7 each. The total amount of money earned from ticket sales equals $395. The number of children was 5 more than twice the number of adults. Work out how many adults attended the movie show.

Let the number od adults be x
Number of children $= 2x + 5$
$Solve\ (7x + 4(2x + 5) = 395, x)$
$\therefore x = 25$
25 adults attended the show.

9. Al, Ben and Connor are three friends and run a small business in the city. At the end of the financial year, they made a profit of $42 000. Sharing the profits, Ben earns twice as much as Al and Connor earns $2000 more than Al. Determine how much each partner received.

Suppose Al receives x
Ben will receive $2x$ and
Connor receives $x + 2000$.
$Solve\ (x + 2x + x + 2000 = 42000, x)$
$\therefore x = 10000$
Al receives 10000, Ben 20000 and Connor receives $12000.

10. Tickets for a school play were priced at $5, $8 and $10. The number of $5 tickets sold was twice the number of $8 tickets. The number of $10 tickets sold was 80 more than the number of $8 tickets. The number of $8 tickets sold was x.
(a) Find an expression, in terms of x, for the total sum of money received from the sale of the tickets.

$5(2x) + 8x + 10(x + 80)$

(b) Given that $4160 was received from the sale of the tickets, form an equation in x. Solve this equation and state the number of $8 tickets that were sold.

$Solve\ (5(2x) + 8x + 10(x + 80) = 4160, x)$
$\therefore x = 120$

CHAPTER 7 USING EQUATIONS TO SOLVE PROBLEMS

7D SOLVING EQUATIONS USING RATIOS

Ratios can be used to solve simple algebraic equations. It can also be used to solve similar shapes problems. The examples that follow attempt to show how ratios can be used efficiently to work out unknowns in equations.

EXAMPLES

Find the value of the pronumeral in each case.

(a) $x : 2 = 5 : 1$

SOLUTION

Express the ratios as fractions

$$\frac{x}{2} = \frac{5}{1}$$

Cross multiply, we have

$$x = 10$$

(b) $3 : y = 18 : 9$

SOLUTION

Express the ratios as fractions

$$\frac{3}{y} = \frac{18}{9}$$

Cross multiply, we have

$$18y = 27$$
$$\therefore y = 1.5$$

(c) Use ratios to find the height of Tom.

SOLUTION

In $\triangle ABC$ and $\triangle ADE$,

$\angle CAB = \angle DAE$ (Common angle)

$\angle CBA = \angle DEA = 90°$

Hence $\angle ACB = \angle ADE$ (sum of angles in a triangle = 180°)

$\therefore \triangle ABC \sim \triangle ADE$

Using ratios, we have

$x : 20 = 4 : 45$

Express the ratios as fractions

$$\frac{x}{20} = \frac{4}{45}$$

Cross multiply, we have

$45x = 80 \therefore x = 1.78m$

Hence Tom is 1.78m tall.

EXERCISE 7D

Find the value of the pronumeral in each case.

1. $x : 6 = 1 : 3$

$$\frac{x}{6} = \frac{1}{3}$$
$$3x = 6$$
$$x = 2$$

2. $2 : y = 36 : 3$

$$\frac{2}{y} = \frac{36}{3}$$
$$36y = 6$$
$$y = \frac{1}{6}$$

3. $x : 18 = 1 : 9$

$$\frac{x}{18} = \frac{1}{9}$$
$$9x = 18$$
$$x = 2$$

4. $2.5 : x = 12 : 60$

$$\frac{2.5}{x} = \frac{12}{60}$$
$$12x = 150$$
$$x = 12.5$$

5. $5 : y = 7 : 8$

$$\frac{5}{y} = \frac{7}{8}$$
$$7y = 40$$
$$y = \frac{40}{7}$$

6. $x : 3 = 6 : 1.5$

$$\frac{x}{3} = \frac{6}{1.5}$$
$$1.5x = 18$$
$$x = 12$$

7. The distance run by Peter and Pan is in the ratio 4 : 5. The distance run by Peter is 14km. Find the distance run by Pan.

$4 : 5 = 14 : x$

$$\frac{4}{5} = \frac{14}{x}$$
$$4x = 70$$
$$x = 17.5$$

Pan runs 17.5 km.

8. The ratio of boys to girls in a Year 11 Mathematics class is 3:4. If the number of boys is 9, find the number of girls.

$3 : 4 = 9 : x$

$$\frac{3}{4} = \frac{9}{x}$$
$$3x = 36$$
$$x = 12$$

$\therefore$ ***There are 12 girls.***

9. Given that $w : 6 = 2 : 5$, find the value of w.

$$\frac{w}{6} = \frac{2}{5}$$
$$5w = 12$$
$$w = 2.4$$

10. Given that $5x = 3y$, find the ratio $x : y$.

$$\frac{x}{y} = \frac{3}{5}$$
$$x : y = 3 : 5$$

CHAPTER 7 USING EQUATIONS TO SOLVE PROBLEMS : SOLUTIONS

11. Determine the height of the tree using ratios given that $\triangle ABC \sim \triangle ADE$.

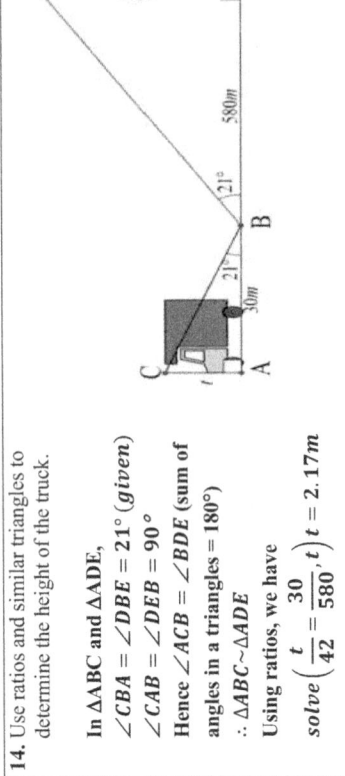

$$Solve\left(\frac{2.4}{22.4} = \frac{1.75}{h}, h\right)$$
$$h = 16.3\ m$$

12. Use ratios to determine the height of the coconut tree given that the two given triangles are similar.

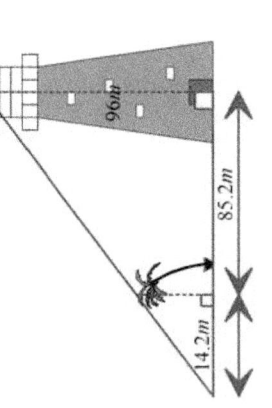

$$Solve\left(\frac{h}{96} = \frac{14.2}{99.4}, h\right)$$
$$h = 13.71\ m$$
The tree is 13.71 m high.

13. The diagram shows a 2.5m high sailing boat and a huge ship. Using the measurements determine the height of the ship.

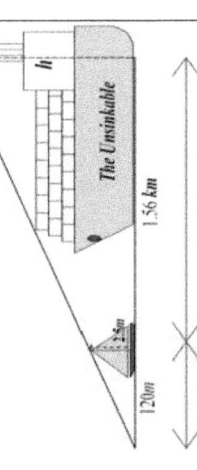

$$1.56\ km = 1560\ m$$
$$Solve\left(\frac{h}{2.5} = \frac{1680}{120}, h\right)$$
$$h = 35\ m$$
The ship is 35 m high.

14. Use ratios and similar triangles to determine the height of the truck.

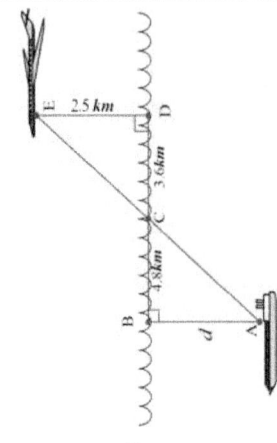

In $\triangle ABC$ and $\triangle ADE$,
$\angle CBA = \angle DBE = 21°\ (given)$
$\angle CAB = \angle DEB = 90°$
Hence $\angle ACB = \angle BDE$ (sum of angles in a triangles = $180°$)
$\therefore \triangle ABC \sim \triangle ADE$
Using ratios, we have
$$solve\left(\frac{t}{42} = \frac{30}{580}, t\right) t = 2.17\ m$$

15. Find the depth of the submarine under water level.

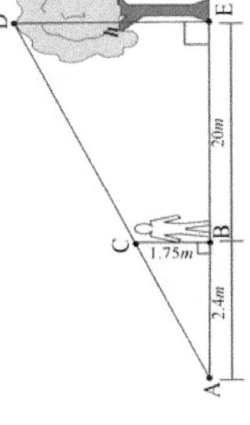

In $\triangle ABC$ and $\triangle CDE$,
$\angle ACB = \angle ECD\ (vertically\ opp.)$
$\angle ABC = \angle CDE = 90°\ (given)$
Hence $\angle BAC = \angle CED$ (sum of angles in a triangles = $180°$)
$\therefore \triangle ABC \sim \triangle CDE$
Using ratios, we have
$$solve\left(\frac{d}{2.5} = \frac{4.8}{3.6}, d\right) d = 3.33\ km$$

CHAPTER 8

LINEAR RELATIONSHIPS

8A GRADIENT

Consider the four diagrams below, all showing ladders held in a different direction and angle.

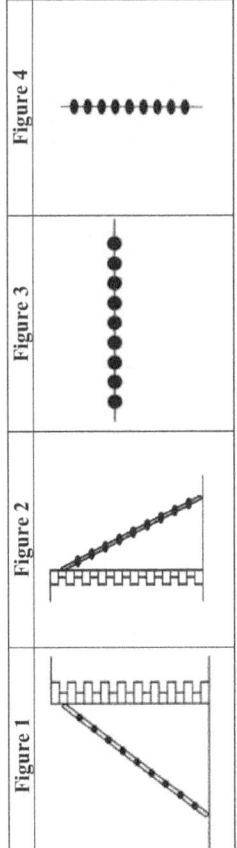

Figure 1	Figure 2	Figure 3	Figure 4

Comparing Figures 1 and 2, obviously Figure 2 seems steeper as it has a greater slope. The slope of the ladders represents the gradient of the ladder. Note that,

- ❖ Figure 1 has a positive gradient, sloping upward from left to right.
- ❖ Figure 2 has a negative gradient sloping downwards from left to right.
- ❖ Figure 3 has no slope as it is horizontal. The gradient is zero.
- ❖ Figure 4 being vertical, we say the gradient is undefined.

EXERCISE 8A

For each of the lines below, determine whether it has a positive, negative, undefined or zero gradient.

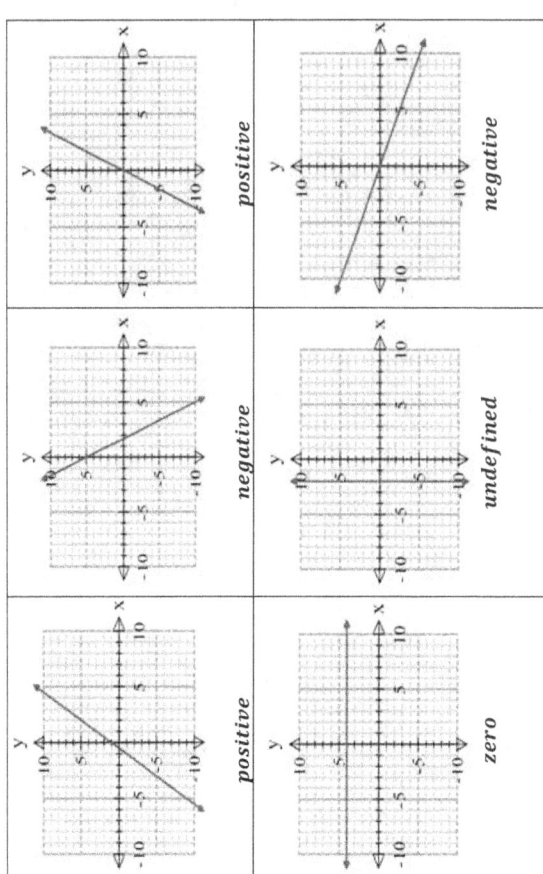

positive	negative	positive
zero	undefined	negative

8B DETERMINING GRADIENT OF A LINE FROM A GRAPH

As we have seen earlier, the slope of the ladder represents the gradient. But how do we calculate the gradient?

Gradient is defined as the ratio of the vertical increase (rise) to the horizontal increase (run).

Gradient of AB = $\frac{RISE}{RUN} = \frac{length\ of\ BC}{length\ of\ AC}$

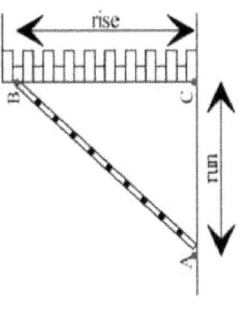

EXAMPLE

Determine the gradient of the following straight lines.

1.
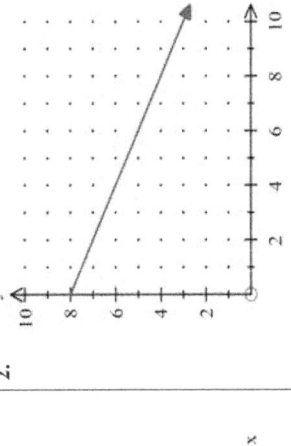

2.
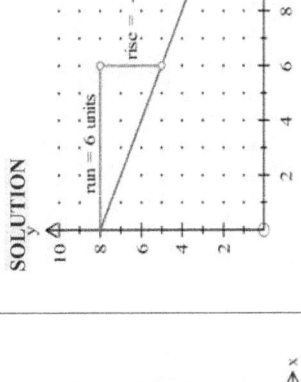

SOLUTION

1.

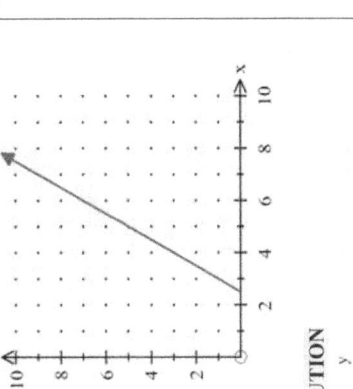

To determine the gradient, we can draw any size triangles as far the points lie exactly on the given line.

Gradient = $\frac{rise}{run} = \frac{4}{2} = 2$.

Now, a gradient of 2 implies every one unit we move to the right we have to move two units up to reach the line again.

SOLUTION

As we can see for each unit we move right, the graph is not rising as such, it is going down. This is the reason the rise is negative and thereby giving a negative gradient as a result.

Gradient = $\frac{rise}{run} = \frac{-3}{6} = -\frac{1}{2}$.

Now, a gradient of -0.5 implies every one unit we move to the right we have to move down 0.5 unit to touch the graph again.

EXERCISE 8B

Determine the gradient (m) of the following straight lines.

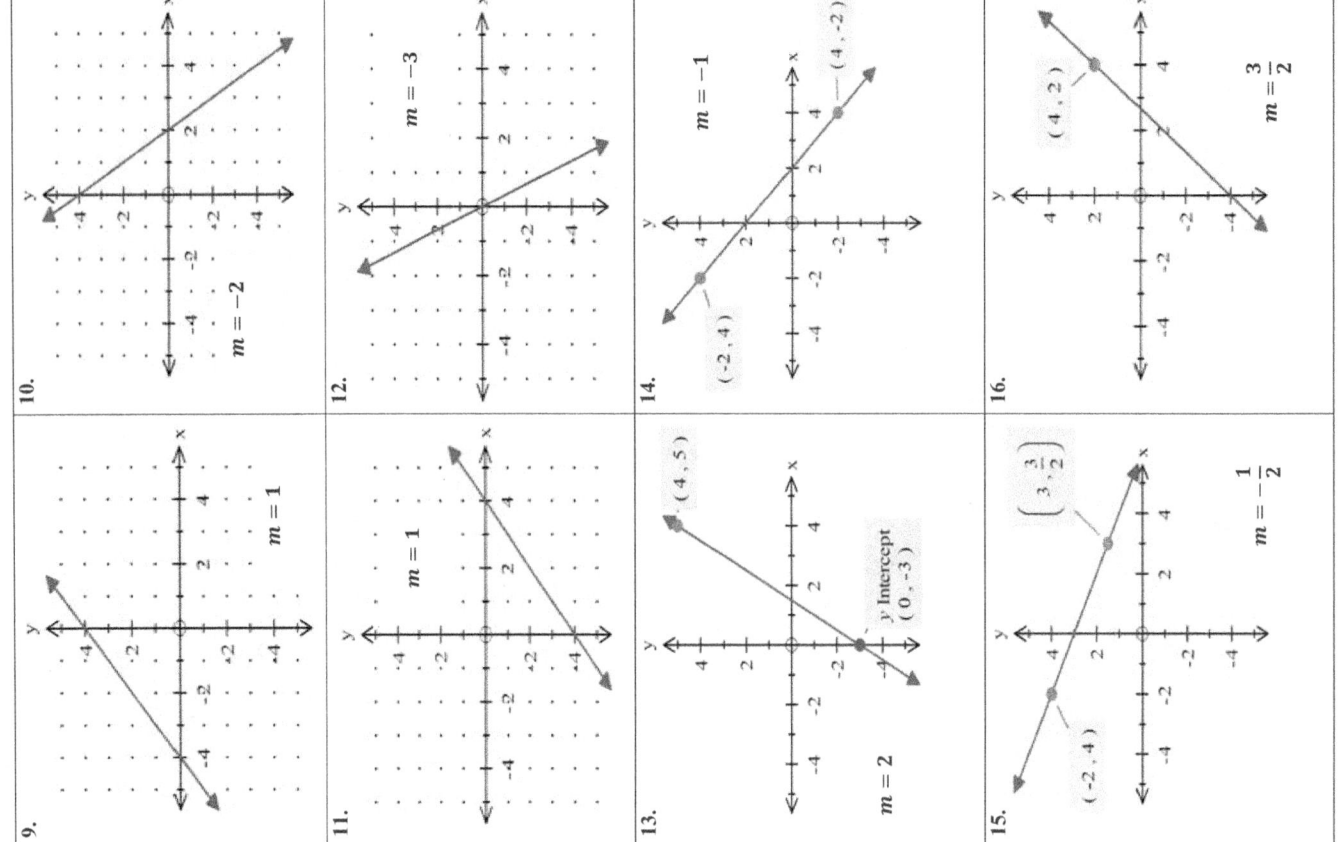

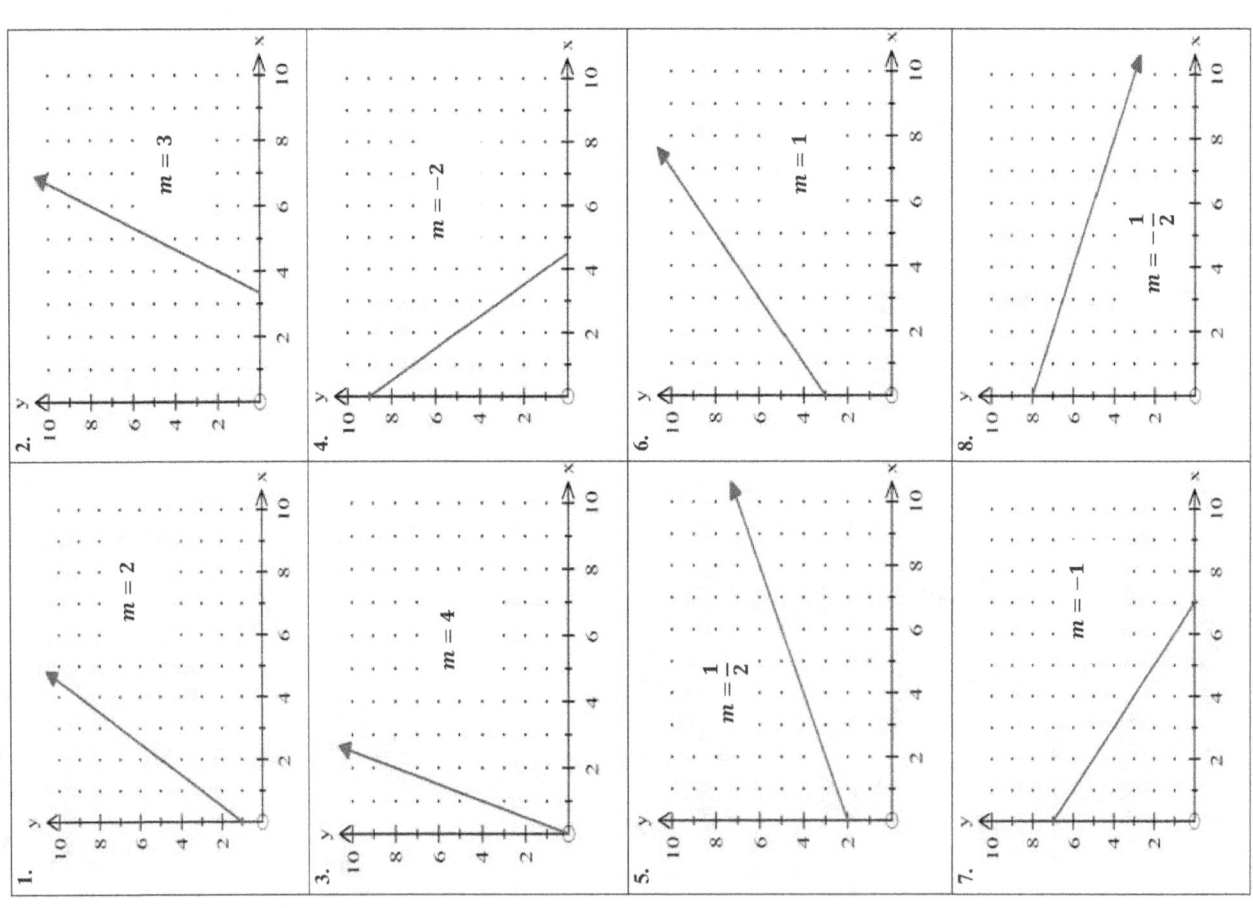

CHAPTER 8 : LINEAR RELATIONSHIPS: SOLUTIONS

8C DETERMINING GRADIENT GIVEN TWO POINTS

Now we are going to derive the general rule when we have to find the gradient of a line segment joining two points.

To find the gradient (**m**) of a line joining two points $A(x_1, y_1)$ and $B(x_2, y_2)$, use the formula

$$m = \frac{y_2 - y_1}{x_2 - x_1}$$

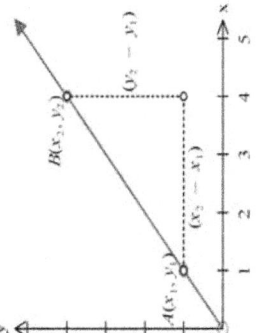

EXERCISE 8C

Find the gradient (m) of the line joining the points.

Points	Gradient (m)	Points	Gradient (m)
1. (2,3) and (4, 7)	$m = \frac{7-3}{4-2} = \frac{4}{2} = 2$	**2.** (0,2) and (3, 5)	$m = \frac{5-2}{3-0} = \frac{3}{3} = 1$
3. (1, -4) and (3, 6)	$m = \frac{6-(-4)}{3-1} = \frac{10}{2} = 5$	**4.** (2, -3) and (1, 0)	$m = \frac{0-(-3)}{1-2}$ $= \frac{3}{-1} = -3$
5. (1,2) and (5, 6)	$m = \frac{6-2}{5-1} = \frac{4}{4} = 1$	**6.** (1,2) and (8, 10)	$m = \frac{10-2}{8-1} = \frac{8}{7}$
7. (1, 3) and (3, 9)	$m = \frac{9-3}{3-1} = \frac{6}{2} = 3$	**8.** (5, 4) and (3, -2)	$m = \frac{-2-4}{3-5}$ $= \frac{-6}{-2} = 3$
9. (-1,6) and (3, 10)	$m = \frac{10-6}{3-(-1)}$ $= \frac{4}{4} = 1$	**10.** (0,2) and (3, 5)	$m = \frac{5-2}{3-0} = \frac{3}{3} = 1$
11. (-2, 10) and (3, 10)	$m = \frac{10-10}{3-(-2)}$ $= \frac{0}{5} = 0$	**12.** (2, 4) and (2, 8)	$m = \frac{8-4}{2-2}$ $= \frac{4}{0}$ (undefined)

8D FINDING GRADIENT AND Y-INTERCEPT FROM AN EQUATION

EXAMPLES

Express each of the following in the form $y = mx + c$, where necessary. Hence state the gradient (**m**) and the y-intercept (**c**).

1. $y = 3x + 5$
Compare this equation with $y = mx + c$.
Clearly $m = 3$ and $c = 5$
So, $y = 3x + 5$ has a gradient of 3 and a y-intercept of 5.

2. $y = 10 - 2x$
This equation can be re-written as
$y = -2x + 10$
Compare this equation with $y = mx + c$.
Clearly $m = -2$ and $c = 10$
So, $y = 10 - 2x$ has a gradient of -2 and a y-intercept of 10.

3. $2y = 8x + 1$
Remember to make y the subject of formula before stating the value of m and c.
Since the coefficient of y is 2, we divide each term by 2 and get
$\frac{2y}{2} = \frac{8x}{2} + \frac{1}{2} \therefore y = 4x + 0.5$
Compare this equation with $y = mx + c$.
Clearly $m = 4$ and $c = 0.5$.
So, $2y = 8x + 1$ has a gradient of 4 and a y-intercept of 0.5.

4. $3y + 6x = 2$
Making y the subject of formula, we have
$3y = -6x + 2$
$\frac{3y}{3} = \frac{-6x}{3} + \frac{2}{3} \therefore y = -2x + \frac{2}{3}$
Compare this equation with $y = mx + c$.
Clearly $m = -2$ and $c = \frac{2}{3}$
So, $3y + 6x = 2$ has a gradient of -2 and a y-intercept of $\frac{2}{3}$.

USING TECHNOLOGY

We can also make use of technology to make y the subject of the formula as shown in the examples below.

Use the following steps on your CAS:

- Main
- Action
- Advanced
- Solve

5. For the equation, $4y - 12x - 15 = 0$ state the gradient and the y-intercept.
SOLUTION

$Solve\ (4y - 12x - 15 = 0, y)$
$\{y = 3x + 3.75\}$

$m = 3\ and\ c = 3.75$

6. For the equation, $3x + 2.5y + 9 = 0$ state the gradient and the y-intercept.
SOLUTION

$Solve\ (3x + 2.5y + 9 = 0, y)$
$\{y = -1.2x - 3.6\}$

$m = -1.2\ and\ c = -3.6$

CHAPTER 8 : LINEAR RELATIONSHIPS:SOLUTIONS

EXERCISE 8D

For each of the following equations, state the gradient (m) and the y-intercept (c).

Equation (RULE)	Gradient (m)	y-intercept (c)	Equation (RULE)	Gradient (m)	y-intercept (c)
1. $y = 2x + 5$	2	5	2. $y = x + 6$	1	6
3. $y = -3x + 7$	-3	7	4. $y = 8 - 5x$	-5	8
5. $y = 5 - x$	-1	5	6. $y = -x + 9$	-1	9
7. $y = 0.5x + 1$	0.5	1	8. $y = 4 - 0.75x$	-0.75	4
9. $y = -2$	0	-2	10. $y = 3x$	3	0
11. $y = 5$	0	5	12. $y = -7x$	-7	0
13. $y = 4x + 10$	2	5	14. $2y = x + 8$	0.5	4
Divide by 2 $y = 2x + 5$			$y = 0.5x + 4$		
15. $3y = 9x$	3	0	16. $4y = 2x$	0.5	0
Divide by 3 $y = 3x$			*Divide by 4* $y = 0.5x$		
17. $2y = 6x - 1$	3	-0.5	18. $3y = 4x + 3$	$\frac{4}{3}$	1
Divide by 2 $y = 3x - 0.5$			*Divide by 3* $y = \frac{4}{3}x + 1$		
19. $y - 2x = 10$	2	10	20. $y + 5x = 4$	-5	4
$y = 2x + 10$			$y = -5x + 4$		
21. $y - 0.5x - 2 = 0$	0.5	2	22. $y + 2x - 7 = 0$	-2	7
$y = 0.5x + 2$			$y = -2x + 7$		
23. $5y - 2x = 15$	$\frac{2}{5}$	3	24. $5y - 10x = 4$	2	$\frac{4}{5}$
$5y = 2x + 15$ $y = \frac{2}{5}x + 3$			$5y = 10x + 4$ $y = 2x + \frac{4}{5}$		

MATHEMATICS APPLICATIONS UNIT 2

8E GIVEN A RULE HOW TO TABULATE

By using simple substitution learnt from Chapter 1 in Unit 1, we should be able to find a set of values of y for corresponding values of x.

EXERCISE 8E

Complete the following tables without using a calculator.

x	-2	-1	0	1	2	5
$y = 2x$	-4	**-2**	0	**2**	**4**	10

x	-3	-2	0	1	4	10
$y = 3x$	-9	**-6**	0	**3**	12	**30**

x	0	1	2	3	4	5
$y = 7 - 2x$	7	**5**	**3**	1	**-1**	**-3**

Complete the following tables using technology.

The following steps must be used while using the tabulating capabilities of your calculator.

- Menu
- Graphs & Table
- Insert the equation e.g $y_1 = 8 - 3x$
- Check the box
- Tap on the tabulate icon (second from top left hand corner)
- Tap on the icon to set the correct domain (values of x).

x	-2	-1	1	2	5	10
$y = 8 - 3x$	14	11	5	2	**-7**	**-22**

x	2	3	5	8	10	20
$y = 5x - 2$	8	**13**	**23**	38	**48**	**98**

x	-2	3	0	1	3	5
$y = 3x - 5$	-11	**-8**	**-5**	**-2**	**4**	10

x	2	3	5	8	10	20
$y = 4x + 1$	9	13	**21**	33	**41**	81

8F LINEAR OR NOT?

We have seen that a linear function can be written in the form $y = mx + c$.

Examples of linear functions are:

$$y = 4x, \quad y = 5 - 3x, \quad y = \frac{x}{2}, \quad \frac{x}{4} + \frac{y}{5} = 2$$

For pairs of values of x and y as shown below, we can say it is linear if the first differences in the y-values are same.

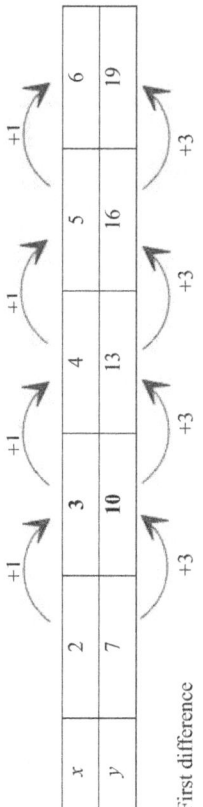

x	2	3	4	5	6
y	7	10	13	16	19

First difference +3 +3 +3 +3

As we can see the first differences in the y-values is constant and increasing by 3.

What is the rule then?

To figure out the rule, we need to work out the differences in the x-values as well. The differences in the y-values divided by the differences in the x-values gives the gradient.

Here the gradient is $m = \frac{3}{1} = 3$

To work out the value of c, we can use any pair of values from the table and substitute in the equation $y = 3x + c$.

Say we choose the point $(3, 10)$ shown bold in the table. Substitute $x = 3$ and $y = 10$. We can then figure out the value of the y-intercept c.

$$10 = 3(3) + c \therefore c = 1$$

Hence the rule is $y = 3x + 1$.

USING TECHNOLOGY

To find the rule using your calculator, apply the following steps

- Menu
- Statistics
- Enter the x-values in List 1 and the y-values in List 2.
- Calc
- Linear Reg
- OK
- $y = ax + b$ appears, replace a and b in the equation which is the rule.

EXERCISE 8F

1. State which of the following are linear equations.

Function	Linear or Not	Function	Linear or Not
$y = 2x - 1$	YES	$y = 7 + 4x$	YES
$y = \frac{3}{4}x$	YES	$y^2 = 2x$	NO
$y = x^2 + 4$	NO	$y = \frac{5}{x}$	NO
$\frac{x}{2} = y - 4$	YES	$y = 0.25x$	YES
$xy = 10$	NO	$y = 25 - 0.2x$	YES

2. Investigate which of the following points will lie on a straight line if plotted. If yes, find the rule without a calculator.

(a)

x	1	2	3	4	5	6
y	3	5	7	9	11	13

First difference   2   2   2   2   2

It is Linear and the rule is $y = 2x + 1$

(b)

x	-1	0	1	2	3	4
y	7	10	16	19	31	46

First difference   3   6   3   12   15

Since the first differences are not same, it is not linear.

(c)

x	3	5	7	9	11	13
y	10	18	26	34	42	50

First difference   8   8   8   8   8

It is Linear and the rule is $y = 4x - 2$

CHAPTER 8 : LINEAR RELATIONSHIPS:SOLUTIONS

(d)

x	1	2	3	4	5	6
y	10	5	0	-5	-10	-15

First difference   -5   -5   -5   -5   -5

It is Linear and the rule is $y = -5x + 15$

(e)

x	10	12	14	16	18	20
y	10	20	40	80	160	320

First difference   10   20   40   80   160

Since the first differences are not same, it is not linear.

(f)

x	5	6	7	8	9
y	-10	-13	-16	-19	-22

First difference   -3   -3   -3   -3

It is Linear and the rule is $y = -3x + 5$

3. Find the rule of the following linear functions using technology.

(a)

x	4	5	6	7	8	9
y	-2	-5	-8	-11	-14	-17

$y = -3x + 10$

(b)

x	2	4	6	8	10	12
y	5	10	15	20	25	30

$y = 2.5x$

(c)

x	-1	2	5	8	11
y	10	16	22	28	34

$y = 2x + 12$

MATHEMATICS APPLICATIONS UNIT 2

8G   EQUATION OF LINE : GIVEN GRADIENT AND Y-INTERCEPT

The equation of a line is given by $y = mx + c$, where m is the gradient and c is the y-intercept.
To find the equation of a line we need

1. the gradient
2. the y-intercept

EXAMPLES

1. Find the equation of the line having gradient 5 and crossing the y-axis at 3.
   SOLUTION
   $$y = 5x + 3$$

2. Find the equation of the line having gradient -6 and crossing the y-axis at (0,4).
   SOLUTION
   $$y = -6x + 4$$

EXERCISE 8G

1. Find the equation of the line having gradient 7 and crossing the y-axis at 2.
   $$y = 7x + 2$$

2. Find the equation of the line having gradient 3 and crossing the y-axis at -2.
   $$y = 3x - 2$$

3. Find the equation of the line having slope -1 and y-intercept 6.
   $$y = -x + 6$$

4. Find the equation of the line having gradient 9 and crossing the y-axis at 7.
   $$y = 9x + 7$$

5. Find the equation of the line having slope -2 and y-intercept 10.
   $$y = -2x + 10$$

6. Find the equation of the line having slope -5 and y-intercept 1.
   $$y = -5x + 1$$

7. Find the equation of the line having gradient 1 and crossing the y-axis at (0,5).
   $$y = x + 5$$

8. Find the equation of the line having gradient 4 and crossing the y-axis at (0,-3).
   $$y = 4x - 3$$

CHAPTER 8 : LINEAR RELATIONSHIPS:SOLUTIONS

8H EQUATION OF A LINE: GRADIENT GIVEN AND A POINT

EXAMPLES

1. Find the equation of the line having gradient 2 and passing through A(3,5).

SOLUTION
Let $y = mx + c$ be the equation of the line
Here $m = 2$, $x = 3$ and $y = 5$
Substitute these values in the equation $y = mx + c$ to determine the value of c
$$5 = 2(3) + c$$
$$5 = 6 + c$$
$$c = -1$$
$$\therefore y = 2x - 1$$

2. Find the equation of the line having gradient −0.5 and passing through A(−10,7).

SOLUTION
Let $y = mx + c$ be the equation of the line
Here $m = -0.5$, $x = -10$ and $y = 7$
Substitute these values in the equation $y = mx + c$ to determine the value of c
$$7 = -0.5(10) + c$$
$$7 = -5 + c$$
$$c = 12$$
$$\therefore y = -0.5x + 12$$

EXERCISE 8H

1. Find the equation of the line having gradient 3 and passing through (5,7).

$$y = mx + c$$
$$7 = 3(5) + c$$
$$7 = 15 + c$$
$$\therefore c = -8$$
$$y = 3x - 8$$

2. Find the equation of the line having gradient 4 and passing through (1,2).

$$y = mx + c$$
$$2 = 4(1) + c$$
$$2 = 4 + c$$
$$\therefore c = -2$$
$$y = 4x - 2$$

3. Find the equation of the line having slope 6 and passing through (−2,5).

$$y = mx + c$$
$$5 = 6(-2) + c$$
$$5 = -12 + c$$
$$\therefore c = 17$$
$$y = 6x + 17$$

4. Find the equation of the line having gradient −4 and passing through (1,5).

$$y = mx + c$$
$$5 = -4(1) + c$$
$$5 = -4 + c$$
$$\therefore c = 9$$
$$y = -4x + 9$$

5. Find the equation of the line having gradient -5 and passing through (−2,3).

$$y = mx + c$$
$$3 = -5(-2) + c$$
$$3 = 10 + c$$
$$\therefore c = -7$$
$$y = -5x - 7$$

MATHEMATICS APPLICATIONS UNIT 2

8I EQUATION OF A LINE PASSING THROUGH TWO POINTS

To find the equation of a line passing through two points we need to find

1. the gradient by using $\frac{y_2 - y_1}{x_2 - x_1}$
2. the y-intercept, by choosing any one of the two given points.

EXAMPLE

Find the equation of the line passing through A(3,1) and B(5,9)

Solution
Since the gradient is not given, we use the formula $m = \frac{y_2 - y_1}{x_2 - x_1}$ to find m.

$$m = \frac{9-1}{5-3} = \frac{8}{2} = 4$$

From the two points A and B, we can choose any point as the final answer would be same.
Let us choose A (3,1) to find the y-intercept c.
Let $y = mx + c$ be the equation of the line

$$1 = 4(3) + c$$
$$1 = 12 + c$$
$$c = -11$$
$$y = 4x - 11$$

EXERCISE 8I

1. Find the equation of the line passing through A(2,5) and B(6,4).

$$m = \frac{4-5}{6-2} = \frac{1}{-4} = -\frac{1}{4}$$
$$y = mx + c$$
$$5 = -\frac{1}{4} \times 2 + c$$
$$5 = -0.5 + c$$
$$c = 5.5$$
$$y = -\frac{1}{4}x + 5.5$$

2. Find the equation of the line passing through A(−2,3) and B(1,9).

$$m = \frac{9-3}{1-(-2)} = \frac{6}{3} = 2$$
$$y = mx + c$$
Using A, $3 = 2(-2) + c$
$$3 = -4 + c$$
$$c = 7$$
$$y = 2x + 7$$

3. Find the equation of the line passing through P(0,2) and B(2,8).

$$m = \frac{8-2}{2-0} = \frac{6}{2} = 3$$
$$y = mx + c$$
$$\therefore y = 3x + 2$$

as P is the y-intercept

4. Find the equation of the line passing through A(−3,1) and B(1,5).

$$m = \frac{5-1}{1-(-3)} = \frac{4}{4} = 1$$
$$y = mx + c$$
Using A, $1 = 1(-3) + c$
$$c = 4$$
$$y = x + 4$$

8J DOES THE POINT LIE ON THE LINE?

To show a point lies on a given line,
- replace the x and y values in the equation of the line
- we have to obtain the constant value

EXAMPLES

1. Show that the point A(2,5) lies on the line with equation $4x + y = 13$.

SOLUTION

Replace $x = 2$ and $y = 5$ in the equation. We have to obtain the right hand side number.
$4(2) + (5) = 13$ **(shown)**

2. Does the point B(3,-2) lie on the line with equation $2x + y = 10$?

SOLUTION

Replace $x = 3$ and $y = -2$ in the equation.
$2(3) + (-2) = 4 \neq 10$
Therefore B does not lie on the line.

3. The point A(p,3) lies on the line $4x - 2y = 2$, find the value of p.

SOLUTION

Replace x by p and y by 3 in the given equation and solve for p.
$4(p) - 2(3) = 2$
$4p - 6 = 2$
$4p = 8$
$\therefore p = 2$

EXERCISE 8J

1. Show that the point (4,3) lies on the line with equation $3x + 2y = 18$.

$3(4) + 2(3) = 12 + 6 = 18 \ (RHS)$

2. Show that the point (5,-1) lies on the line with equation $x + 2y = 3$.

$5 + 2(-1) = 5 - 2 = 3 \ (RHS)$

3. Does the point (2,5) lie on the line with equation $3x - y = 1$?

$3(2) - 5 = 6 - 5 = 1 \ (RHS)$
YES.

4. The point A(t,4) lies on the line $4x - 2y = 12$, find the value of t.

$4(t) - 2(4) = 12$
$4t - 8 = 12$
$4t = 20$
$t = 5$

5. The point A(4,m) lies on the line $5x + 2y = 6$, find the value of m.

$5(4) + 2m = 6$
$2m = -14$
$m = -7$

6. The point A(q,4) lies on the line $3x - 5y = -2$, find the value of q.

$3q - 5(4) = -2$
$3q = 18$
$q = 6$

---

5. Find the equation of the line passing through A(3,11) and B(5,15).

$m = \dfrac{15 - 11}{5 - 3} = \dfrac{4}{2} = 2$
$y = mx + c$
$11 = 2 \times 3 + c$
$11 = 6 + c$
$c = 5$
$y = 2x + 5$

6. Find the equation of the line passing through A(-2,16) and B(-1,13).

$m = \dfrac{13 - 16}{-1 - (-2)} = \dfrac{-3}{1} = -3$
$y = mx + c$
$16 = -3 \times -2 + c$
$16 = 6 + c$
$c = 10$
$y = -3x + 10$

7. Find the equation of the line passing through P(4,10) and B(6,11).

$m = \dfrac{11 - 10}{6 - 4} = \dfrac{1}{2} = 2$
$y = mx + c$
$10 = \dfrac{1}{2} \times 4 + c$
$10 = 2 + c$
$c = 8$
$y = \dfrac{1}{2}x + 8$

8. Find the equation of the line passing through A(0,-3) and B(-1,-8).

$m = \dfrac{-8 - (-3)}{-1 - 0} = \dfrac{-5}{-1} = 5$
$y = mx + c$
$y = 5x - 3$
as A(0,-3) *is the y – intercept.*

9. Find the equation of the line passing through P(-5,-5) and B(2,16).

$m = \dfrac{16 - (-5)}{2 - (-5)} = \dfrac{21}{7} = 3$
$y = mx + c$
$16 = 3 \times 2 + c$
$16 = 6 + c$
$c = 10$
$y = 3x + 10$

10. Find the equation of the line passing through A(4,6) and B(10,3).

$m = \dfrac{3 - 6}{10 - 4} = \dfrac{-3}{6} = -\dfrac{1}{2}$
$y = mx + c$
$6 = -\dfrac{1}{2} \times 4 + c$
$6 = -2 + c$
$c = 8$
$y = -\dfrac{1}{2}x + 8$

CHAPTER 8 : LINEAR RELATIONSHIPS: SOLUTIONS

8K HOW TO SKETCH A LINE WITHOUT A CALCULATOR?

To sketch a line, we can use one of the methods shown below. Note that each method works best depending the way the equation is presented to us. In some cases, we can make use of use of a table of values and in other cases we might be better off using the intercept-method.

EXAMPLES

1. Sketch the line $y = 2x + 3$.
SOLUTION: METHOD 1
We make a table of values choosing a few values of x. Always choose numbers that you would be able to work out mentally.
Note: two pairs of values are enough.

x	0	1	2
$y = 2x + 3$	3	5	7

Now plot these pairs of values on the Cartesian plane to obtain the sketch of the line as shown.

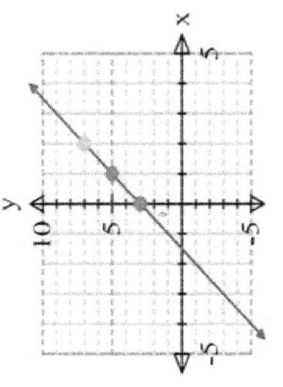

2. Sketch the line $y = 3x − 5$
SOLUTION: METHOD 2
Clearly the y-intercept is -5.
First, we plot the point (0,-5) as we know that the line passes through this point.
Now since the gradient is 3, for each unit we move right move 3 units up.
Plot another couple points just to make sure no error is left behind.
Now draw a smooth line passing through the y-intercept and the other two points.

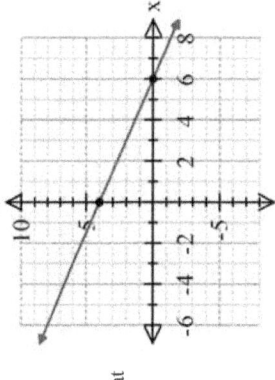

3. Sketch the line $2x + 3y = 12$
SOLUTION
This type of equation would be easier to sketch using the intercept method.
When $x = 0$ (hide $2x$ with your finger)
You'll see $3y = 12$
   $\therefore y = 4$
Similarly, when $y = 0$ (hide $3y$ with your finger)
Only $2x = 12$ will be visible
$\therefore x = 6$
Hence the line crosses the y-axis at 4 and the x-axis at 6,
producing the sketch as shown.

MATHEMATICS APPLICATIONS UNIT 2

EXERCISE 8K

1. Sketch the line $y = x + 3$.

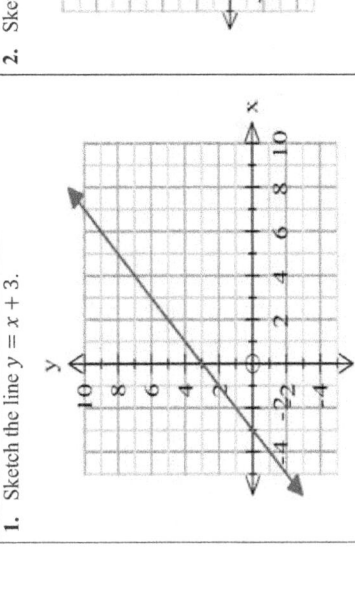

2. Sketch the line $y = 3x − 1$.

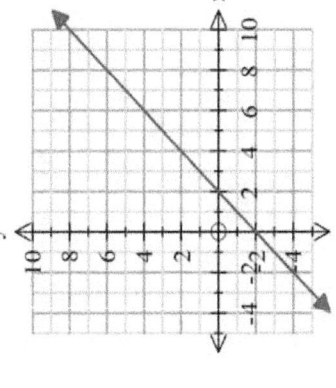

3. Sketch the line $y = 4 − x$.

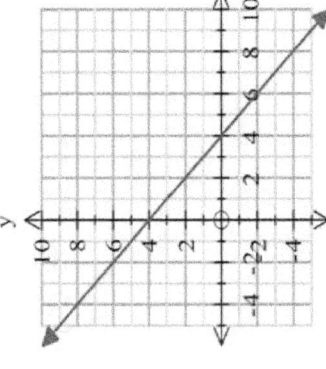

4. Sketch the line $y = x − 2$.

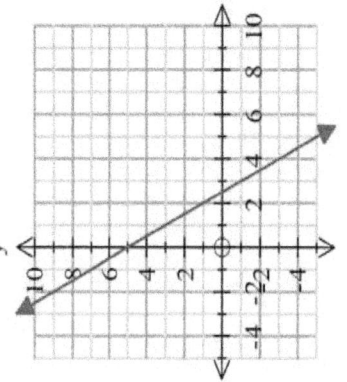

5. Sketch the line $y = 5 − 2x$.

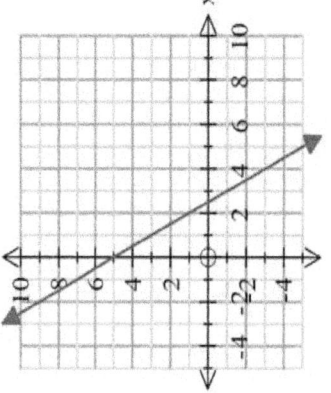

6. Sketch the line $y = 3x + 2$.

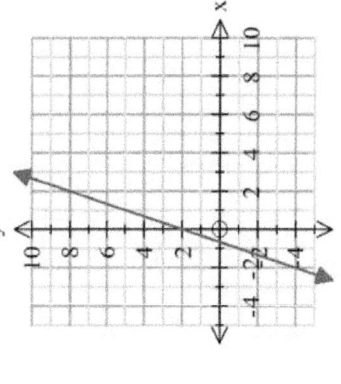

CHAPTER 8 : LINEAR RELATIONSHIPS: SOLUTIONS

7. Sketch the line $y = 8 - 2x$.

8. Sketch the line $y = 5x + 2$.

9. Sketch the line $2x + y = 8$.

10. Sketch the line $x - 3y = 9$.

11. Sketch the line $2x + 3y = 12$.

12. Sketch the line $2y - x = 4$.

MATHEMATICS APPLICATIONS UNIT 2

8L   HORIZONTAL AND VERTICAL LINES

Consider the sketch on the right.
It is a vertical line passing through the points $(5,-4), (5,1)$ and $(5,7)$.

Note that all the three points have x-coordinate 5.
Hence the equation of the line is $x = 5$.

Vertical lines are of the type $x = a$, where a is any number.

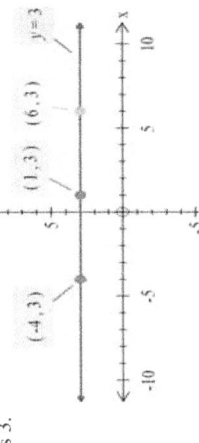

Similarly, in the diagram on the right it is a horizontal line passing through the points $(-4,3), (1,3)$ and $(6,3)$.

Note again that all the three points have y-coordinate as 3.
Hence the equation of the line is $y = 3$.

Horizontal lines are of the type $y = b$, where b is any number.

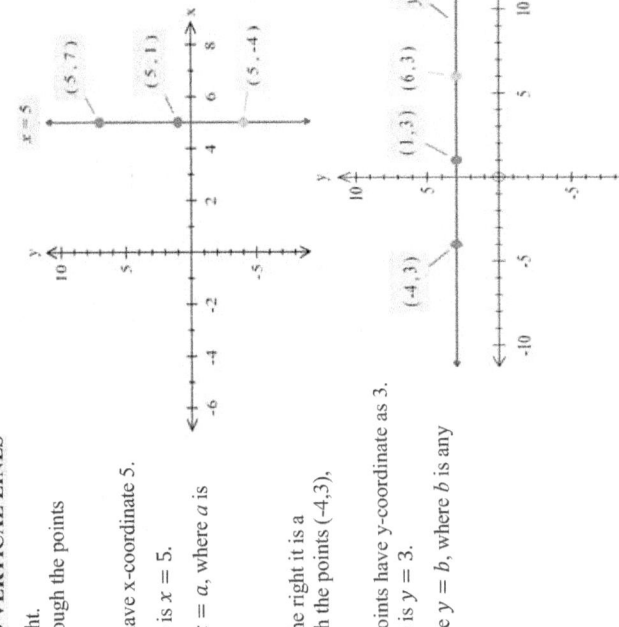

EXAMPLES

1. On the same set of axes sketch the line $x = 4, x = -7,$ and $y = 8$.

SOLUTION

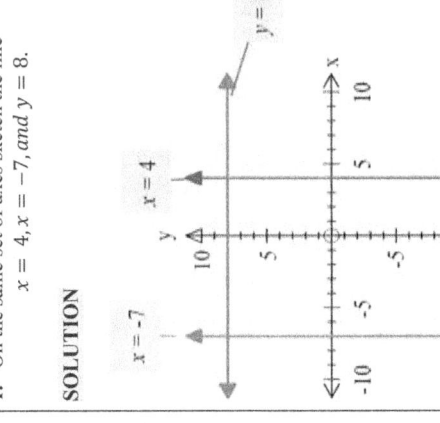

2. State the equation of each of the lines labelled A, B and C.

SOLUTION

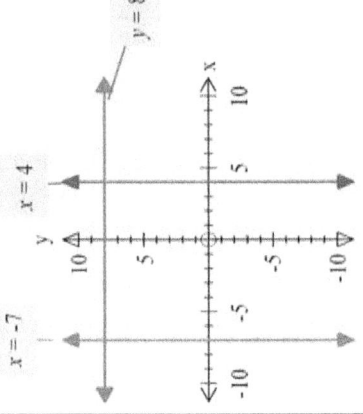

$A : x = -5$
$B : x = 8$
$C : y = -6$

CHAPTER 8 : LINEAR RELATIONSHIPS: SOLUTIONS

EXERCISE 8L

1. On the same set of axes sketch the line $y = 4, y = -1,$ and $x = 2$.

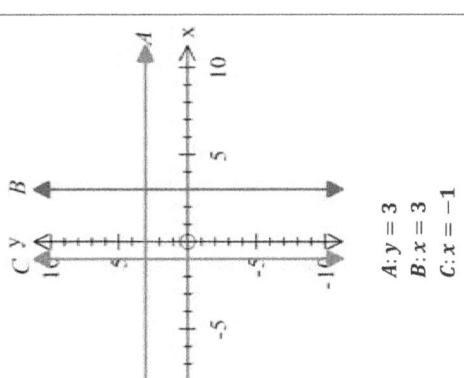

2. State the equation of each of the lines labelled A, B and C.

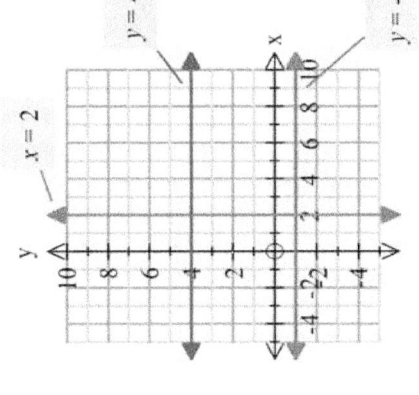

$A: y = 3$
$B: x = 3$
$C: x = -1$

3. State the equation of each of the lines labelled A, B and C.

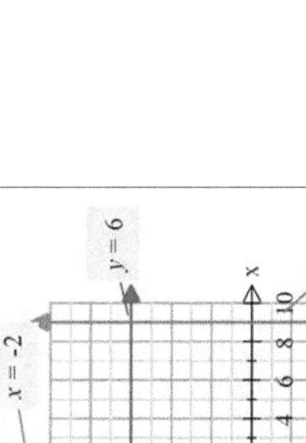

$A: y = 7$
$B: x = 5$
$C: x = -4$

4. On the same set of axes sketch the line $x = 9, x = -2,$ and $y = 6$.

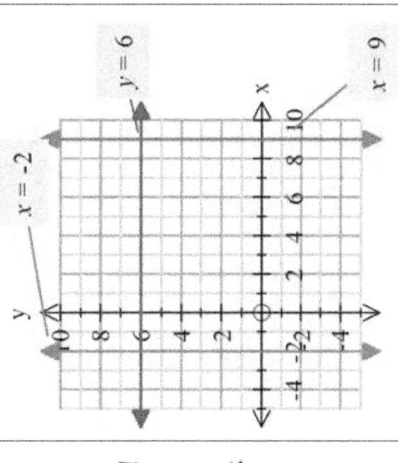

MATHEMATICS APPLICATIONS UNIT 2

8M   SKETCHING LINES USING TECHNOLOGY

To sketch a line or even a few lines, use the following steps on your calculator

- Menu
- Graph & Table
- Insert the equations
- check the box
- click on the graph icon (1st on top left)

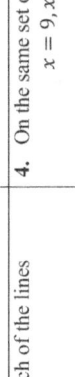

EXAMPLES

Sketch the following using technology.

(a) $y = 3x + 5$
(b) $y = 4 - 3x$
(c) $y = 0.5x + 2$

SOLUTION

Your calculator must display a similar screen version to have the correct graphs.

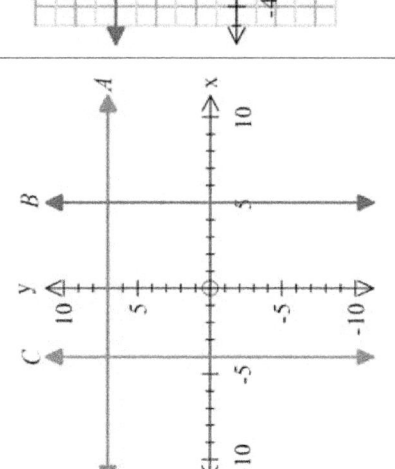

CLASS ACTIVITY

Sketch the following lines using technology. You don't have to graph them. Just practice!

1. $y = 2x + 9$
2. $y = 10 - 2x$
3. $y = 3x + 9$
4. $y = 0.25x + 4$
5. $y = 6 - 3x$
6. $y = x + 6$
7. $x + y = 8$
8. $2x + y = 8$
9. $x + 2y = 10$

CHAPTER 8 : LINEAR RELATIONSHIPS: SOLUTIONS

8N   HOW TO FIND THE EQUATION OF A LINE GIVEN THE SKETCH?

EXAMPLES

1. Find the equation of the line.

2. Find the equation of the line.

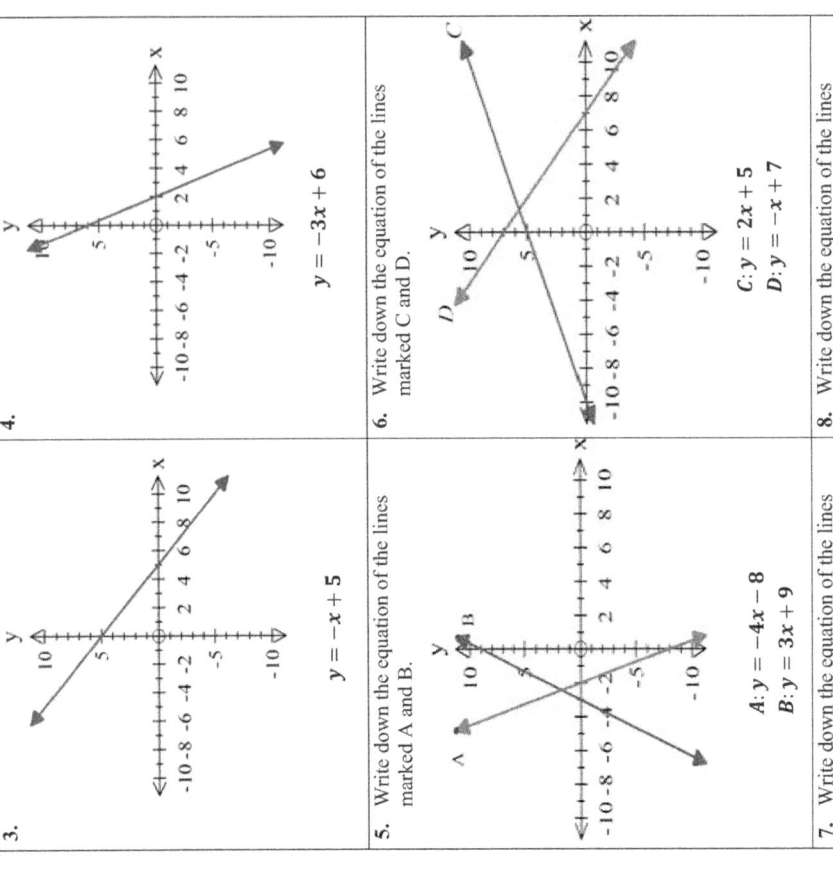

SOLUTION

This line has a y-intercept of 2.

The gradient $= \frac{rise}{run} = \frac{1}{1} = 1$

So the equation is $y = x + 2$.

SOLUTION

This line has a y-intercept of 6.

The gradient $= \frac{rise}{run} = \frac{-6}{2} = -3$

So the equation is $y = -3x + 6$.

EXERCISE 8N

Find the equations of the following lines.

1.

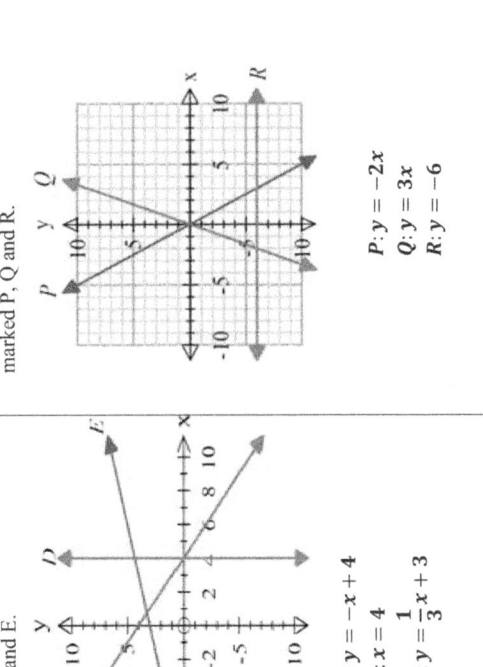

$y = 2x + 8$

2.

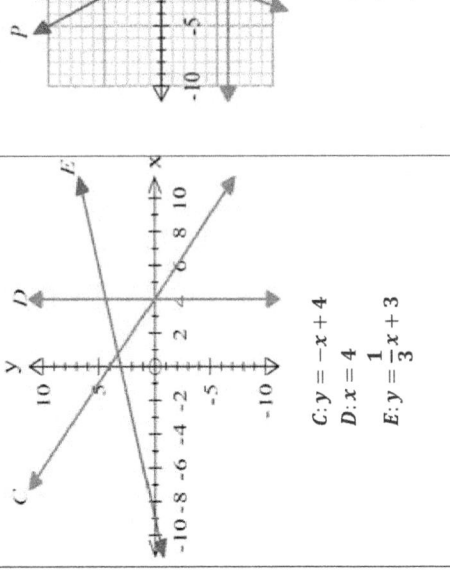

$y = x + 4$

3.

$y = -x + 5$

4.

$y = -3x + 6$

5. Write down the equation of the lines marked A and B.

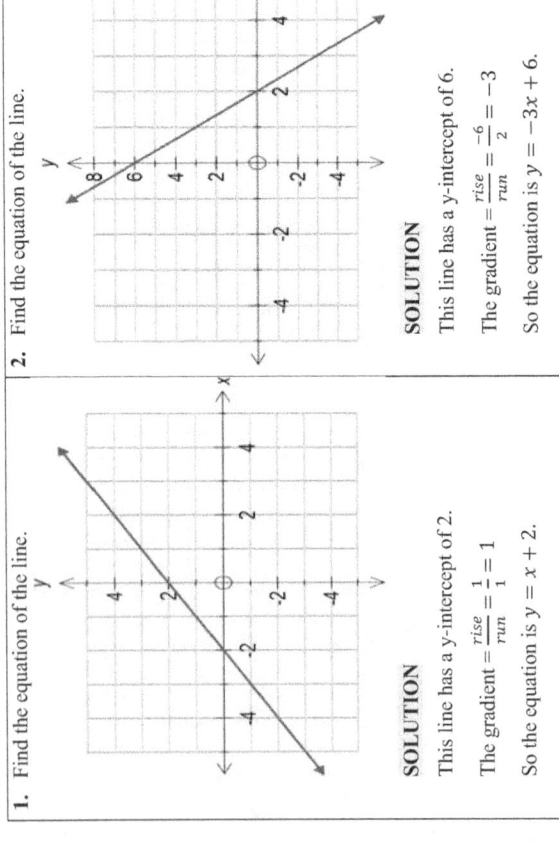

$A: y = -4x - 8$
$B: y = 3x + 9$

6. Write down the equation of the lines marked C and D.

$C: y = 2x + 5$
$D: y = -x + 7$

7. Write down the equation of the lines marked C, D and E.

$C: y = -x + 4$
$D: x = 4$
$E: y = \frac{1}{3}x + 3$

8. Write down the equation of the lines marked P, Q and R.

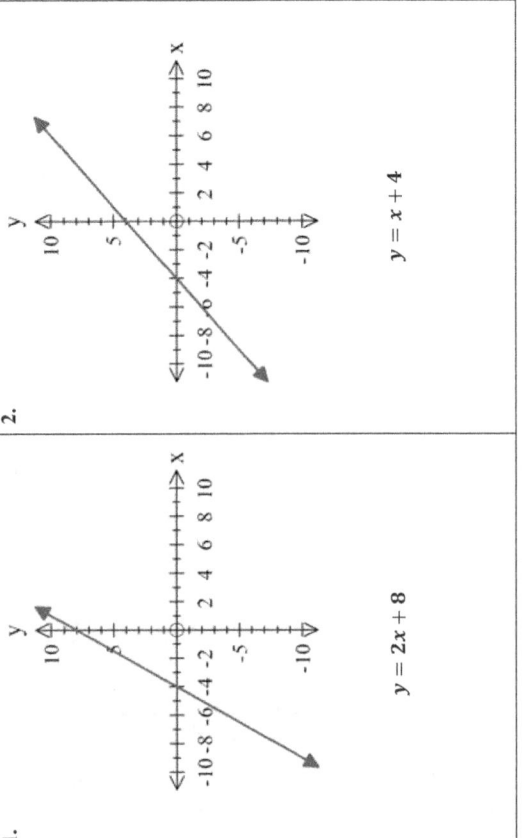

$P: y = -2x$
$Q: y = 3x$
$R: y = -6$

CHAPTER 8 : LINEAR RELATIONSHIPS: SOLUTIONS

8O APPLICATIONS

In the last section of the chapter, an attempt has been made to familiarize the reader with all the concepts learnt so far and put them into practice by solving real world problems.

EXAMPLE

Tony is a professional electrician and charges an hourly rate of $60 per hour but no callout fee.

(a) Complete the table of values below to show the cost of having Tony complete jobs of varying lengths.

Time (t hours)	0	1	2.5	4	8
Cost ($C)	0	60	150	240	480

(b) On the axes below, plot the cost of Tony completing a job of length t hours. The cost of Grant, another electrician, has already been plotted.

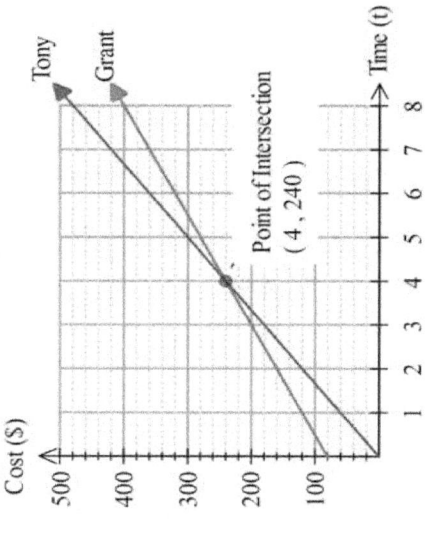

(c) Write a rule to calculate the cost of employing Tony for any length of time (t).

$$C = 60t$$

(d) Write a rule to calculate the cost of employing Grant for any length of time (t).

$$C = 80 + 40t$$

(f) You are keen to pay as little money as possible. For what interval of time would you employ Tony instead of Grant?

When t = 4, the cost of employing both electrician is the same ($240).
For any time less than 4 hours Tony is cheaper

EXERCISE 8O

1. A mobile phone company charges a flat rate of $30 per month and an additional charge of 9c for each minute of service.

   (a) Write a linear equation for the monthly charge ($C) based upon the number of minutes (t) of service each month.

   $$C(t) = 30 + 0.09t$$

   (b) What will be the charge for 90 minutes of service?

   $$C(t) = 30 + 0.09(90) = \$38.10$$

   (c) Johanna can only afford a $45 of phone bill each month. How long can she afford to talk on the phone each month?

   $$30 + 0.09t = 45$$
   $$t \approx 167 \text{ minutes}$$

2. A small garment company makes cheap brand T shirts and the profit function P, of the company follows a linear model as shown below.

   $$P = 2.50N - 2000$$

   where N is the number of T shirts sold.

   (a) What does the figure 2000 represent?

   It is the fixed cost of running the business.

   (b) What does 2.50 mean in the above context?

   It is the selling price per T Shirt

   (c) Calculate the profit of the company if they sell 1500 T shirts.

   $$P = 2.50(1500) - 2000 = \$1750$$

   (d) How many T shirts must the company sell to break even (no profit no loss)?

   $$0 = 2.50N - 2000$$
   $$\therefore N = 800$$

   The company must sell 800 T shirts to break even.

CHAPTER 8 : LINEAR RELATIONSHIPS:SOLUTIONS

3. A cab company charges $3.80 boarding rate and an additional $1.10 for every kilometre.

   (a) Write a linear equation for the charge ($C) based upon the number of kilometres (d) of travel.

   $$C(t) = 3.80 + 1.10d$$

   (b) What will be the charge for 40 km of journey?

   $$C(t) = 3.80 + 1.10(40) = \$47.80$$

   (c) A random passenger paid $50 for catching a cab to his work place. What distance did the cab cover?

   $$Solve\ (3.80 + 1.10d = 50, d)$$
   $$d = 42\ km$$

   (d) Mishka is a tourist and want to do some sight-seeing around the city. She has a $100 note. How many kilometres would she be able to travel in a cab?

   $$Solve\ (3.80 + 1.10d = 100, d)$$
   $$d = 87.5\ km$$

4. A toy company makes little car toys. The Cost function $C, of the company follows a linear model as shown below.

   $$C = 6.80N + 1250$$

   where N is the number of car toys produced.

   (a) What does the figure 1250 represent?

   It is the fixed cost of running the business.

   (b) What does 6.80 mean in the above context?

   It is the cost per car

   (c) Calculate the cost of the company if they produce 3000 cars.

   $$C = 6.80(3000) + 1250 = \$21650$$

   (d) The company sells each car for $10. How many cars must the company sell to break even?

   $$Solve\ (6.80N + 1250 = 10N, N)$$
   $$N \approx 391$$

MATHEMATICS APPLICATIONS UNIT 2

5. Michael is a plumber who charges $40 call out fee plus $35 per hour of labour.

   (a) Complete the table below to show the cost of having Michael complete jobs of varying lengths.

Time (t hours)	0	2	3.5	6	10
Cost ($C)	40	110	162.50	250	390

   (b) On the axes below, plot the cost of Michael completing a job in t hours. The cost of Simpson, another plumber, has already been plotted.

   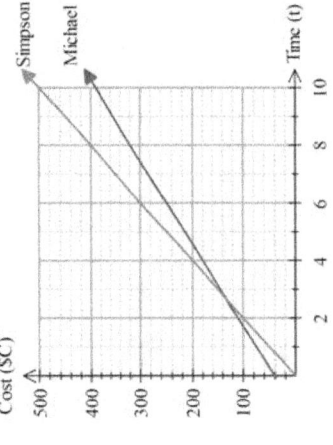

   (c) Calculate the hourly cost of employing Simpson.

   $50

   (d) Use your graph to determine the number of hours of labour for which the both plumbers would cost the same.

   $$\approx 2.67\ hours$$

6. A white water rafting company rents kayaks at $12 per person for their guides fee and $8 dollars an hour to rent a kayak. Write an equation representing the cost per person, $C, of renting a kayak for x hours and work out the cost of renting a kayak for 5 hours?

   $$C = 12 + 8x$$
   $$C = 12 + 8(5) = \$52$$

CHAPTER 8 : LINEAR RELATIONSHIPS:SOLUTIONS

7. A sequence of shapes is made of squares to form the shapes as shown below.

   Shape Number 1   Shape Number 2   Shape Number 3

    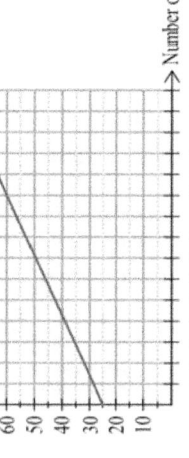

The table of results is shown below.

Shape Number (n)	1	2	3	4	5
Number of squares (s)	5	9	13	17	21

(a) Plot the above data on the axes below.

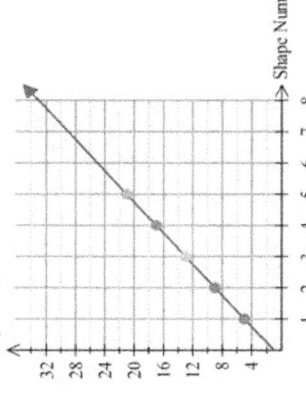

(b) Write a rule linking s and n, where s and n are as defined in the table.

$$S = 4n + 1$$

(c) Determine the number of squares required for Shape Number 10.

$$S = 4(10) + 1 = 41\ squares$$

(d) Justify that the point (15,61) lies on the line that would pass through the points plotted in (a).

$$4(15) + 1 = 60 + 1 = 61$$

(e) Mary has 80 squares. What is the biggest Shape Number she would be able to make?

$$Solve\ (4n + 1 = 80, n)$$
$$n = 19.75$$
$$Biggest\ shape\ number = 19$$

MATHEMATICS APPLICATIONS UNIT 2

8. Asian Motors will rent-a-car for $35 plus $0.20 per kilometre travelled.

(a) Write a function expressing rental cost ($C) as a function of the kilometres travelled (d).

$$C = 35 + 0.20d$$

(b) Find the cost of renting a car to travel 800 km.

$$C = 35 + 0.20(800) = \$195$$

(e) John's bill was $51.40. How far did he drive?

$$Solve\ (35 + 0.20d = 51.40, d)$$
$$d = 82\ km$$

9. Dream Video Store charges a $25 membership fee and $3.50 for each DVD rented.

(a) Write a function expressing rental cost ($C) as a function of the number of DVDs (n) rented.

$$C = 25 + 3.50n$$

(b) Find the cost of renting 62 DVD's.

$$C = 25 + 3.50(62) = \$242$$

(c) On the axes below, plot the cost of renting DVD's.

(d) If a new member paid the store $193 in the last 6 months, how many DVDs were rented?

$$solve\ (25 + 3.50n = 193, n)$$
$$n = 48$$

CHAPTER 9

PIECEWISE FUNCTIONS

9A GRAPHING PIECEWISE FUNCTIONS

In chapter 8 we have learnt about lines and how they can be sketched by long hand and by using technology. In many real-life problems, however, functions are represented by a combination of equations, each corresponding to a part of the domain. Such functions are called piecewise functions.

For example, the piecewise function given by

$$f(x) = \begin{cases} x + 2 & \text{for } x < 3 \\ 2x - 1 & \text{for } x \geq 3 \end{cases}$$

is defined by two equations. One equation gives the values of $f(x)$ when x is less than 3 and the other equation gives the values of $f(x)$ when x is greater than or equal to 3.

If we sketch the above piecewise function, we have two lines as shown.

To the left of $x = 3$, we have the graph of $y = x + 2$ and to the right and including $x = 3$ the graph is given by $y = 2x - 1$.

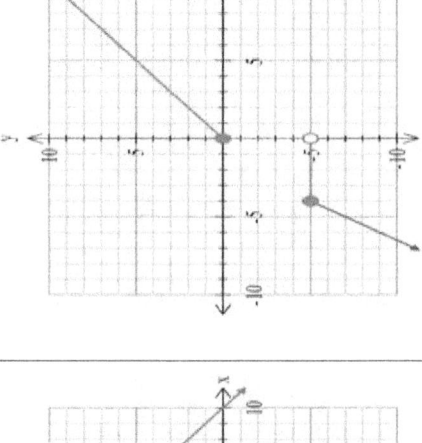

EXAMPLE

Graph the following piecewise function.

$$f(x) = \begin{cases} x - 4 & \text{for } x \leq 2 \\ 2 & \text{for } 2 < x < 5 \\ -x & \text{for } x \geq 5 \end{cases}$$

SOLUTION

This piecewise function will consist of 3 different lines.
The open ended circles in the horizontal lines indicate that 2 and 5 are not included in the domain.

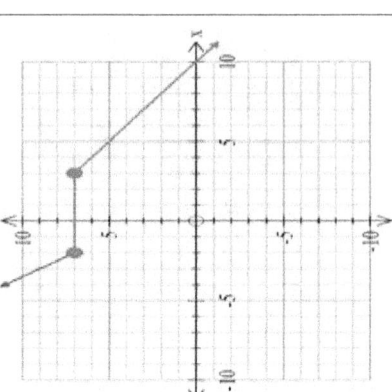

EXERCISE 9A

Graph the following piecewise functions.

1. $f(x) = \begin{cases} x + 3 & \text{for } x < 1 \\ 3x + 1 & \text{for } x \geq 1 \end{cases}$

2. $f(x) = \begin{cases} 3x - 1 & \text{for } x < -2 \\ x - 5 & \text{for } x \geq -2 \end{cases}$

3. $f(x) = \begin{cases} -2x + 3 & \text{for } x \leq -2 \\ 7 & \text{for } -2 < x \leq 3 \\ -x + 10 & \text{for } x > 3 \end{cases}$

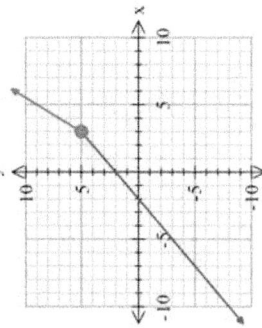

4. $f(x) = \begin{cases} 2x + 3 & \text{for } x \leq -4 \\ -5 & \text{for } -4 < x < 0 \\ x & \text{for } x \geq 0 \end{cases}$

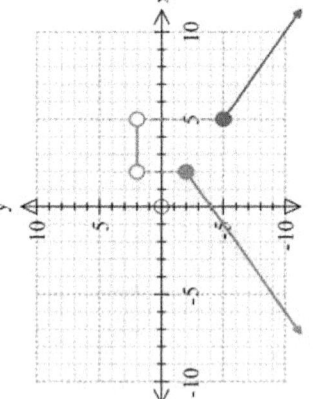

CHAPTER 9 : PIECEWISE FUNCTIONS SOLUTIONS

9B   DETERMINING EQUATIONS GIVEN GRAPHS OF PIECEWISE FUNCTIONS

The diagram shows a piecewise function.
Fill in the missing equations for each given domain (values of x).

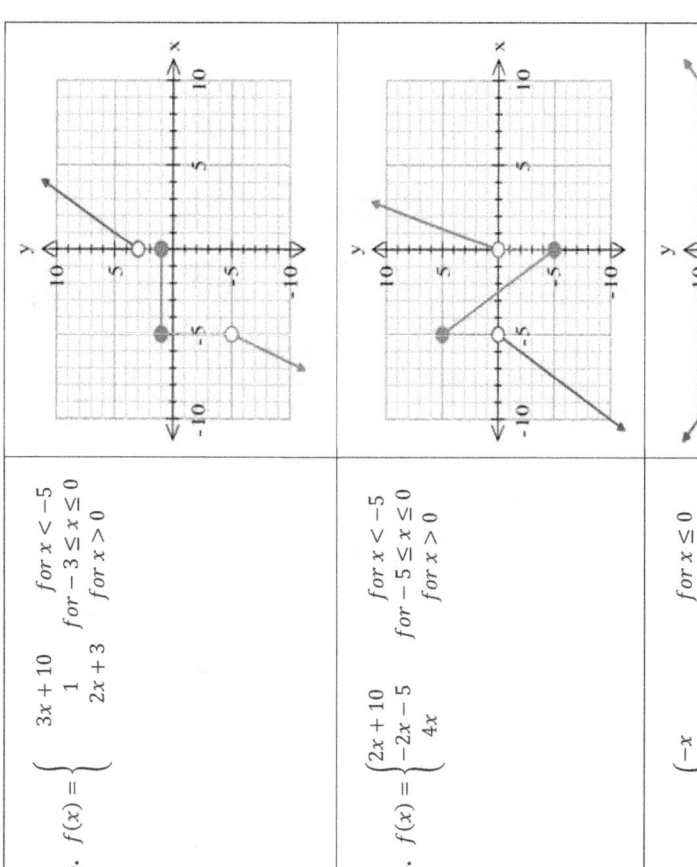

$$f(x) = \begin{cases} & \text{for } x \leq 0 \\ & \text{for } 0 < x < 3 \\ & \text{for } x \geq 3 \end{cases}$$

SOLUTION

For $x \leq 0$, the line has a gradient of -1 and a y-intercept of 4.
It's equation is $y = -x + 4$.

For $0 < x < 3$, it is a horizontal line with equation $y = 6$.
For $x \geq 3$, the line has a slope of 1 and if produced will cross the y-axis at -3. Hence its equation is $y = x - 3$.
Now we can define our whole piecewise function as under:

$$f(x) = \begin{cases} -x + 4 & \text{for } x \leq 0 \\ 6 & \text{for } 0 < x < 3 \\ x - 3 & \text{for } x \geq 3 \end{cases}$$

EXERCISE 9B

Complete the following piecewise function given their graphs.

1. $f(x) = \begin{cases} x & \text{for } x \geq 0 \\ -x & \text{for } x < 0 \end{cases}$	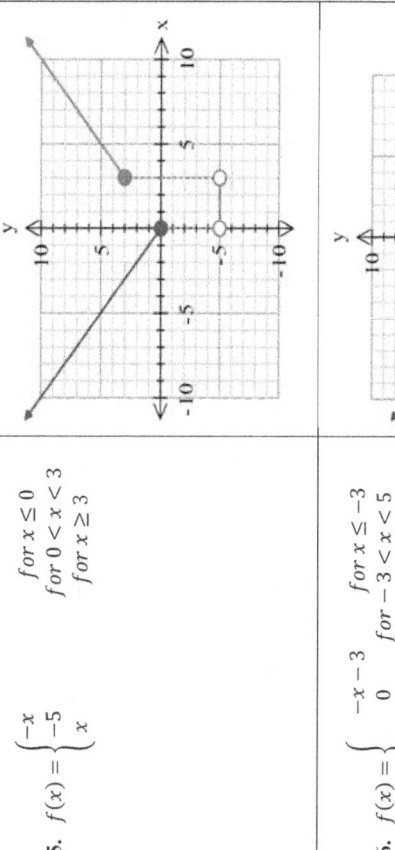
2. $f(x) = \begin{cases} 8 - x & \text{for } x \geq 0 \\ x + 5 & \text{for } x < 0 \end{cases}$	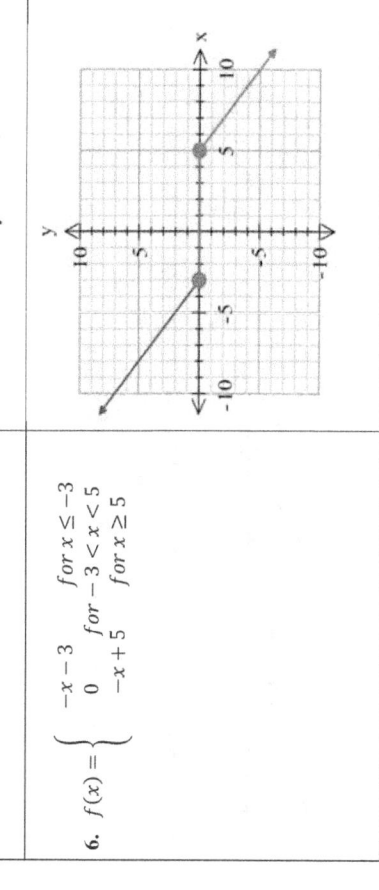

3. $f(x) = \begin{cases} 3x + 10 & \text{for } x < -5 \\ 1 & \text{for } -3 \leq x \leq 0 \\ 2x + 3 & \text{for } x > 0 \end{cases}$	
4. $f(x) = \begin{cases} 2x + 10 & \text{for } x < -5 \\ -2x - 5 & \text{for } -5 \leq x \leq 0 \\ 4x & \text{for } x > 0 \end{cases}$	
5. $f(x) = \begin{cases} -x & \text{for } x \leq 0 \\ -5 & \text{for } 0 < x < 3 \\ x & \text{for } x \geq 3 \end{cases}$	
6. $f(x) = \begin{cases} -x - 3 & \text{for } x \leq -3 \\ 0 & \text{for } -3 < x < 5 \\ -x + 5 & \text{for } x \geq 5 \end{cases}$	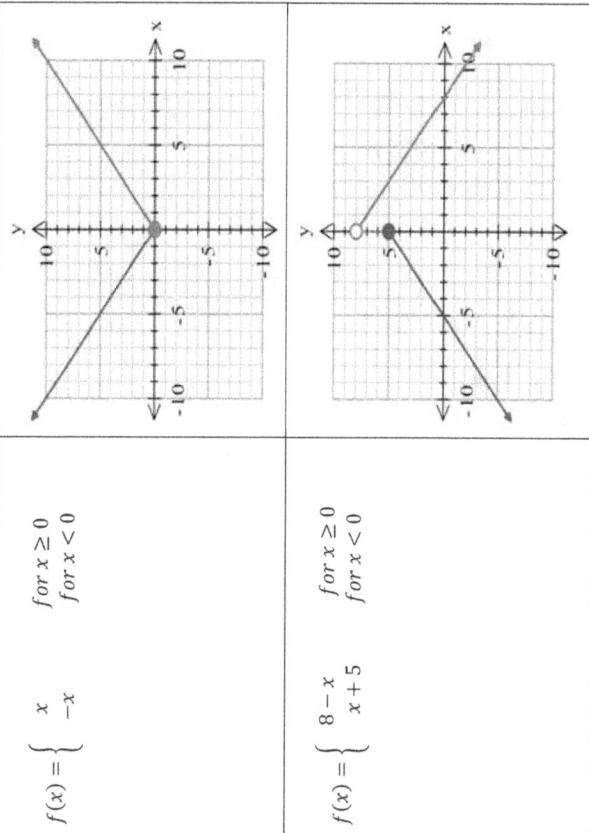

CHAPTER 9 : PIECEWISE FUNCTIONS

9C USING TECHNOLOGY TO GRAPH PIECEWISE FUNCTIONS

In this part of the chapter, we are going to make use of the capabilities of our calculators to graph linear piecewise functions. The steps have been outlined below.

- Menu
- Graph & table
- Keyboard
- Select the symbol $\begin{cases} \blacksquare \\ \square \end{cases}$, $\begin{cases} \square \\ \square \end{cases}$
- In the first column, insert all the equations you wish to graph. Remember do not type $y =$, just the expressions involving x. If we have more equations just tap on the same symbol mentioned above and it will create more rows.
- In the second column, insert the domain (the range of values of x).
- Tick the box
- Tap on the graph icon (first on top left hand corner)

EXERCISE 9C

Use your calculator to graph the following piecewise functions.

1. $f(x) = \begin{cases} x + 4 & \text{for } x < -4 \\ 2x + 8 & \text{for } x \geq -4 \end{cases}$

2. $f(x) = \begin{cases} -x + 5 & \text{for } x < -2 \\ 3x + 1 & \text{for } x \geq -2 \end{cases}$

3. $f(x) = \begin{cases} -x + 2 & \text{for } x \leq 0 \\ 4 & \text{for } 0 < x < 3 \\ x - 5 & \text{for } x \geq 3 \end{cases}$

4. $f(x) = \begin{cases} -2x + 4 & \text{for } x \leq 2 \\ 4 & \text{for } 2 < x < 5 \\ x - 3 & \text{for } x \geq 5 \end{cases}$

9D APPLICATIONS

There are some real-life practical examples for studying piecewise linear functions. For example, it can be used to determine the tax payable by an individual. In other cases, we can use piecewise linear functions to calculate the electricity bills, commission paid to a salesman and so on.

EXAMPLE 1

The diagram on the right shows the distance time graph of a paper boy named Charlie. He uses his motorbike to deliver newspapers in the surrounding suburbs and part of the country as well.

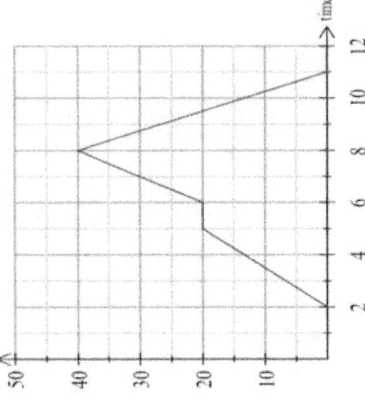

(a) When did Charlie start his job?

2 a.m.

(b) Charlie stopped and had breakfast on his journey. At what time did he stop and for how long?

5 a.m. , for 1 hour

(c) Calculate Charlie's speed on the way home.

$$speed = \frac{distance}{time} = \frac{40}{3} = 13.\dot{3} \; km/h$$

EXAMPLE 2

Home Care Electric Company has this rate for a family home:

- Monthly service charge $35.50
- First 100 kWh : $1.25 per kWh
- Over 100 kWh : $1.50 per kWh

Write down a piecewise model for the monthly charge C(x) as a function where x is the number of kWh.

SOLUTION

$$C(x) = \begin{cases} 35.50 + 1.25x & \text{for } 0 \leq x \leq 100 \\ 35.50 + 1.25 \times 100 + 1.50(x - 100) & x > 100 \end{cases}$$

which simplifies to

$$C(x) = \begin{cases} 35.50 + 1.25x & \text{for } 0 \leq x \leq 100 \\ 160.50 + 1.50(x - 100) & x > 100 \end{cases}$$

CHAPTER 9 : PIECEWISE FUNCTIONS SOLUTIONS

EXERCISE 9D

1. The diagram shows the round trip journey of a delivery truck travelling from Town A to Town B.

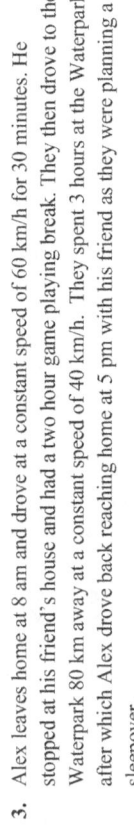

(a) At what time did the truck depart?

1 pm

(b) Calculate the speed of the truck on the first leg of its journey.

$$\frac{60}{3} = 20 \ km/h$$

(c) Kevin, the truck driver stopped for lunch on the way to Town B. For how long did he stop for?

2 hours

(d) On the way back Kevin drove at a constant speed. Trucks are only allowed a maximum speed of 40 km/h. There is a speed camera detector on the way from Town B to Town A. Do you think that Kevin's truck must have received a flash while passing the speed camera? Explain.

$$speed \ on \ way \ back = \frac{100}{2} = 50 \ km/h.$$

Kevin was certainly speeding and went over the 40 km/h limit.

2. The graph on the right shows the round trip journey of a car and a van between Towns P and Q.

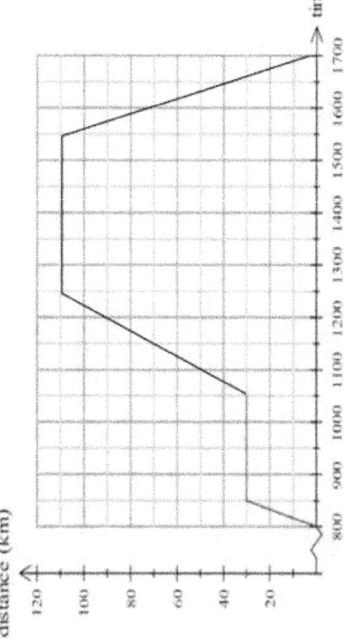

(a) At what time did the van start its journey?

9.30 am

(b) Explain the motion of the car between 10 a.m to 11 a.m.

It stopped.

(c) Estimate the time when the van first overtook the car.

$$\approx 09:35$$

(d) Determine the average speed of the car over its entire journey.

$$\frac{200}{4} = 50 \ km/h$$

MATHEMATICS APPLICATIONS UNIT 2

3. Alex leaves home at 8 am and drove at a constant speed of 60 km/h for 30 minutes. He stopped at his friend's house and had a two hour game playing break. They then drove to the Waterpark 80 km away at a constant speed of 40 km/h. They spent 3 hours at the Waterpark after which Alex drove back reaching home at 5 pm with his friend as they were planning a sleepover.

(a) Complete the distance-time graph on the axes below.

(b) Calculate the return journey's speed.

$$\frac{110}{1.5} = 73.33 \ km/h$$

4. Relax Islands taxes the first $30 000 of an individual's income at a rate of 12%, and all income over $30 000 is taxed at 25%.

(a) Pinocchio makes $18 000 and Geppetto makes $40 000. How much is each taxed?

$$Pinocchio: 0.12 \times 18000 = \$2160$$
$$Geppetto: 0.12 \times 30000 + 0.25 \times 10000 = \$6100$$

(b) Write a piecewise function T that specifies the total tax on an income of x dollars.

$$T(x) = \begin{cases} 0.12x & for \ x \leq 30 \ 000 \\ 0.12 \times 30000 + 0.25(x - 30 \ 000) & x > 30 \ 000 \end{cases}$$

5. Tina is employed as a sales person and is paid commission as follows:

First $ 20000 → 4 %

Amount exceeding $ 20000 → $800 + 3% of each $1 over $20000

Write a piecewise function C that specifies the total commission on an income of x dollars.

$$C(x) = \begin{cases} 0.04x & for \ x \leq 20 \ 000 \\ 800 + 0.03 \times (x - 20 \ 000) & x > 20 \ 000 \end{cases}$$

CHAPTER 9 : PIECEWISE FUNCTIONS SOLUTIONS

6. The diagram below shows the tax system in an American State. As we can see the tax system is progressive meaning the more you earn the more tax you pay.

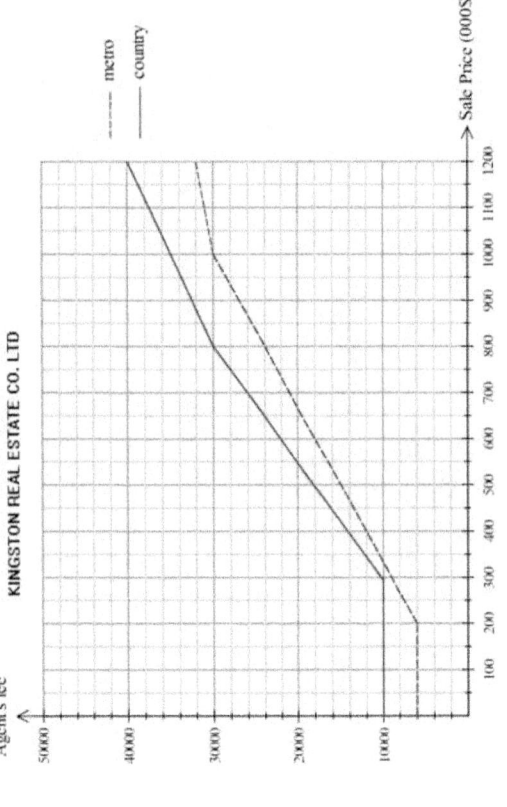

Use the graph to estimate the tax payable by
(a) An accountant with a yearly income of $80000.

$14 000

(b) A teacher with a yearly income of $64000.

$9 000

(c) A soccer player with a yearly income of $98000.

$21 000

Angela paid $12000 in tax last calendar year,
(d) What was her taxable income?

$75 000

Jimmy is a quantity surveyor and earns $100 000 a year. However, because his wife went on maternity leave and was no longer working, Jimmy decided to declare two incomes of $50 000 each to the tax office instead. How much can he save in tax by doing so?

Tax paid on $100 000 = $22 000
Tax paid on $50 000 = $6 000
savings = 22000 − (2 × 6000) = $10 000.

MATHEMATICS APPLICATIONS UNIT 2

7. When it comes to selling a property, Kingston Real Estate's agents charge a commission fee. How much fees someone pays depends on the location of the property. The graph below shows the amount of fees paid in two different areas: metro and country.

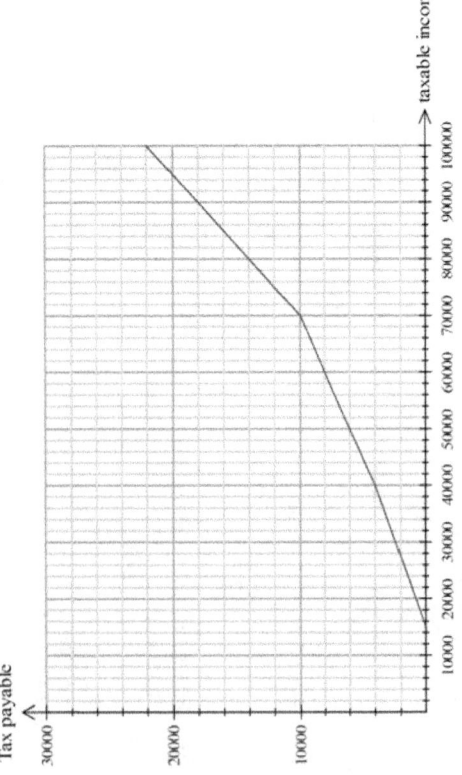

Use the graph to estimate the fees payable on a property sold for
(a) $700 000 in the country.

$26 000

(b) $600 000 in the metro.

$18 000

Angela paid an agent fee $22000 for having her house sold by Kingston Real Estate.
(c) Is her house located in the metro or the country?

Not enough information.
She can own a $600 000 in the country or a $725 000 house in the metro.

CHAPTER 10

TRIGONOMETRY FOR RIGHT TRIANGLES

10A   HYPOTENUSE, ADJACENT AND OPPOSITE

Trigonometry for right-angled triangles is the study of the relationship between the side lengths and its angles. There are three ratios that are used to compare the different lengths of the sides of a right-angled triangle. These ratios are the sine, cosine and tangent ratios commonly named as the trigonometric ratios. The reference angle, usually θ, determines the name of the sides as shown below.

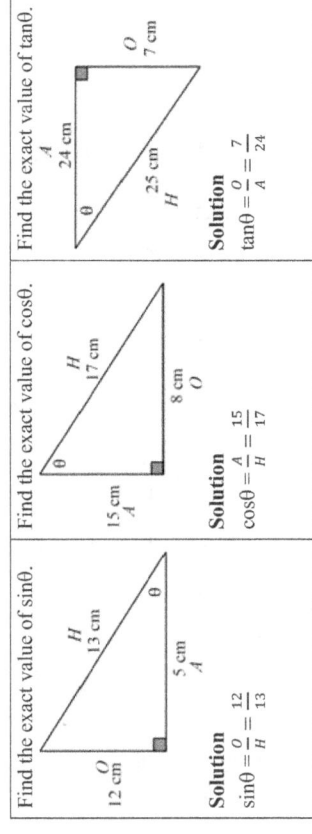

- The hypotenuse is the longest side and faces the right-angle sign.
- The opposite side is the side facing the reference angle θ.
- The adjacent side is the side next to the reference angle θ.

CLASS ACTIVITY 1

In the following triangles label the hypotenuse (H), the opposite (O) and the adjacent (A).

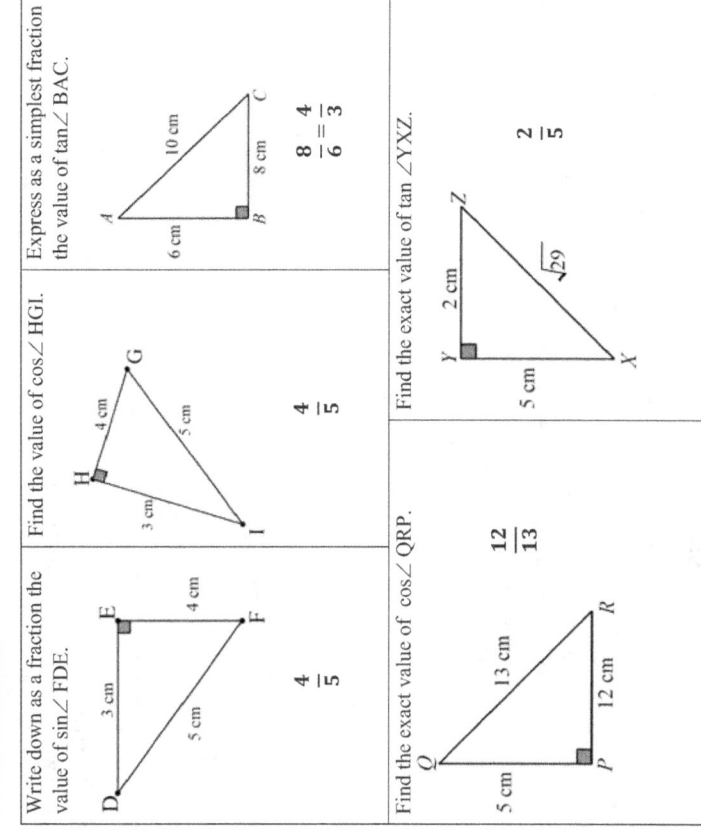

10B   TRIGONOMETRIC RATIOS

As seen earlier, there are three trigonometric ratios: sine (sin), cosine (cos) and tangent (tan).
A simple way of remembering the trigonometric ratios is by using the acronym SOH CAH TOA.

$$\sin\theta = \frac{Opposite}{Hypotenuse}, \quad \cos\theta = \frac{Adjacent}{Hypotenuse}, \quad \tan\theta = \frac{opposite}{adjacent}$$

EXAMPLES

Find the exact value of $\sin\theta$.

Solution
$\sin\theta = \frac{O}{H} = \frac{12}{13}$

Find the exact value of $\cos\theta$.

Solution
$\cos\theta = \frac{A}{H} = \frac{15}{17}$

Find the exact value of $\tan\theta$.

Solution
$\tan\theta = \frac{O}{A} = \frac{7}{24}$

CLASS ACTIVITY

Write down as a fraction the value of $\sin\angle FDE$.

$\frac{4}{5}$

Find the value of $\cos\angle HGI$.

$\frac{4}{5}$

Express as a simplest fraction the value of $\tan\angle BAC$.

$\frac{8}{6} = \frac{4}{3}$

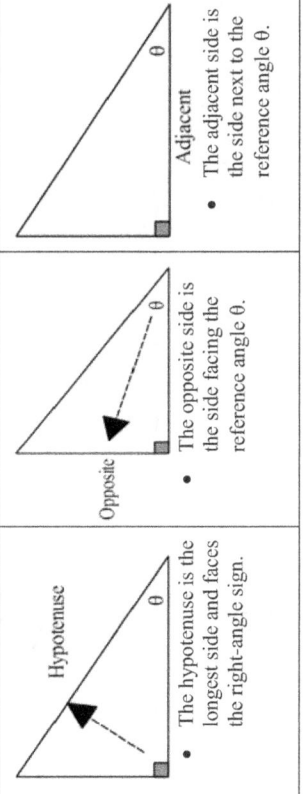

Find the exact value of $\cos\angle QRP$.

$\frac{12}{13}$

Find the exact value of $\tan\angle YXZ$.

$\frac{2}{5}$

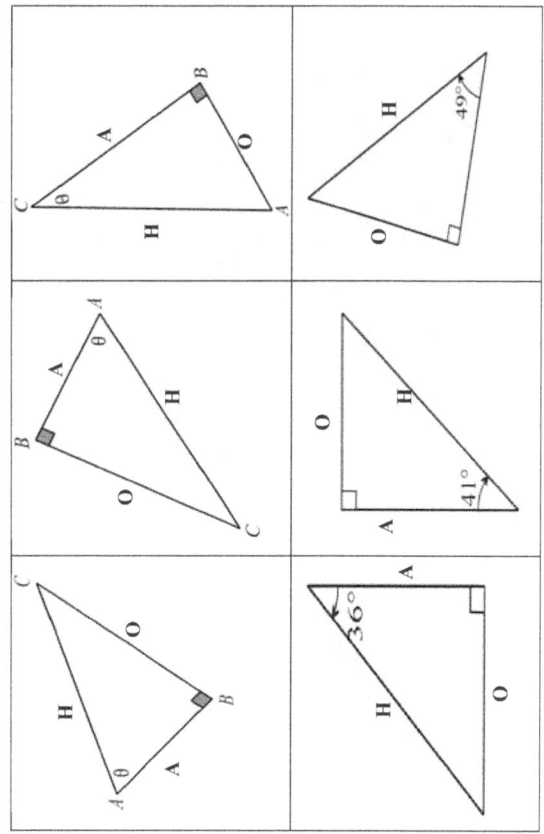

CHAPTER 10 : TRIGONOMETRY FOR RIGHT TRIANGLES: SOLUTIONS

10C   USING THE CALCULATOR

CLASS ACTIVITY

Use your calculator to find the value of the pronumerals in each of the following cases, giving your answers to one decimal place.

1. $x = 20 \times sin36°$ $= 11.8$	2. $y = 24 \times cos45°$ 17.0	3. $x = 13 \times tan28°$ 6.9
4. $x = 32 \times sin51°$ 24.9	5. $y = 32 \times cos63°$ 14.5	6. $z = 15 \times tan54°$ 20.6
7. $x = \frac{10}{sin27°}$ 22.0	8. $x = \frac{12}{cos46°}$ 17.3	9. $y = \frac{9}{tan30°}$ 15.6
10. $sin30° = \frac{x}{10.6}$ (x in the numerator, always multiply) $x = 10.6 \times sin30°$ $= 5.3$	11. $tan60° = \frac{x}{8.6}$ 14.9	12. $cos45° = \frac{x}{16.5}$ 11.7
13. $sin40° = \frac{25}{x}$ (x in the denominator, divide) $x = \frac{25}{sin40°}$ $= 38.9$	14. $tan49° = \frac{12}{x}$ 10.4	15. $cos35° = \frac{14.7}{y}$ 17.9
16. $sin52° = \frac{x}{9.8}$ 7.7	17. $tan32° = \frac{25}{x}$ 40.0	18. $cos60° = \frac{x}{18}$ 9.0
19. $sin63° = \frac{22}{x}$ 24.7	20. $tan78° = \frac{x}{10.2}$ 48.0	21. $cos49° = \frac{25.4}{y}$ 38.7

10D   FINDING SIDE LENGTHS

To be able to find the lengths of the sides of a right-angled triangle, we have to identify the correct trigonometric ratio to be used. The following steps might be helpful.

- If the triangle is not given, draw one showing all known information.
- Label the sides with the letters O, A and H. (remember to label only two sides, leave the blank side unlabelled)
- Use SOH-CAH-TOA to decide which trigonometric ratio to be used.
- Form an equation and solve for the unknown.

EXAMPLES

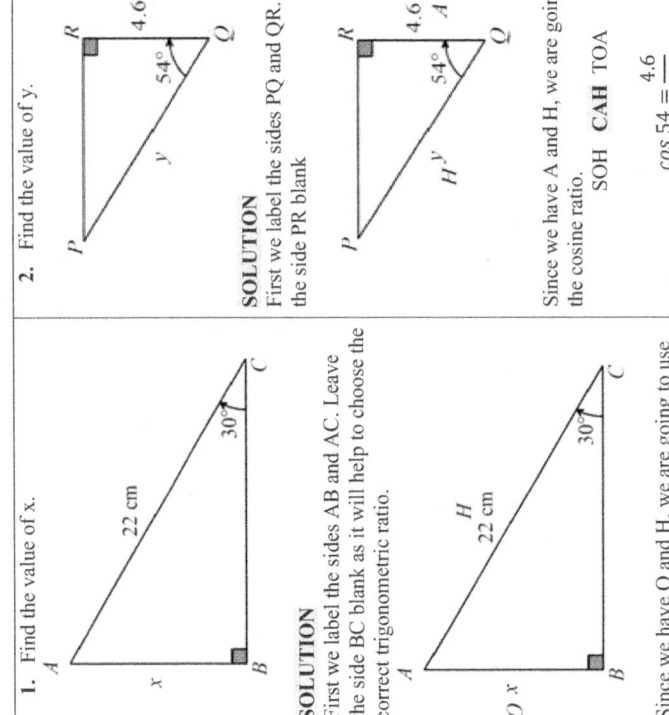

1. Find the value of x.

SOLUTION
First we label the sides AB and AC. Leave the side BC blank as it will help to choose the correct trigonometric ratio.

Since we have O and H, we are going to use the sine ratio.   **SOH**   CAH   TOA

$$sin30° = \frac{x}{22}$$

If the unknown is in the numerator we have to multiply

$$\therefore x = 22 \times sin30 = 11\ cm$$

2. Find the value of y.

SOLUTION
First we label the sides PQ and QR. Leave the side PR blank

Since we have A and H, we are going to use the cosine ratio.   SOH   **CAH**   TOA

$$cos\ 54 = \frac{4.6}{y}$$

If the unknown is in the denominator we have to divide

$$\therefore y = \frac{4.6}{cos54} = 7.83\ cm$$

EXERCISE 10D

Find the value of each variable, correct to two decimal places where applicable.

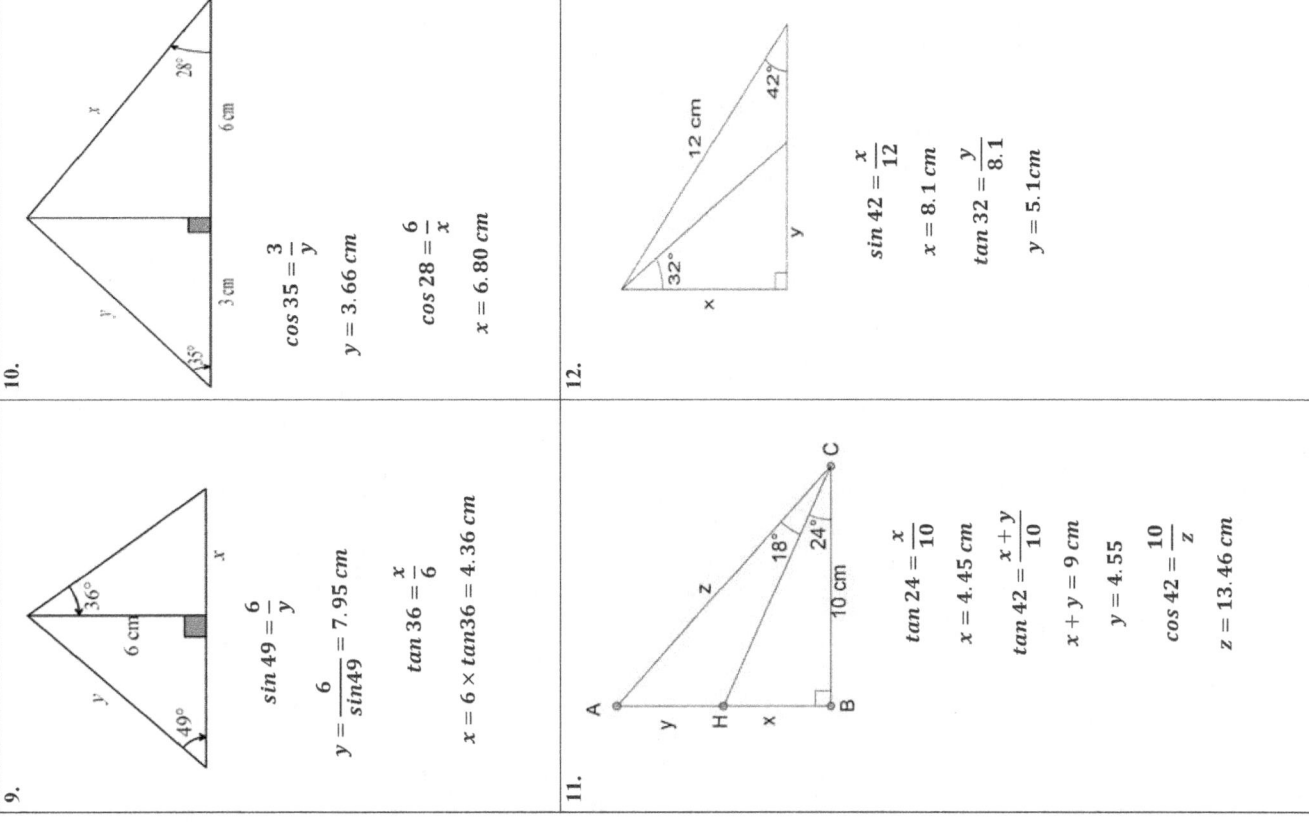

1. $\sin 48 = \dfrac{x}{9.8}$
   $x = 9.8 \times \sin 48 = 7.28\ cm$

2. $\cos 28 = \dfrac{x}{15.2}$
   $x = 13.42\ cm$

3. $\cos 45 = \dfrac{y}{10.6}$
   $x = 7.50\ cm$

4. $\tan 32 = \dfrac{5.6}{x}$
   $x = \dfrac{5.6}{\tan 32} = 8.96\ cm$

5. $\tan 40 = \dfrac{8.3}{x}$
   $x = \dfrac{8.3}{\tan 40} = 9.89\ cm$

6. $\sin 42 = \dfrac{7.5}{x}$
   $x = \dfrac{7.5}{\sin 42} = 11.21\ cm$

7. $\tan 40 = \dfrac{6.25}{x}$
   $x = \dfrac{6.25}{\tan 40} = 7.45\ cm$

8. $\sin 32 = \dfrac{7.1}{x}$
   $x = \dfrac{7.1}{\sin 32} = 13.40\ cm$

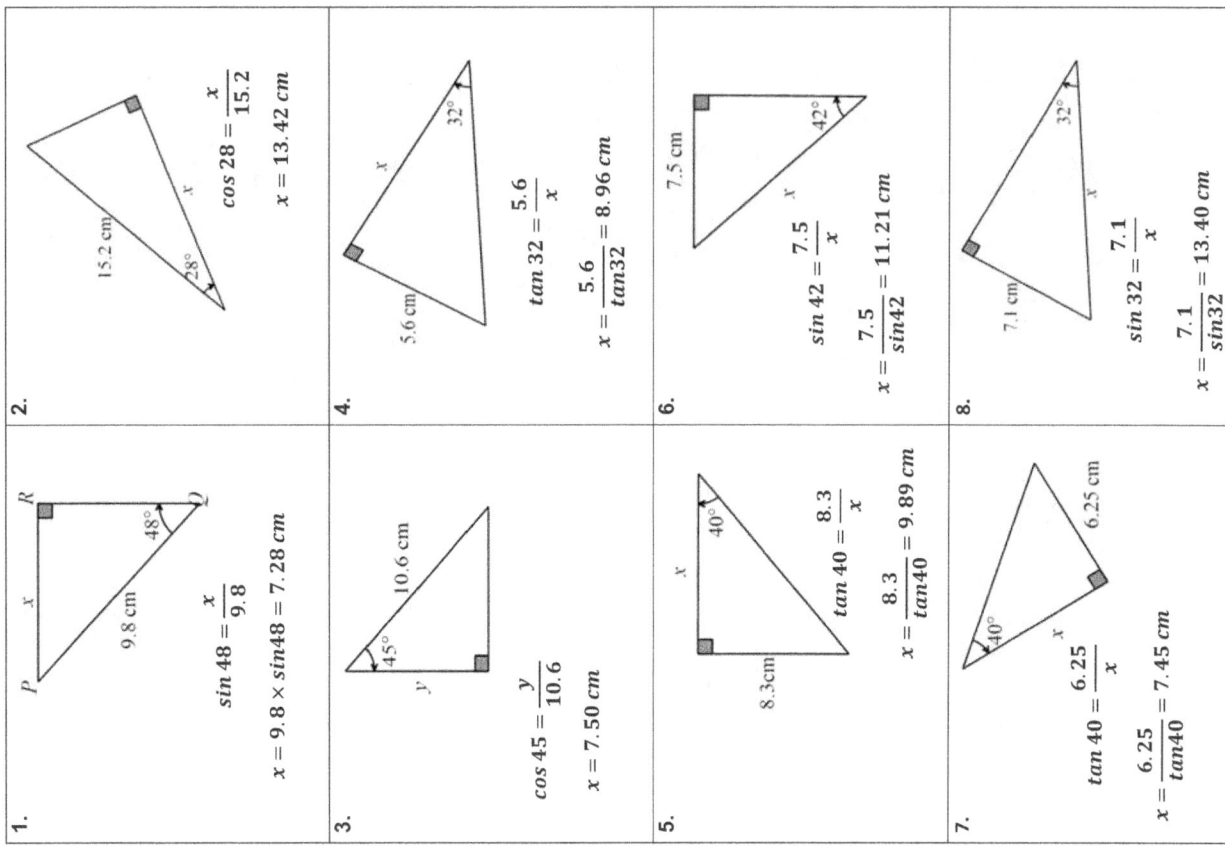

9. $\sin 49 = \dfrac{6}{y}$
   $y = \dfrac{6}{\sin 49} = 7.95\ cm$
   $\tan 36 = \dfrac{x}{6}$
   $x = 6 \times \tan 36 = 4.36\ cm$

10. $\cos 35 = \dfrac{3}{y}$
    $y = 3.66\ cm$
    $\cos 28 = \dfrac{6}{x}$
    $x = 6.80\ cm$

11. $\tan 24 = \dfrac{x}{10}$
    $x = 4.45\ cm$
    $\tan 42 = \dfrac{x+y}{10}$
    $x + y = 9\ cm$
    $y = 4.55$
    $\cos 42 = \dfrac{10}{z}$
    $z = 13.46\ cm$

12. $\sin 42 = \dfrac{x}{12}$
    $x = 8.1\ cm$
    $\tan 32 = \dfrac{y}{8.1}$
    $y = 5.1\ cm$

CHAPTER 10 : TRIGONOMETRY FOR RIGHT TRIANGLES: SOLUTIONS

10E FINDING MISSING ANGLES

To be able to find the missing angle in a right angled triangle, at least two of the sides must be known. Consider the triangle ABC on the right. To determine the size of angle BAC (labelled as x in the diagram), we follow similar steps as seen previously in 10D.

- Label the sides with the letters O, A and H.
- Use SOH-CAH-TOA to decide which trigonometric ratio to be used.
- Form an equation and solve for the unknown.
- Use inverse trigonometry to find x. ($sin^{-1}, cos^{-1}, tan^{-1}$)

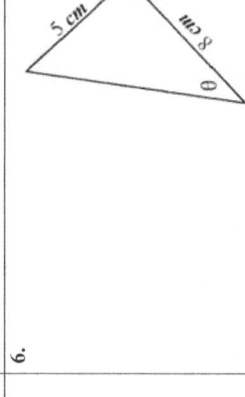

EXAMPLES

1. Find the size of angle BCA (θ) correct to the nearest degree.

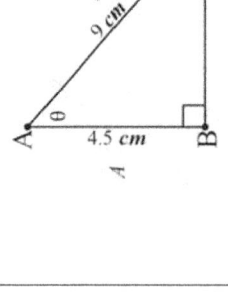

SOLUTION
Using SOH CAH TOA
$$\sin \theta = \frac{3.6}{6}$$
$$\theta = \sin^{-1}\left(\frac{3.6}{6}\right)$$
$$\theta = 37°$$

2. Find the value of θ in the diagram below.

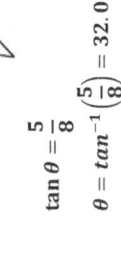

SOLUTION
Using SOH CAH TOA
$$\cos \theta = \frac{4.5}{9}$$
$$\theta = \cos^{-1}\left(\frac{4.5}{9}\right)$$
$$\theta = 60°$$

3. Find the angle that the ladder makes with the wall.

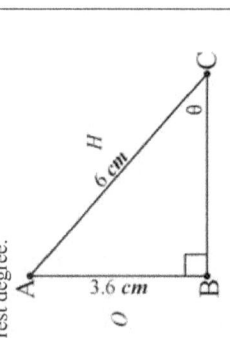

SOLUTION
Using SOH CAH TOA
$$\tan x = \frac{5}{10}$$
$$x = \tan^{-1}\left(\frac{5}{10}\right)$$
$$\therefore x = 26.6°$$
$$x = \tan^{-1}\left(\frac{5}{10}\right)$$
$$\theta = 26.6°$$

MATHEMATICS APPLICATIONS UNIT 2

EXERCISE 10E

Find the value of θ in each of the following triangles.

1.

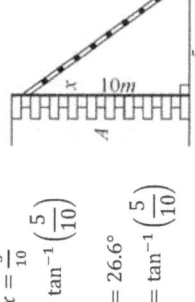

$$\sin \theta = \frac{2.3}{4.6}$$
$$\theta = \sin^{-1}\left(\frac{2.3}{4.6}\right) = 30°$$

2.

$$\cos \theta = \frac{6.1}{9.4}$$
$$\theta = \cos^{-1}\left(\frac{6.1}{9.4}\right) = 49.54°$$

3.

$$\tan \theta = \frac{6.3}{5.2}$$
$$\theta = \tan^{-1}\left(\frac{6.3}{5.2}\right) = 50.46°$$

4.

$$\cos \theta = \frac{3.9}{10}$$
$$\theta = \cos^{-1}\left(\frac{3.9}{10}\right) = 67.05°$$

5.

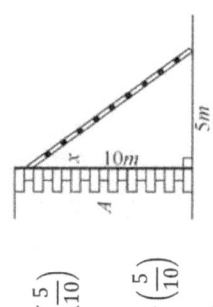

$$\sin \theta = \frac{5.6}{11.4}$$
$$\theta = \sin^{-1}\left(\frac{5.6}{11.4}\right) = 29.42°$$

6.

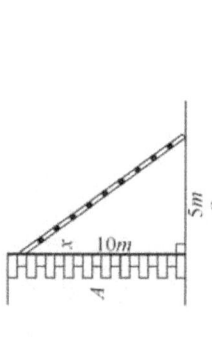

$$\tan \theta = \frac{5}{8}$$
$$\theta = \tan^{-1}\left(\frac{5}{8}\right) = 32.01°$$

CHAPTER 10 : TRIGONOMETRY FOR RIGHT TRIANGLES: SOLUTIONS

7. In triangle ABC, AB = 10.4cm, AC = 6 cm and ∠ACB = 90°. Find ∠ABC.

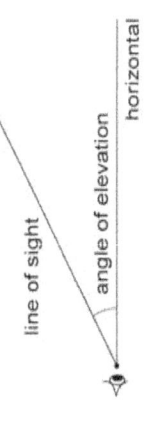

$\sin \theta = \dfrac{6}{10.4}$

$\theta = sin^{-1}\left(\dfrac{6}{10.4}\right) = 35.23°$

8. In triangle PQR, PQ = 10cm, QR = 6.8 cm and ∠PQR = 90°. Find ∠PRQ.

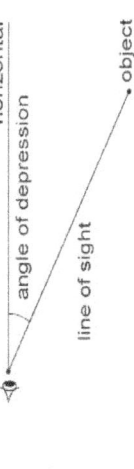

$\tan \theta = \dfrac{10}{6.8}$

$\theta = tan^{-1}\left(\dfrac{10}{6.8}\right) = 55.78°$

9. Determine the size of angle FDG.

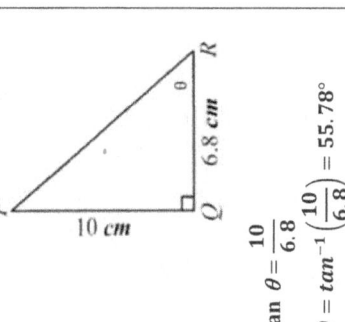

$\tan \theta = \dfrac{5.8}{12.6}$

$\theta = tan^{-1}\left(\dfrac{5.8}{12.6}\right) = 24.72°$

$\tan x = \dfrac{8.3}{12.6}$

$\theta = tan^{-1}\left(\dfrac{8.3}{12.6}\right) = 33.37°$

∠FDG = 33.37° − 24.72° = 8.65°

10. In the diagram AB = 14.6 cm, AD = 15 cm and CD = 10.2 cm. Find the difference between the size of angles x and y.

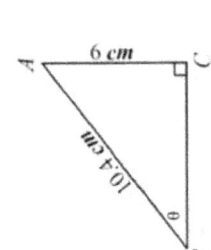

$\cos x = \dfrac{10.2}{15}$

$\theta = cos^{-1}\left(\dfrac{10.2}{15}\right) = 47.17°$

$\tan 47.17 = \dfrac{AC}{10.2}$

AC = 11 cm

$\sin y = \dfrac{11}{14.6}$

$\theta = sin^{-1}\left(\dfrac{11}{14.6}\right) = 48.89°$

$x - y = 48.49 - 47.17 = 1.72°$

MATHEMATICS APPLICATIONS UNIT 2

10F   APPLICATIONS : ANGLE OF ELEVATION AND DEPRESSION

The **angle of elevation** of an object as seen by an observer is the angle between the horizontal and the line of sight from observer's eye.

However, if the object is below the level of the observer, then the angle between the horizontal and the observer's line of sight is called the **angle of depression**.

EXAMPLES

1. Tony standing at the top of a cliff sights a ship 650m at sea as shown. The angle of depression of the ship from Toni is 43°. If Tony is 1.82 m, find the height of the cliff, to the nearest metre.

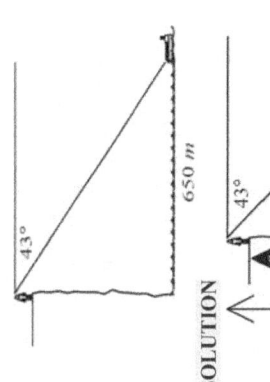

SOLUTION

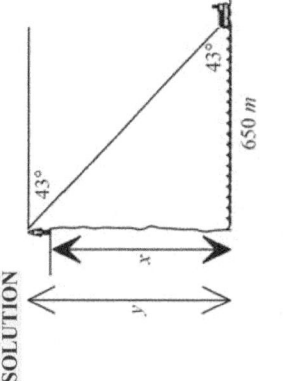 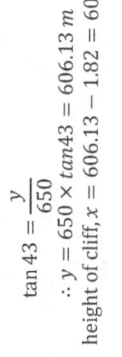

$\tan 43 = \dfrac{y}{650}$

∴ $y = 650 \times tan43 = 606.13 \; m$

height of cliff, $x = 606.13 - 1.82 = 604 \; m$

2. Alex is 1.72m tall. He sights a bird at the top of the tree. The angle of elevation of the top of the tree from Alex's eye-level is 33°. Given that Alex is 32m away from the tree, calculate the height of the tree.

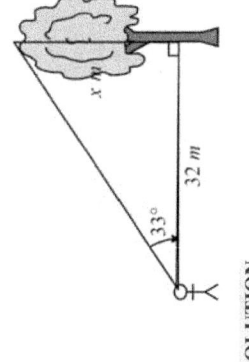

SOLUTION

$\tan 33 = \dfrac{x}{32}$

∴ $x = 32 \times tan33 = 20.78 \; m$

height of tree = 20.78 + 1.72 = 22.50 m

CHAPTER 10 : TRIGONOMETRY FOR RIGHT TRIANGLES: SOLUTIONS

EXERCISE 10F

1. The sun is at an angle of elevation of 51°. A tree casts a shadow 26 metres long on the ground. How tall is the tree?

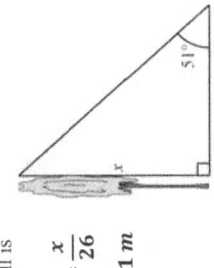

$$\tan 51 = \frac{x}{26}$$
$$x = 32.11\ m$$

2. From the top of a lighthouse 82 m high, the angle of depression of a boat is 44°. Find the distance from the boat to the foot of the lighthouse.

$$\tan 44 = \frac{82}{x}$$
$$x = 84.91\ m$$

3. At a point on the ground 48 m from the foot of a tree, the angle of elevation to the top of the tree is 35°. Find the height of the tree.

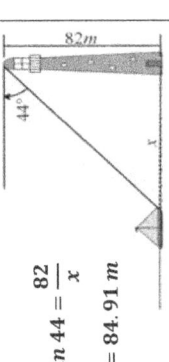

$$\tan 35 = \frac{x}{48}$$
$$x = 33.61\ m$$

4. A 13 m ladder leans against a wall so that the base of the ladder is 5m from the base of the wall. Label the diagram with correct measurements and find the ladder's angle of elevation?

$$\cos\theta = \frac{5}{13}$$
$$\theta = 67.38°$$

5. A ship is on the surface of the water, and its radar detects a submarine at a distance of 98m, at an angle of depression of 25°. Determine how deep underwater is the submarine.

$$\sin 25 = \frac{x}{98}$$
$$x = 41.42\ m$$

6. A dog, who is 15 metres from the base of a light post, spots a bird at the top of the post at an angle of elevation of 39°. Find the direct-line distance between the dog and the bird.

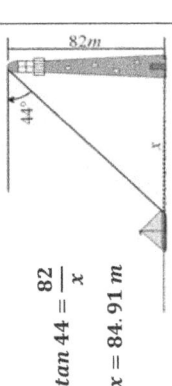

$$\cos 39 = \frac{15}{x}$$
$$x = 19.30\ m$$

7. The angle of elevation from a car to St George Cathedral is 29°. If the church is 38 m high. Determine the distance of the car from the church.

$$\tan 29 = \frac{38}{x}$$
$$x = 68.55\ m$$

8. A person living on the 11th floor of a hotel sights the top and bottom of an apartment 140 m. away. The angle of elevation for the top of the apartment is 34° and the angle of depression for the base of the building is 45°. Find the height of the apartment?

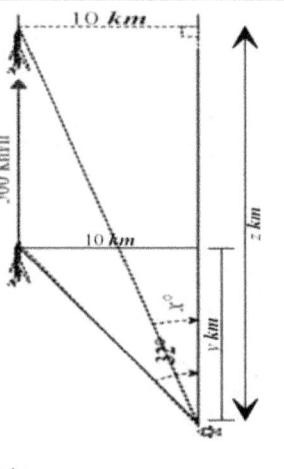

$$\tan 34 = \frac{x}{140}$$
$$x = 94.43\ m$$

$$\tan 45 = \frac{y}{140}$$
$$y = 140\ m$$

height of the apartment $= x + y$
$= 140 + 94.43 = 234.43\ m$

9. An aircraft is travelling horizontally at a contant speed of 500km/h at a height of 10000 m above level ground. At 10 am, Ally sees the aircraft at an angle of elevation 32°. Determine the angle of elevation at 10.30 am.

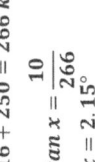

$$\tan 32 = \frac{10}{y}$$
$$y = 16\ km$$

$$z = 16 + 250 = 266\ km$$

$$\tan x = \frac{10}{266}$$
$$x = 2.15°$$

10. The angle of elevation of the top of a tree from a statue's feet is 46° as shown. If the tree casts a shadow of 24m,

(a) Determine the height of the tree.

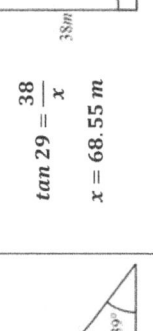

$$\tan 46 = \frac{x}{24}$$
$$x = 24 \times \tan 46 = 24.85\ m$$

(b) If a statue casts a shadow of 4 m, find the statue 's height, assuming the two triangles are similar.

$$solve\left(\frac{4}{24} = \frac{x}{24.85}, x\right)$$
$$x = 4.14\ m$$
The statue is 4.14 m high.

CHAPTER 10 : TRIGONOMETRY FOR RIGHT TRIANGLES

10G BEARINGS

Bearings are a measure of direction, with North taken as being the starting point.
When we travel North, our bearing is 000°, and this is usually represented as a vertical line pointing up.
When we travel in any other direction, our *bearing* is measured **clockwise** from North.
Bearings are also written as 3 digits. Example 059°, 007° and so on.

Consider the diagrams below:

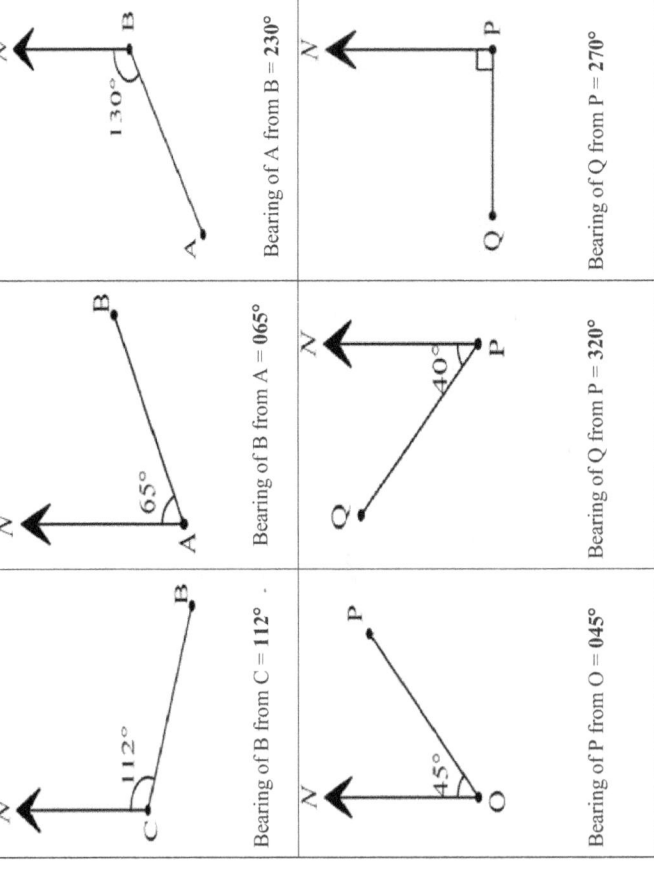

FIGURE 1	FIGURE 2

In Figure 1, if someone walks on the direction shown by the arrow, we say the bearing is 106°.

Similarly, in Figure 2 the bearing is 215°.

EXAMPLE 1
An aeroplane flies from Porto to Rio on a bearing of 065°.
On what bearing should the pilot fly, to return to Porto from Rio?

SOLUTION
First, we find the supplementary angle = 180° − 65° = 115°
Bearing to return to Porto from Rio = 360° − 115° = 245°

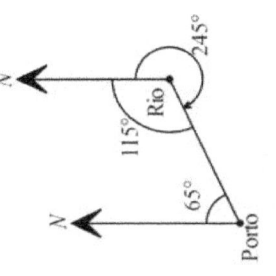

EXAMPLE 2
The diagram shows the positions of three towns R, S and T.
Find the bearing of R from S.

SOLUTION

180° − 57° = 123°

360° − 131° − 123° = 106°

Bearing R from S = 131° + 106° = 237°

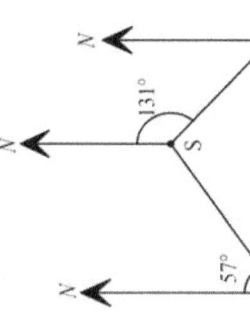

EXERCISE 10G

1. Complete the following by filling in the blanks.

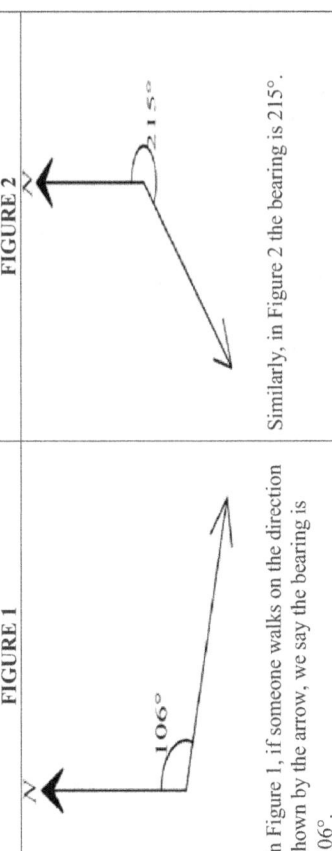

Bearing of B from C = **112°**	Bearing of B from A = **065°**	Bearing of A from B = **230°**
Bearing of P from O = **045°**	Bearing of Q from P = **320°**	Bearing of Q from P = **270°**

2. Study the diagram carefully then answer the questions on the right.

(a) The bearing of R from P = **040°**

(b) The bearing of S from P = **100°**

(c) The bearing of Q from P = **305°**

(d) The bearing of P from Q = **125°**

(e) The bearing of P from R = **220°**

(f) The bearing of P from S = **280°**

CHAPTER 10 : TRIGONOMETRY FOR RIGHT TRIANGLES: SOLUTIONS

3. Study the diagram carefully then answer the questions on the right.

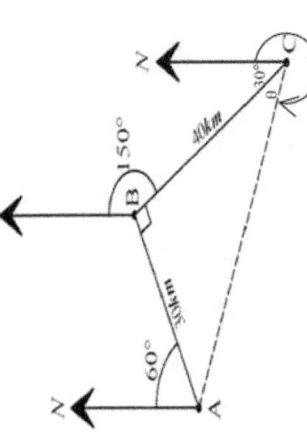

(a) The bearing of R from P = **060°**

(b) The bearing of S from P = **130°**

(c) The bearing of Q from P = **220°**

(d) The bearing of P from Q = **040°**

(e) The bearing of P from R = **240°**

(f) The bearing of P from S = **310°**

4. The map of an holiday island has been shown below.

What is the bearing of

(a) Waterfall from the lighthouse,

090°

(b) Beach from the lighthouse,

135°

(c) Lighthouse from the church.

135°

(d) The lighthouse from the village.

0°

MATHEMATICS APPLICATIONS UNIT 2

5. A ship sets sail from a harbour A and travels 30 km to a harbour B on a bearing of 060°. It stops for a few hours, then sails to a harbour C on a bearing of 150°, travelling a distance of 40 km.

(a) Draw a diagram to illustrate the above situation.

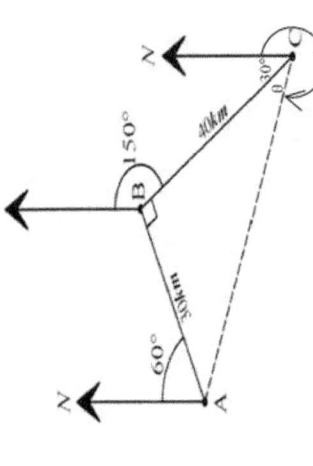

(b) Hence calculate the distance A to C.

$$\sqrt{30^2 + 40^2} = 50\ km$$

(c) On what bearing must the ship set sail to go back to harbour A?

$$tan\ \theta = \frac{30}{40}$$

$$\theta = 36.9°$$
$$bearing = 360 - (30 + 36.9)$$

6. Given that
A is North of B.
C is South-East of B.
C is on a bearing of 160 ° from A.

Find the bearing of:

(a) A from B

0°

(b) A from C

340°

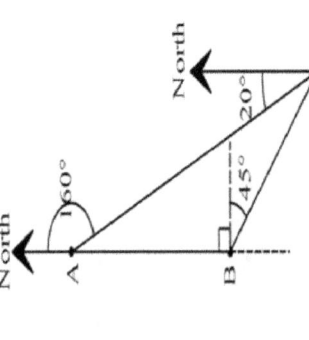

CHAPTER 11

TRIGONOMETRY FOR NON-RIGHT ANGLED TRIANGLES

11A   AREA OF TRIANGLE : BASE AND HEIGHT KNOWN

If the base and the height of a particular triangle are known, then the area can be calculated by the formula

$$Area = \frac{1}{2} \times base \times height$$

EXERCISE 11A

Find the area of the following triangles.

1.
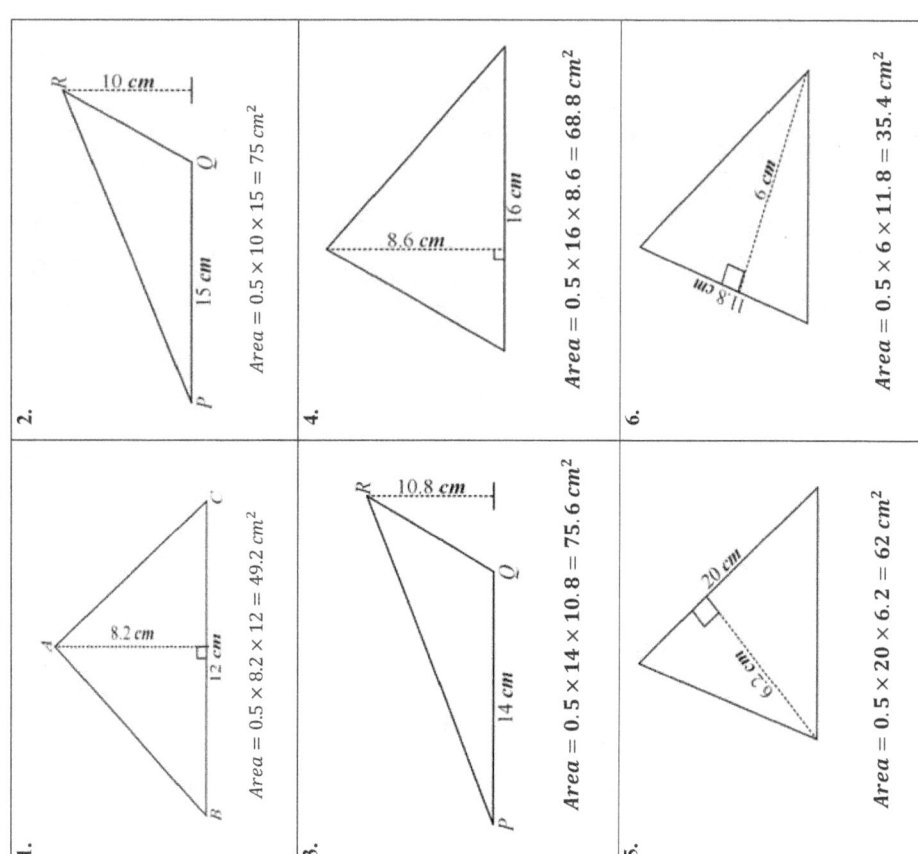

$Area = 0.5 \times 8.2 \times 12 = 49.2\ cm^2$

2.

$Area = 0.5 \times 10 \times 15 = 75\ cm^2$

3.

$Area = 0.5 \times 14 \times 10.8 = 75.6\ cm^2$

4.

$Area = 0.5 \times 16 \times 8.6 = 68.8\ cm^2$

5.

$Area = 0.5 \times 20 \times 6.2 = 62\ cm^2$

6.

$Area = 0.5 \times 6 \times 11.8 = 35.4\ cm^2$

11B   AREA OF TRIANGLE : HEIGHT UNKNOWN

If the base and the height of a particular triangle are unknown, then the area can be calculated by the formula

$$Area = \frac{1}{2} \times a \times b \times sinC,$$

where the C is the angle included between the two sides being used.

EXAMPLES

Find the area of the following triangles.

1.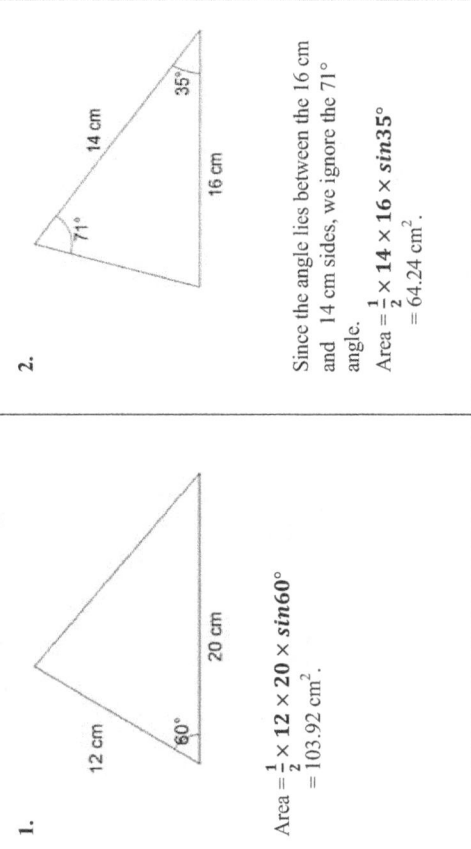

$Area = \frac{1}{2} \times 12 \times 20 \times sin60°$
$= 103.92\ cm^2$.

2.

Since the angle lies between the 16 cm and 14 cm sides, we ignore the 71° angle.
$Area = \frac{1}{2} \times 14 \times 16 \times sin35°$
$= 64.24\ cm^2$.

3. Given that the area of the triangle is 60 cm², find the value of x.

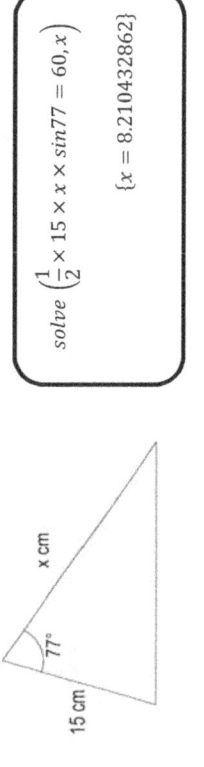

$\frac{1}{2} \times 15 \times x \times sin77° = 60$
Use solve facility on calculator as shown on the right.
∴ $x = 8.21\ cm$.

$solve\left(\frac{1}{2} \times 15 \times x \times sin77 = 60, x\right)$
$\{x = 8.210432862\}$

CHAPTER 11 : TRIGONOMETRY FOR NON-RIGHT ANGLED TRIANGLES: SOLUTIONS

EXERCISE 11B

In questions 1-4, find the area of the triangles.

1.

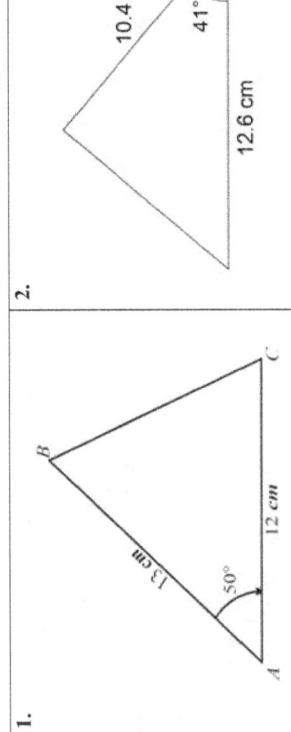

$Area = \frac{1}{2} \times 12 \times 13 \times sin 50$
$= 59.75 \; cm^2$

2.

10.4 cm, 12.6 cm, 41°

$Area = \frac{1}{2} \times 12.6 \times 10.4 \times sin 41$
$= 42.98 \; cm^2$

3.

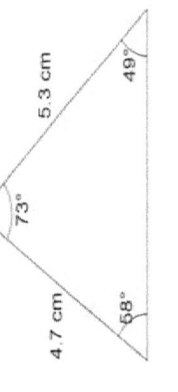

5.7 cm, 6 cm, 75°, 38°

$Area = \frac{1}{2} \times 6 \times 5.7 \times sin 38$
$= 10.53 \; cm^2$

4.

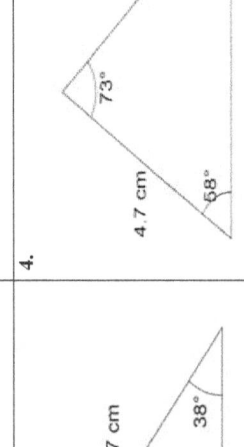

4.7 cm, 5.3 cm, 73°, 58°, 49°

$Area = \frac{1}{2} \times 4.7 \times 5.3 \times sin 73$
$= 11.91 \; cm^2$

5. Given that the area of the triangle is 12 cm², find the value of x.

5.6 cm, x cm, 72°

$Solve(\frac{1}{2} \times 5.6 \times x \times sin 72 = 12, x)$
$= 4.48 \; cm$

6. If the area of the triangle is 15 cm², find the value of x.

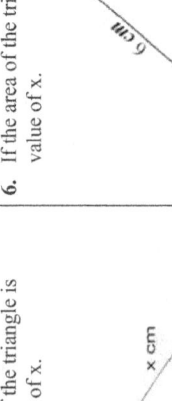

6 cm, x cm, 51°

$Solve(\frac{1}{2} \times 6 \times x \times sin 51 = 15, x)$
$= 6.43 \; cm$

MATHEMATICS APPLICATIONS UNIT 2

11C   AREA OF TRIANGLE: HERON'S FORMULA

If the lengths of the three sides of a triangle are known, a very quick method for calculating the area of the triangle is to use the Heron's formula which states that

$$Area = \sqrt{s(s-a)(s-b)(s-c)},$$

where a, b and c are the lengths of the sides of a triangle and $s = \frac{a+b+c}{2}$.

EXAMPLES

Find the area of the following triangles using the Heron's formula.

1.

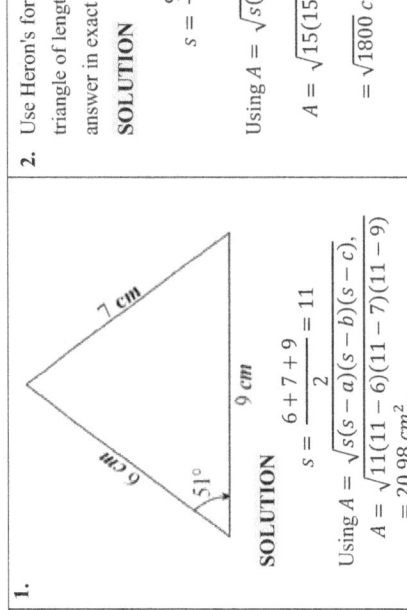

6 cm, 7 cm, 9 cm, 51°

SOLUTION

$s = \frac{6+7+9}{2} = 11$

Using $A = \sqrt{s(s-a)(s-b)(s-c)}$,
$A = \sqrt{11(11-6)(11-7)(11-9)}$
$= 20.98 \; cm^2$.

2. Use Heron's formula to find the area of a triangle of lengths 9, 10 and 11. Give your answer in exact form.

SOLUTION

$s = \frac{9+10+11}{2} = 15$

Using $A = \sqrt{s(s-a)(s-b)(s-c)}$,
$A = \sqrt{15(15-9)(15-10)(15-11)}$
$= \sqrt{1800} \; cm^2$.

3. A level block of land is triangular in shape. (Shown as triangle ABC in the diagram below). Because of its shape, The Real Estate Company decides to plant trees on the block instead of selling it. Trees can be obtained locally from a nursery for $4.65 each. It is ideal that one tree be planted every 3.6 m² of land. How many trees are required to fill the area and what will be the total cost to fill the block of land with the trees?

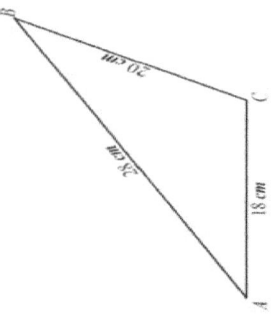

SOLUTION

$s = \frac{18+28+20}{2} = 33$

$\therefore s = 33$

Using $A = \sqrt{s(s-a)(s-b)(s-c)}$,
$A = \sqrt{33(33-18)(33-28)(33-20)}$
$= 179.4 \; m^2$.
Number of trees $= 179.4 \div 3.6 \approx 49 \; trees$
Cost $= 49 \times 4.65 = \$227.85$

CHAPTER 11 : TRIGONOMETRY FOR NON-RIGHT ANGLED TRIANGLES: SOLUTIONS

EXERCISE 11C

1. Use Heron's formula to find the area of the triangle.

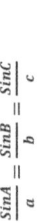

(Triangle with sides 12 cm, 13 cm, 15 cm; vertices B, A, C)

$$s = \frac{12 + 13 + 15}{2} = 20$$

$$A = \sqrt{20(20-12)(20-13)(20-15)}$$

$$= 74.83\ cm^2.$$

2. Find the area of the triangle.

(Triangle with sides 9 cm, 11 cm, 13 cm; vertices A, B, C)

$$s = \frac{9 + 11 + 13}{2} = 16.5$$

$$\sqrt{16.5(16.5-9)(16.5-11)(16.5-13)}$$

$$= 48.81\ cm^2.$$

3. Find the area of a triangular playground, to the nearest square metre, with sides of length 8 m, 10 m and 14 m.

$$s = \frac{8 + 10 + 14}{2} = 16$$

$$A = \sqrt{16(16-8)(16-10)(16-14)}$$

$$= 39\ m^2.$$

4. Find the triangular area, to the nearest square metre, enclosed by three pieces of fencing 50 m, 60 m and 75 m long.

$$s = \frac{50 + 60 + 75}{2} = 92.5$$

$$\sqrt{92.5(92.5-50)(92.5-60)(92.5-75)}$$

$$= 1495\ m^2.$$

5. Find the area of a triangle with side lengths 11 m, 13 m and 20 m.

$$s = \frac{11 + 13 + 20}{2} = 22$$

$$A = \sqrt{22(22-11)(22-13)(22-20)}$$

$$= 66\ m^2.$$

6. Use Heron's formula to find the area of a triangle of lengths 3 cm, 7cm and 8cm. Give your answer in **exact form**.

$$s = \frac{3 + 7 + 8}{2} = 9$$

$$A = \sqrt{9(9-3)(9-7)(9-8)}$$

$$= 6\sqrt{3}\ cm^2.$$

MATHEMATICS APPLICATIONS UNIT 2

11D – THE SINE RULE : CALCULATOR ASSUMED

When do we use the sine rule?
1. For non-right angled triangles
2. At least two angles involved.

The sine rule is given by $\quad \dfrac{SinA}{a} = \dfrac{SinB}{b} = \dfrac{SinC}{c}$

EXAMPLES

1. Find the value of x, correct to 2 decimal places.

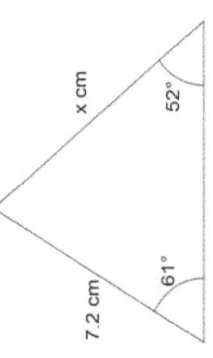

(Triangle with 7.2 cm on left side, angle 61°, angle 52°, side x cm)

SOLUTION

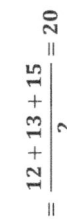

Always use a cross as shown to know which side is divided by which one.

Use the solve facility on your calculator (Main, Action, Advanced, Solve)

$$solve\left(\frac{sin(61)}{x} = \frac{sin(52)}{7.2}, x\right)$$

$$\{x = 7.991340045\}$$

$$x = 7.99\ cm$$

2. Find the value of x.

(Triangle with sides 12 cm, 15 cm, angle 41°, angle x)

SOLUTION

$$solve\left(\frac{sin(41)}{15} = \frac{sin(x)}{12}, x\right)$$

$$\{x = 360.\,constn(1) + 148.342041,$$
$$x = 360.\,constn(2) + 31.6579589\overline{6}$$

$$x = 31.7°\ or\ 148.3°$$

3. Find the value of x:

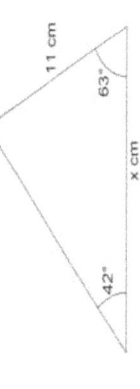

(Triangle with 11 cm, 63°, 42°, side x cm)

SOLUTION

The angle opposite side x can be found as
$x = 180 - (42 + 63) = 75°$

$$solve\left(\frac{sin(42)}{11} = \frac{sin(75)}{x}, x\right)$$

$$\{x = 15.87908846\}$$

$$x = 15.9\ cm$$

CHAPTER 11 : TRIGONOMETRY FOR NON-RIGHT ANGLED TRIANGLES: SOLUTIONS

EXERCISE 11D

Use the solve facility on your calculator to find the value of x in each of the following cases.

1. $\dfrac{\sin(60)}{x} = \dfrac{\sin(45)}{7}$

   $x = 8.57$

2. $\dfrac{\sin(x)}{9.8} = \dfrac{\sin(61)}{10.4}$

   $x = 55.5°$

3. $\dfrac{\sin(x)}{28} = \dfrac{\sin(38)}{20}$

   $x = 120.5°$ or $x = 59.5°$

4. $\dfrac{\sin(57)}{12} = \dfrac{\sin(49)}{x}$

   $x = 10.8$

Find the value of x in each of the following triangles.

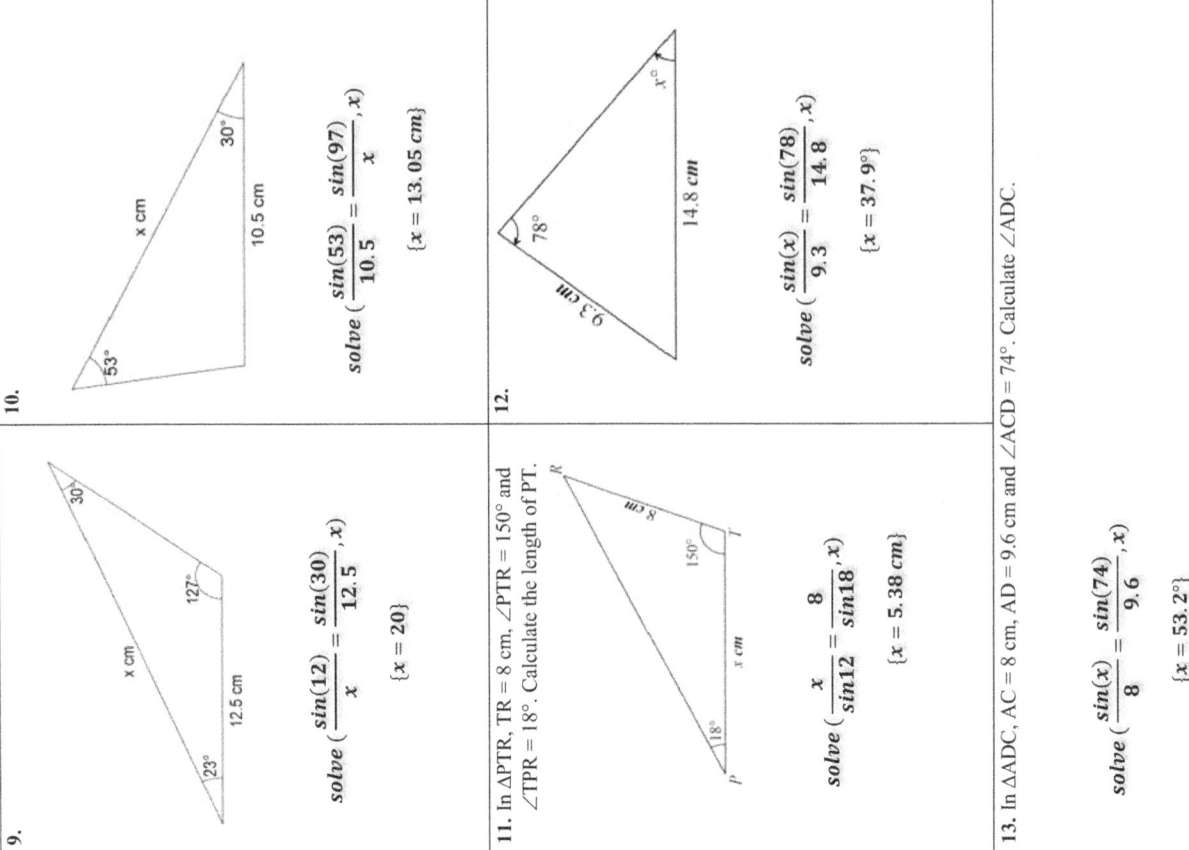

5. $solve\left(\dfrac{\sin(77)}{x} = \dfrac{\sin(45)}{10}, x\right)$

   $\{x = 13.78\}$

6. $solve\left(\dfrac{\sin(69)}{x} = \dfrac{\sin(72)}{8.9}, x\right)$

   $\{x = 8.74\}$

7. $solve\left(\dfrac{\sin(x)}{5} = \dfrac{\sin(82)}{8}, x\right)$

   $\{x = 38.2°\}$

8. $solve\left(\dfrac{\sin(x)}{6.5} = \dfrac{\sin(54)}{9}, x\right)$

   $\{x = 35.75°\}$

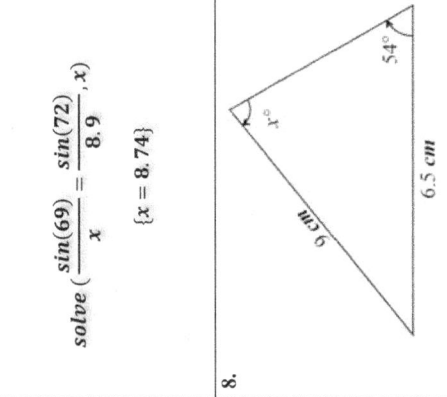

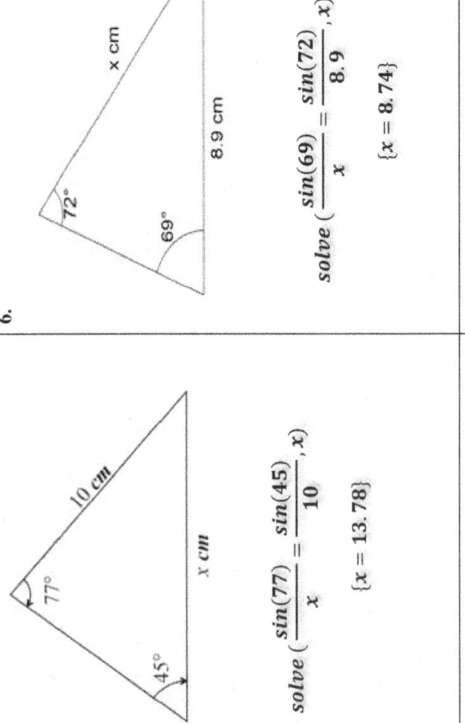

9. $solve\left(\dfrac{\sin(12)}{x} = \dfrac{\sin(30)}{12.5}, x\right)$

   $\{x = 20\}$

10. $solve\left(\dfrac{\sin(53)}{10.5} = \dfrac{\sin(97)}{x}, x\right)$

    $\{x = 13.05\ cm\}$

11. In $\triangle$PTR, TR = 8 cm, $\angle$PTR = 150° and $\angle$TPR = 18°. Calculate the length of PT.

    $solve\left(\dfrac{x}{\sin 12} = \dfrac{8}{\sin 18}, x\right)$

    $\{x = 5.38\ cm\}$

12. $solve\left(\dfrac{\sin(x)}{9.3} = \dfrac{\sin(78)}{14.8}, x\right)$

    $\{x = 37.9°\}$

13. In $\triangle$ADC, AC = 8 cm, AD = 9.6 cm and $\angle$ACD = 74°. Calculate $\angle$ADC.

    $solve\left(\dfrac{\sin(x)}{8} = \dfrac{\sin(74)}{9.6}, x\right)$

    $\{x = 53.2°\}$

CHAPTER 11 : TRIGONOMETRY FOR NON-RIGHT ANGLED TRIANGLES: SOLUTIONS

11E THE SINE RULE : NON CALCULATOR

In this part of the chapter, we are still going to use the sine rule to solve triangles but restrict the use of calculators. The examples below will definitely help the reader to understand how such problems can be tackled.

EXAMPLES

1.

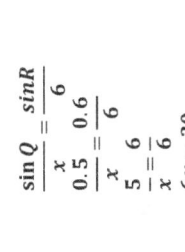

In $\triangle PQR$, $QR = 12$ cm, $\sin Q = 0.3$ and $\sin \angle P = 0.5$. Calculate the length of PR.

SOLUTION

$$\frac{\sin P}{12} = \frac{\sin Q}{x}$$

$$\frac{0.5}{12} = \frac{0.3}{x}$$

Multiply both sides by 10

$$\frac{5}{12} = \frac{3}{x}$$

Cross multiply $5x = 36$

$$\therefore x = \frac{36}{5} = 7.2 \text{ cm}$$

2.

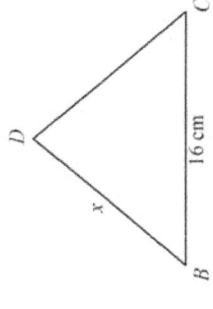

In the above triangle, $AB = 16$ cm, $BC = 10$ cm and $\sin A = 0.25$. Calculate the value of $\sin C$.

SOLUTION

$$\frac{\sin A}{10} = \frac{\sin C}{16}$$

$$\frac{0.25}{10} = \frac{x}{16}$$

Cross multiply $10x = 4$
$\therefore x = 0.4$
$\sin C = 0.4$

---

3. In $\triangle ABC$, $BC = 20$ cm, $\sin A = 0.4$ and $\sin B = 0.2$. Calculate the length of AC.

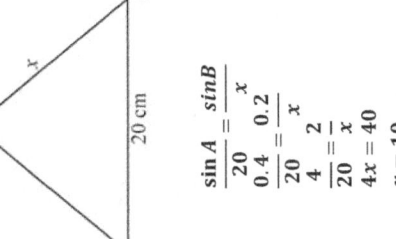

$$\frac{\sin A}{20} = \frac{\sin B}{x}$$
$$\frac{0.4}{20} = \frac{0.2}{x}$$
$$\frac{4}{20} = \frac{2}{x}$$
$$4x = 40$$
$$x = 10$$

4. In triangle PQR, the length of the side PQ is 6 cm, $\sin R = 0.6$ and $\sin Q = 0.5$. Determine the exact length of the side PR.

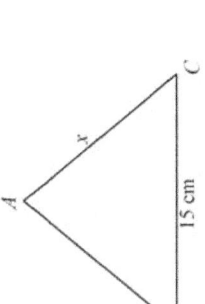

$$\frac{\sin Q}{x} = \frac{\sin R}{6}$$
$$\frac{0.5}{x} = \frac{0.6}{6}$$
$$\frac{5}{x} = \frac{6}{6}$$
$$6x = 30$$
$$x = 5$$

EXERCISE 11E

1. In $\triangle ABC$ below, $BC = 10$ cm, $\sin A = 0.5$ and $\sin B = 0.4$. Calculate the length of AC.

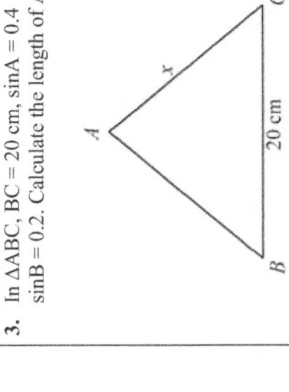

$$\frac{\sin A}{10} = \frac{\sin B}{x}$$
$$\frac{0.5}{10} = \frac{0.4}{x}$$
$$0.5x = 4$$
$$x = 8$$

2. In $\triangle DEF$ below, $DF = 8$ cm, $\sin E = 0.2$ and $\sin D = 0.5$. Calculate the length of EF.

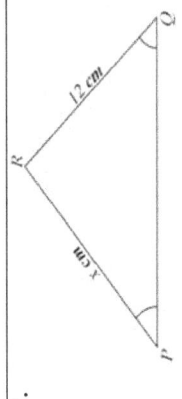

$$\frac{\sin D}{x} = \frac{\sin E}{8}$$
$$\frac{0.5}{x} = \frac{0.2}{8}$$
$$0.2x = 4$$
$$x = 20$$

5. In $\triangle ABC$, $BC = 15$ cm, $\sin A = 0.3$ and $\sin B = 0.5$. Calculate the length of AC.

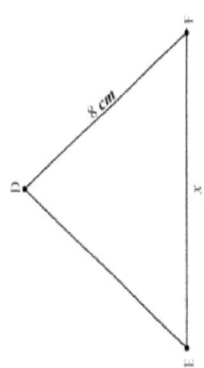

$$\frac{\sin A}{15} = \frac{\sin B}{x}$$
$$\frac{0.3}{15} = \frac{0.5}{x}$$
$$\frac{3}{15} = \frac{5}{x}$$
$$3x = 45$$
$$x = 15$$

6. In triangle BCD, the length of the side BC is 16 cm, $\sin D = 0.8$ and $\sin C = 0.4$. Determine the exact length of the side BD.

$$\frac{\sin D}{16} = \frac{\sin C}{x}$$
$$\frac{0.8}{16} = \frac{0.4}{x}$$
$$\frac{8}{16} = \frac{4}{x}$$
$$8x = 64$$
$$x = 8$$

CHAPTER 11 : TRIGONOMETRY FOR NON-RIGHT ANGLED TRIANGLES: SOLUTIONS

11F THE COSINE RULE (I) – FINDING SIDES

When do we use the cosine rule?

1. For non-right angled triangles
2. At most one angle involved.

The cosine rule is given by $a^2 = b^2 + c^2 - 2bc\,CosA$, where a is the side facing the angle.

EXAMPLES

1. Find the value of x.

SOLUTION

$x^2 = 10^2 + 12^2 - 2 \times 10 \times 12 \times \cos 53°$

Use the solve facility to find x as follows.

(Main, Action, Advanced, Solve)

solve $(x^2 = 10^2 + 12^2 - 2 \times 10 \times 12 \times \cos(53), x)$

$\{x = -9.978195951,$
$x = 9.978195951\}$

We reject the negative value of x as the length of a triangle cannot be less than zero.
Hence $x = 9.98\ cm$

2. Find the value of x in the triangle ABC.

SOLUTION

Use the solve facility on your calculator,

solve $(15^2 = x^2 + 10^2 - 2 \times 10 \times x \times Cos58, x)$

$(x = -7.073415921, x = 17.67180121)$

Here again, we reject the negative answer.

$x = 17.7\ cm$

EXERCISE 11F

Find the value of x in each of the following using the solve facility in your calculator.

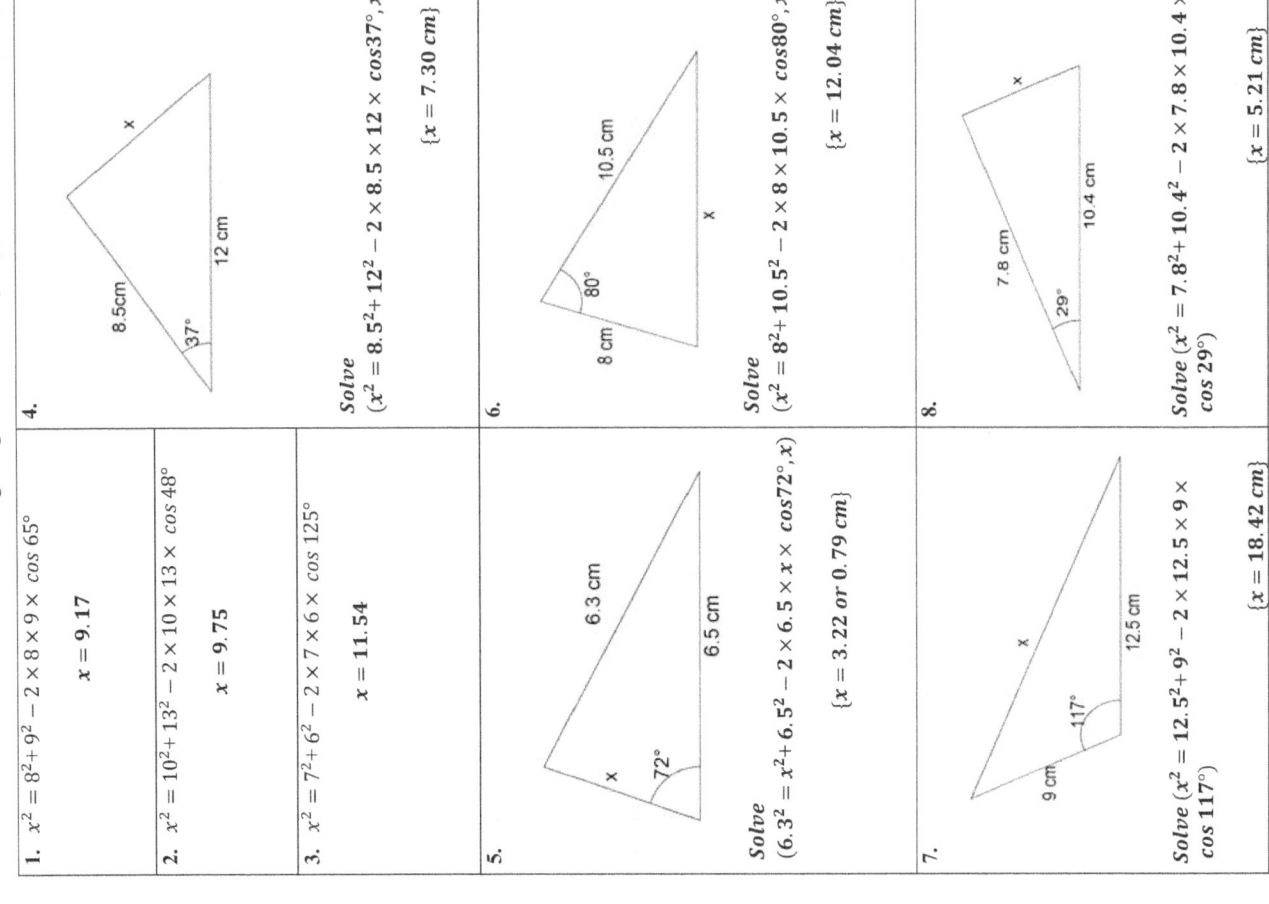

1. $x^2 = 8^2 + 9^2 - 2 \times 8 \times 9 \times \cos 65°$

$x = 9.17$

2. $x^2 = 10^2 + 13^2 - 2 \times 10 \times 13 \times \cos 48°$

$x = 9.75$

3. $x^2 = 7^2 + 6^2 - 2 \times 7 \times 6 \times \cos 125°$

$x = 11.54$

4.

Solve $(x^2 = 8.5^2 + 12^2 - 2 \times 8.5 \times 12 \times cos37°, x)$

$\{x = 7.30\ cm\}$

5.

Solve $(6.3^2 = x^2 + 6.5^2 - 2 \times 6.5 \times x \times cos72°, x)$

$\{x = 3.22\ or\ 0.79\ cm\}$

6.

Solve $(x^2 = 8^2 + 10.5^2 - 2 \times 8 \times 10.5 \times cos80°, x)$

$\{x = 12.04\ cm\}$

7.

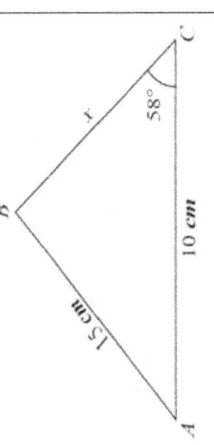

Solve $(x^2 = 12.5^2 + 9^2 - 2 \times 12.5 \times 9 \times cos 117°)$

$\{x = 18.42\ cm\}$

8.

Solve $(x^2 = 7.8^2 + 10.4^2 - 2 \times 7.8 \times 10.4 \times \cos 29°)$

$\{x = 5.21\ cm\}$

CHAPTER 11 : TRIGONOMETRY FOR NON-RIGHT ANGLED TRIANGLES: SOLUTIONS

11G THE COSINE RULE (II) – finding angles

To find angles, using the cosine rule make use of the formula

$$cosA = \frac{b^2+c^2-a^2}{2bc}, \text{ where } a \text{ is the side facing the angle.}$$

EXAMPLE
Find the value of x.

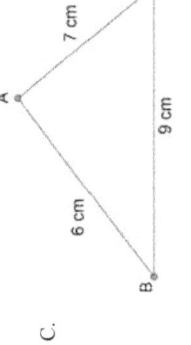

SOLUTION

$Solve\left(cosx = \frac{8^2+13^2-10^2}{2\times 8\times 13}, x\right)$

$x = 50.3°$.

EXERCISE 11G
Find the value of x in each case.

1.

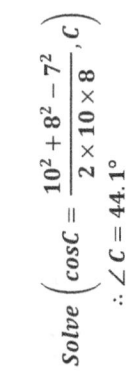

$Solve\left(cosx = \frac{11^2+7^2-8^2}{2\times 11\times 7}, x\right)$

$\{x = 46.5°\}$

2.

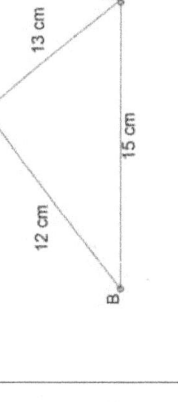

$Solve\left(cosx = \frac{6.5^2+10.5^2-7.2^2}{2\times 6.5\times 10.5}, x\right)$

$\{x = 42.5°\}$

3.

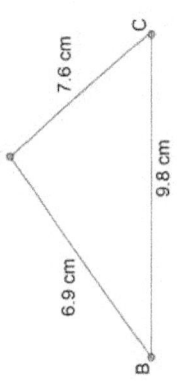

$Solve\left(cosx = \frac{5.4^2+8.5^2-7.6^2}{2\times 5.4\times 8.5}, x\right)$

$\{x = 61.6°\}$

4.

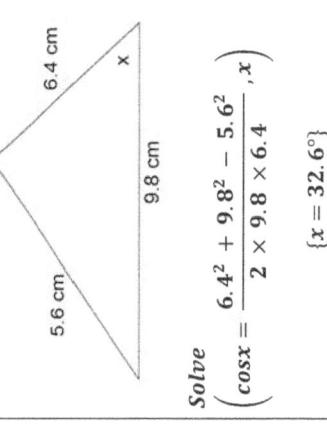

$Solve\left(cosx = \frac{6.4^2+9.8^2-5.6^2}{2\times 9.8\times 6.4}, x\right)$

$\{x = 32.6°\}$

11H THE COSINE RULE (III) – finding the smallest and largest angle

- The smallest angle faces the smallest side.
- The largest angle faces the longest side

EXAMPLE
Find the size of the smallest angle in the given triangle.

SOLUTION
The smallest side being 6 cm, the smallest angle is angle C.

$Solve\left(cos C = \frac{7^2+9^2-6^2}{2\times 7\times 9}, C\right)$

$\therefore \angle C = 41.8°$

EXERCISE 11H

1. Find the size of the largest angle from the triangle.

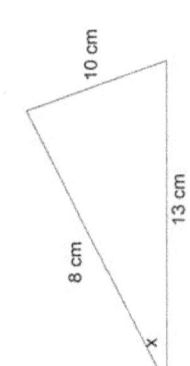

$Solve\left(cosA = \frac{11^2+9^2-13^2}{2\times 11\times 9}, A\right)$

$\therefore \angle A = 80.4°$

2. Find the size of the smallest angle in the given triangle.

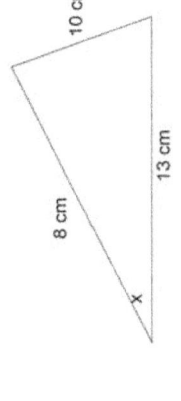

$Solve\left(cosC = \frac{10^2+8^2-7^2}{2\times 10\times 8}, C\right)$

$\therefore \angle C = 44.1°$

3. Find the size of the largest angle from the triangle.

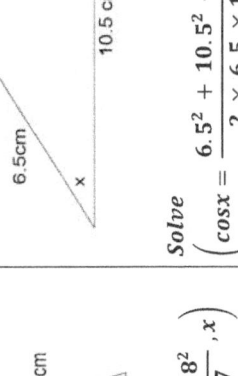

$Solve\left(cosA = \frac{6.9^2+7.6^2-9.8^2}{2\times 6.9\times 7.6}, A\right)$

$\therefore \angle A = 84.9°$

4. Find the size of the largest angle from the triangle.

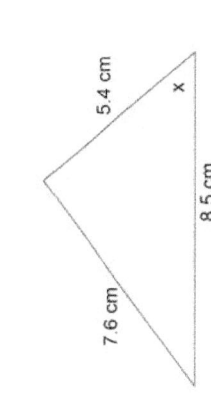

$Solve\left(cosA = \frac{12^2+13^2-15^2}{2\times 12\times 13}, A\right)$

$\therefore \angle A = 73.6°$

CHAPTER 11 : TRIGONOMETRY FOR NON-RIGHT ANGLED TRIANGLES: SOLUTIONS

11.1 SINE RULE, COSINE RULE AND AREA APPLICATIONS

EXAMPLE

The diagram shows a farmer's block of land having the shape of a quadrilateral labelled ABCD. Given that AD = 63m, DC = 75m and BC = 81m. Also $\angle ADC = 56°$ and $\angle ABC = 47°$.

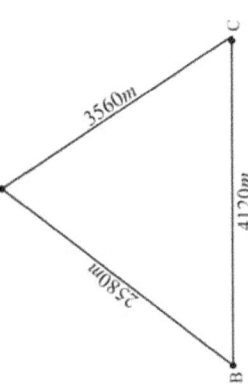

(a) Find the length of AC, correct to the nearest m.

$$AC^2 = 63^2 + 75^2 - 2 \times 63 \times 75 \times \cos 56 = 65.65\ m$$

$$\therefore AC \approx 66\ m$$

(b) Find the area of triangle ADC.

$$A = \frac{1}{2} \times 63 \times 75 \times \sin 56° = 1958.6\ m^2$$

(c) Use the sine rule to determine the size of the acute angle BAC, correct to the nearest degree.

$$\frac{\sin \angle BAC}{81} = \frac{\sin 47}{65.65}$$

$$\angle BAC = 64.47° \approx 64°$$

(d) Find the area of the block ABCD.

$$\angle ACB = 180 - 47 - 64.47 = 68.53°$$

$$\text{Area of } \triangle ABC = \frac{1}{2} \times 81 \times 65.65 \times \sin 68.53° = 2474.3\ m^2$$

Hence area of quadrilateral ABCD = 1958.6 + 2474.3 = 4432.9 m^2

EXERCISE 11.1

1. The diagram below (not drawn to scale) shows a school oval consisting of three walls AB, BC and AC.

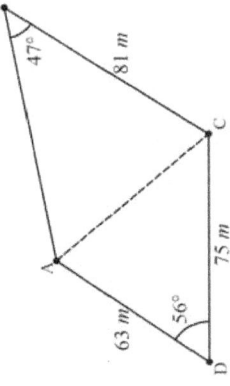

(a) Use trigonometry to determine the size of the angle BAC to the nearest degree.

$$\boldsymbol{Solve\left(cosA = \frac{2580^2 + 3560^2 - 4120^2}{2 \times 2580 \times 3560}, A\right)}$$

$$\{\angle A = 82.63° \approx 83°\}$$

(b) Use the sine rule to determine the size of the angle ABC.

$$\boldsymbol{solve\left(\frac{sinB}{3560} = \frac{sin82.63}{4120}, B\right)}$$

$$\{\angle B = 59.0°\}$$

(c) The section ABC needs to be covered with artificial lawn. The cost of the material is $48 per square metre. Determine the total cost of installing the lawn.

$$\boldsymbol{Area = \frac{1}{2} \times 2580 \times 4120 \times sin 58.97}$$

$$= 4554238.89\ m^2$$

$$\boldsymbol{Cost} = 4554238.89 \times 48$$

$$= \$218603466.5$$

CHAPTER 11 : TRIGONOMETRY FOR NON-RIGHT ANGLED TRIANGLES: SOLUTIONS

2. The diagram below (not drawn to scale) is a survey plan of a new industrial site land *ABCD*.

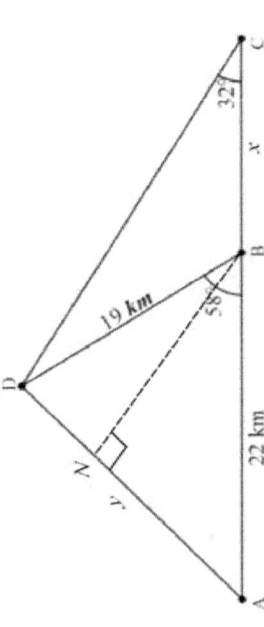

(a) To develop the site a road needs to be constructed along the line segment *AC*. Using trigonometry, calculate the length of this road to the nearest metre.

$$solve\ (x^2 = 660^2 + 740^2 - 2 \times 660 \times 740 \times cos(63), x)$$

$$\{x = 734.67\ m \approx 735m\}$$

(b) Alpha Road Resurfacing Co Ltd has obtained the contract of constructing the road AC at a rate of $1450 per metre. What is the total cost of constructing the road?

$$1450 \times 735 = \$1065750$$

(c) Using trigonometry, determine the size of ∠ABC.

$$solve\ (\frac{sinB}{734.67} = \frac{sin58}{810}, B)$$

$$\{\angle B = 50.3°\}$$

(d) Using trigonometry, determine the area of the site ABCD.

$$Area\ of\ \triangle ADC == \frac{1}{2} \times 660 \times 740 \times sin63$$
$$= 217583.79\ m^2$$
$$\angle ACB = 180 - (58 + 50.3) = 71.7$$
$$Area\ of\ \triangle ABC == \frac{1}{2} \times 734.67 \times 810 \times sin71.7$$
$$= 282493.34\ m^2$$
$$Total\ area = 500077.13m^2$$

3.

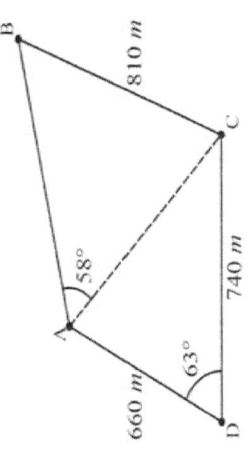

The diagram shows four towns A, B, C and D. ABC is a straight line ABC. AB = 22 km, BD =19 km, ∠ABD = 58° and ∠BCD = 32°. Calculate

(a) The length of BC,

$$\angle DBC = 180 - 58 = 122$$
$$\angle BDC = 180 - (32 + 122) = 26$$
$$solve\ (\frac{sin26}{x} = \frac{sin32}{19}, B)$$
$$\{x = 15.72\ km\}$$

A bridge connects towns A and D.
(b) Determine the length of the bridge AD.

$$solve\ (y^2 = 22^2 + 19^2 - 2 \times 22 \times 19 \times cos(58), y)$$

$$\{y = 20.05\ km\}$$

(c) The area of triangle ABD,

$$Area = \frac{1}{2} \times 22 \times 19 \times sin58$$
$$= 177.24\ km^2$$

(d) The bridge being too busy with the commuters, the government has decided to construct a road from Town B connecting to the bridge AD. Find the shortest distance from B to AD.

Shortest distance = perpendicular distance BN
$$Solve\ (\frac{1}{2} \times 20.05 \times BN = 177.24)$$
$$BN = 17.68\ km$$

CHAPTER 11 : TRIGONOMETRY FOR NON-RIGHT ANGLED TRIANGLES: SOLUTIONS

4. The Dexter's family is planning the front yard of their new house. The diagram below (not drawn to scale) shows the area.

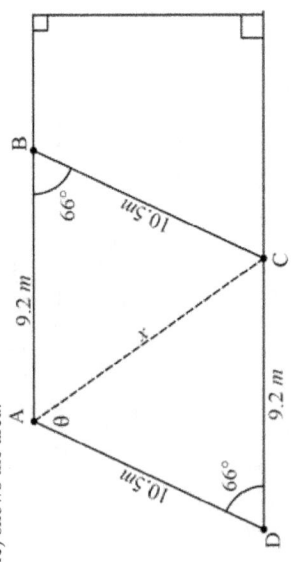

(a) The Dexter's decide to build a limestone wall (one block high) from A to C to split the available space into two tiny little gardens.

(i) Using trigonometry, calculate the length of this wall.

$$solve\ (x^2 = 9.2^2 + 10.5^2 - 2 \times 9.2 \times 10.5 \times \cos(66), x)$$

$$\{x = 10.8\ m\}$$

(ii) Limestone blocks come in 400mm lengths. How many blocks will the Dexter's need to buy?

$$10.8 \times 1000 = 10800 \div 400 = 27$$
They need to buy 27 blocks.

(c) The playground area ACD is to be covered by sand. Using trigonometry, determine

(i) the size of the angle CAD.

$$solve\left(\frac{sin(\theta)}{9.2} = \frac{sin(66)}{10.8}, \theta\right)$$

$$\{\theta = 51.1°\}$$

(ii) the area of the playground ACD.

$$Area = \frac{1}{2} \times 10.5 \times 9.2 \times sin66$$
$$= 44.12\ m^2$$

5. The diagram below shows a sketch of a block of land Lot 91. Find both the **area** and **perimeter** of the block.

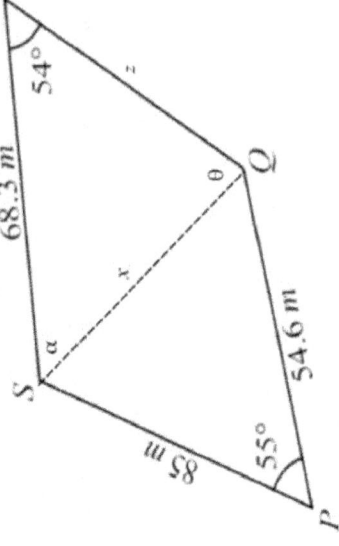

$$solve\ (x^2 = 85^2 + 54.6^2 - 2 \times 85 \times 54.6 \times \cos(55), x)$$

$$\{x = 69.9\ m\}$$

$$solve\left(\frac{sin(\theta)}{68.3} = \frac{sin(54)}{69.9}, \theta\right)$$

$$\{\theta = 52.2°\}$$

$$Area\ of\ \triangle PQS = \frac{1}{2} \times 85 \times 54.6 \times sin55$$
$$= 1900.84\ m^2$$

$$\alpha = 180 - (54 + 52.2) = 73.8$$

$$Area\ of\ \triangle QRS = \frac{1}{2} \times 68.3 \times 69.9 \times sin73.8$$
$$= 2292.30\ m^2$$

$$solve\left(\frac{sin(73.8)}{z} = \frac{sin(54)}{69.9}, z\right)$$

$$\{z = 83.0\ m\}$$

$$Perimeter = 85 + 54.6 + 68.3 + 83.0$$
$$= 290.9$$

$$Area\ of\ PQRS = 2292.30 + 1900.84$$
$$= 4193.14\ m^2$$

CHAPTER 12

SIMULTANEOUS LINEAR EQUATIONS

Simultaneous equations in this chapter are two equations, each containing two unknown letters. We have to use both equations to find the value of the unknown letters through different methods such as elimination, substitution or graphical. Skills needed to tackle this chapter are

- Solving linear equations
- Algebra rules
- Directed numbers

12A   ELIMINATION METHOD

In elimination method, we have to make the coefficients of one the two variables the same by multiplying by a scalar, if applicable.

- If the coefficients have the same sign, either positive or negative, we need to subtract one equation from the other.
- If the coefficients have the different signs, one positive and one negative, we need to add both equations

EXAMPLES

Solve the following pairs of simultaneous equations using the elimination method.

EXAMPLE 1

$$4x + y = 10$$
$$2x + y = 4$$

SOLUTION
First we label both equations
$4x + y = 10$ .... equation 1
$2x + y = 4$ .... equation 2

Clearly the coefficient of y is same in both equations. They both have the same sign, so we subtract to eliminate y.
$4x + y = 10$
$2x + y = 4$
Subtracting we have,
$2x = 6$
$x = 3$
To find the value of y, substitute $x = 3$ in either equation
Using $2x + y = 4$
$2(3) + y = 4$
$6 + y = 4$
$y = -2$

EXAMPLE 2

$$2x + 3y = 9$$
$$3x + y = 10$$

SOLUTION
First we label both equations
$2x + 3y = 9$ .... equation 1
$3x + y = 10$ .... equation 2 ($\times 3$)

Multiply the second equation by 3 to make the coefficient of y same as in equation 1.
$2x + 3y = 9$
$9x + 3y = 30$
Subtracting we have,
$-7x = -21$
$x = 3$
To find the value of y, substitute $x = 3$ in equation 2. (We can also use equation 1.)
$3(3) + y = 10$
$9 + y = 10$
$y = 1$

EXAMPLE 3

$$2a + b = 7$$
$$5a - 2b = 22$$

SOLUTION
First we label both equations
$2a + b = 7$ .... equation 1
$5a - 2b = 22$ .... equation 2

We can make the coefficients of b the same by multiplying equation 1 by 2.
$4a + 2b = 14$
$5a - 2b = 22$
The signs are different, so we add to eliminate b
Adding we have,
$9a = 36$
$a = 4$
To find the value of b, substitute $a = 4$ in either equation
Using $2a + b = 7$
$2(4) + b = 7$
$8 + b = 7$
$b = -1$
Hence $a = 4, b = -1$

EXAMPLE 4

$$3x + 4y = 23$$
$$2x + 3y = 16$$

SOLUTION
First we label both equations
$3x + 4y = 23$ .... equation 1 ($\times 3$)
$2x + 3y = 16$ .... equation 2 ($\times 4$)

In this case neither the coefficients of x are same nor the coefficient of y. We can choose to eliminate one of them, it is a personal choice.
To eliminate x multiply equation 1 by 2 and equation 2 by 3 so that both become $6x$ Or
To eliminate y multiply equation 1 by 3 and equation 2 by 4 so that both become $12y$.
Say we want to eliminate y.
$9x + 12y = 69$
$8x + 12y = 64$
Subtracting we have,
$x = 5$
To find the value of y, substitute $x = 5$ in either equation
Using $2x + 3y = 16$
$2(5) + 3y = 16$
$10 + 3y = 16$
$3y = 6$
$\therefore y = 2$

EXERCISE 12A

Solve the following pairs of simultaneous equations using the elimination method.

1. $5x + y = 9$
   $2x + y = 3$

   Subtracting we have,
   $3x = 6$
   $x = 2$
   Using $2x + y = 3$
   $2(2) + y = 3$
   $4 + y = 3$
   $y = -1$
   Hence $x = 2$ and $y = -1$

2. $x + 2y = 11$
   $x - y = -1$

   Subtracting we have,
   $3y = 12$
   $y = 4$
   Using $x + 2y = 11$
   $x + 2(4) = 11$
   $x + 8 = 11$
   $x = 3$
   Hence $x = 3$ and $y = 4$

CHAPTER 12 : SIMULTANEOUS LINEAR EQUATIONS: SOLUTIONS

3. $3x + y = 5$ ......× 2   equation 1
    $5x + 2y = 8$ ......   equation 2

Multiply equation (1) by 2 to eliminate y.
    $6x + 2y = 10$
    $5x + 2y = 8$

Subtracting we have,
    $x = 2$
Using $3x + y = 5$
    $3(2) + y = 5$
    $6 + y = 5$
    $y = -1$
    $\therefore x = 2, y = -1$

5. $x + 3y = 13$ ......   equation 1
    $2x - y = -2$ ......× 2   equation 2

Multiply equation (2) by 2 to eliminate y.
    $x + 3y = 13$
    $6x - 3y = -6$

Adding we have,
    $7x = 7$
    $x = 1$
Using $x + 3y = 13$
    $1 + 3y = 13$
    $3y = 12$
    $y = 4$
    $\therefore x = 1, y = 4$

7. $3x + 4y = 17$ ......× 2   equation 1
    $2x - 5y = 19$ ......× 3   equation 2

Multiply equation (1) by 2 and equation (2) by 3 to eliminate x.
    $6x + 8y = 34$
    $6x - 15y = 57$

Subtracting we have,
    $23y = -23$
    $y = -1$
Using $3x + 4y = 17$
    $3x + 4(-1) = 17$
    $3x - 4 = 17$
    $3x = 21$
    $x = 7$
    $\therefore x = 7, y = -1$

4. $x + 2y = 11$ ......   equation 1
    $2x - y = -3$ ......× 2   equation 2

Multiply equation (2) by 2 to eliminate y.
    $x + 2y = 11$
    $4x - 2y = -6$

Adding we have,
    $5x = 5$
    $x = 1$
Using $x + 2y = 11$
    $1 + 2y = 11$
    $2y = 10$
    $y = 5$
    $\therefore x = 1, y = 5$

6. $2x + y = 9$ ......× 2   equation 1
    $4x + 3y = 17$ ......   equation 2

Multiply equation (1) by 2 to eliminate x.
    $4x + 2y = 18$
    $4x + 3y = 17$

Subtracting we have,
    $-y = 1$
    $y = -1$
Using $2x + y = 9$
    $2x - 1 = 9$
    $2x = 10$
    $x = 5$
    $\therefore x = 5, y = -1$

8. $2x + y = 7$ ......× 2   equation 1
    $4x - 3y = 19$ ......   equation 2

Multiply equation (1) by 2 to eliminate x.
    $4x + 2y = 14$
    $4x - 3y = 19$

Subtracting we have,
    $5y = -5$
    $y = -1$
Using $2x + y = 7$
    $2x - 1 = 7$
    $2x = 8$
    $x = 4$
    $\therefore x = 4, y = -1$

9. $3x + 2y = 11$ ......× 2   equation 1
    $2x + 5y = 0$ ......× 3   equation 2

Multiply equation (1) by 2 and equation (2) by 3 to eliminate x.
    $6x + 4y = 22$
    $6x + 15y = 0$

Subtracting we have,
    $-11y = 22$
    $y = -2$
Using $3x + 2y = 11$
    $3x + 2(-2) = 11$
    $3x - 4 = 11$
    $3x = 15$
    $x = 5$
    $\therefore x = 5, y = -2$

11. $4x + 5y = 17$ ......   equation 1
    $2x + 3y = 10$ ......× 2   equation 2

Multiply equation (2) by 2 to eliminate x.
    $4x + 5y = 17$
    $4x + 6y = 20$

Subtracting we have,
    $-y = -3$
    $y = 3$
Using $2x + 3y = 10$
    $2x + 3(3) = 10$
    $2x + 9 = 10$
    $2x = 1$
    $x = 0.5$
    $\therefore x = 0.5, y = 3$

13. $2x + 3y = 10$ ......× 5   equation 1
    $5x + 4y = 11$ ......× 2   equation 2

Multiply equation (1) by 2 and equation (2) by 5 to eliminate x.
    $10x + 15y = 50$
    $10x + 8y = 22$

Subtracting we have,
    $7y = 28$
    $y = 4$
Using $2x + 3y = 10$
    $2x + 3(4) = 10$
    $2x + 12 = 10$
    $2x = -2$
    $x = -1$
    $\therefore x = -1, y = 4$

10. $2x + 3y = 11$ ......× 3   equation 1
    $3x + 5y = 17$ ......× 2   equation 2

Multiply equation (1) by 3 and equation (2) by 2 to eliminate x.
    $6x + 9y = 33$
    $6x + 10y = 34$

Subtracting we have,
    $-y = -1$
    $y = 1$
Using $3x + 5y = 17$
    $3x + 5(1) = 17$
    $3x + 5 = 17$
    $3x = 12$
    $x = 4$
    $\therefore x = 4, y = 1$

12. $7x + 2y = 8$ ......× 5   equation 1
    $4x + 5y = -34$ ......× 2   equation 2

Multiply equation (1) by 5 and equation (2) by 2 to eliminate y.
    $35x + 10y = 40$
    $8x + 10y = -68$

Subtracting we have,
    $27x = 108$
    $x = 4$
Using $7x + 2y = 8$
    $7(4) + 2y = 8$
    $28 + 2y = 8$
    $2y = -20$
    $y = -10$
    $\therefore x = 4, y = -10$

14. $3x + 2y = 5$
    $x + 2y = -3$

Subtracting we have,
    $2x = 8$
    $x = 4$
Using $x + 2y = -3$
    $4 + 2y = -3$
    $2y = -7$
    $y = -3.5$
    $\therefore x = 4, y = -3.5$

CHAPTER 12 : SIMULTANEOUS LINEAR EQUATIONS: SOLUTIONS

12B SUBSTITUTION METHOD

The **substitution method** is more valid in case of a pair of simultaneous equations having two unknown and when it is possible to make one of the unknown the subject of the formula. It simply involves putting one of the equations into the other. We can also use the substitution method even if both equations of the linear system are in standard form. We can start by solving one of the equations for one of its variables as illustrated below.

Substitution method can be applied in four simple steps:

Step 1: Solve one of the equations for either x or y as the subject of formula.

Step 2: Substitute the solution from step 1 into the other equation.

Step 3: Solve this new equation.

Step 4: Solve for the second variable.

EXAMPLES

Solve the following simultaneous equations using the substitution method.

1.
$$y = 2x + 1$$
$$2x + 3y = 27$$

SOLUTION
First we label both equations
$y = 2x + 1$ ..... equation 1
$2x + 3y = 27$ ... equation 2

From equation 1, we can see that y is already the subject of the formula
So we replace equation 1 in equation 2 and solve for x first
$$2x + 3(2x + 1) = 27$$
$$2x + 6x + 3 = 27$$
$$8x + 3 = 27$$
$$8x = 24$$
$$\therefore x = 3$$

To find y, replace $x = 3$ in equation 1
$$y = 2(3) + 1 = 7$$

So $x = 3$ and $y = 7$.

2.
$$x - 3y = 8$$
$$2x + 5y = 5$$

SOLUTION
First we label both equations
$x - 3y = 8$ ...... equation 1
$2x + 5y = 5$ ...... equation 2

Rearrange equation 1 to make x the subject
$x = 3y + 8$ ...... equation 3
Replace this in equation 2, we have
$$2(3y + 8) + 5y = 5$$
$$6y + 16 + 5y = 5$$
$$11y + 16 = 5$$
$$11y = -11$$
$$\therefore y = -1$$

To find x replace $y = -1$ in equation 3
$$x = 3(-1) + 8$$
$$x = 5$$

Hence $x = 5, y = -1$

EXERCISE 12B

Solve the following simultaneous equations using the substitution method.

1.
$y = 3x + 2$ .... equation 1
$x + 2y = 11$ .... equation 2

Substitute equation (1) in (2), we have
$x + 2(3x + 2) = 11$
$x + 6x + 4 = 11$
$7x + 4 = 11$
$7x = 7$
$x = 1$
from equation (1) $y = 3(1) + 2 = 5$
$\therefore x = 1, \quad y = 5$

2.
$y = x - 4$ .... equation 1
$5x + y = 32$ .... equation 2

Substitute equation (1) in (2), we have
$5x + (x - 4) = 32$
$6x - 4 = 32$
$6x = 36$
$x = 6$
from equation (1) $y = 6 - 4 = 2$
$\therefore x = 6, \quad y = 2$

3.
$x = 3y + 2$ .... equation 1
$2x + y = 18$ .... equation 2

Substitute equation (1) in (2), we have
$2(3y + 2) + y = 18$
$6y + 4 + y = 18$
$7y + 4 = 18$
$7y = 14$
$y = 2$
from equation (1) $x = 3(2) + 2 = 8$
$\therefore x = 8, \quad y = 2$

4.
$x = y - 3$ .... equation 1
$2x + y = 9$ .... equation 2

Substitute equation (1) in (2), we have
$2(y - 3) + y = 9$
$2y - 6 + y = 9$
$3y - 6 = 9$
$3y = 15$
$y = 5$
from equation (1) $x = 5 - 3 = 2$
$\therefore x = 2, \quad y = 5$

5.
$x + 3y = 13$ .... equation 1
$y = 2x + 2$ .... equation 2

Substitute equation (2) in (1), we have
$x + 3(2x + 2) = 13$
$x + 6x + 6 = 13$
$7x + 6 = 13$
$7x = 7$
$x = 1$
from equation (2) $y = 2(1) + 2 = 4$
$\therefore x = 1, \quad y = 4$

6.
$x = 5y - 3$ .... equation 1
$3x + y = 23$ .... equation 2

Substitute equation (1) in (2), we have
$3(5y - 3) + y = 23$
$15y - 9 + y = 23$
$16y - 9 = 23$
$16y = 32$
$y = 2$
from equation (1) $x = 5(2) - 3 = 7$
$\therefore x = 7, \quad y = 2$

CHAPTER 12 : SIMULTANEOUS LINEAR EQUATIONS: SOLUTIONS

Solve the following simultaneous equations using the substitution method.

7.
$3x + y = 11$ .... equation 1
$5x + 2y = 19$ .... equation 2

From (1) $y = 11 - 3x$ ...... **equation 3**
Substitute equation (3) in (2), we have
$5x + 2(11 - 3x) = 19$
$5x + 22 - 6x = 19$
$-x = -3$
$x = 3$
from equation (3) $y = 11 - 3(3) = 2$
$\therefore x = 3, \quad y = 2$

8.
$x + 3y = 3$ .... equation 1
$2x - 5y = 17$ .... equation 2

From (1) $x = 3 - 3y$ ...... **equation 3**
Substitute equation (3) in (2), we have
$2(3 - 3y) - 5y = 17$
$6 - 6y - 5y = 17$
$-11y = 11$
$y = -1$
from equation (3) $x = 3 - 3(-1) = 6$
$\therefore x = 6, \quad y = -1$

9.
$x = 4y - 3$ .... equation 1
$2x + y = 12$ .... equation 2

Substitute equation (1) in (2), we have
$2(4y - 3) + y = 12$
$8y - 6 + y = 12$
$9y - 6 = 12$
$9y = 18$
$y = 2$
from equation (1) $x = 4(2) - 3 = 5$
$\therefore x = 5, \quad y = 2$

10.
$y = 6 - 3x$ .... equation 1
$2x + 3y = 18$ .... equation 2

Substitute equation (1) in (2), we have
$2x + 3(6 - 3x) = 18$
$2x + 18 - 9x = 18$
$-7x = 0$
$x = 0$
from equation (1) $y = 6 - 3(0) = 6$
$\therefore x = 0, \quad y = 6$

11.
$x - 2y = 3$ .... equation 1
$2x + 3y = 20$ .... equation 2

From (1) $x = 3 + 2y$ ...... **equation 3**
Substitute equation (3) in (2), we have
$2(3 + 2y) + 3y = 20$
$6 + 4y + 3y = 20$
$6 + 7y = 20$
$7y = 14$
$y = 2$
from equation (3) $x = 3 + 2(2) = 7$
$\therefore x = 7, \quad y = 2$

12.
$y = 10 - 4x$ .... equation 1
$2x + 3y = 0$ .... equation 2

Substitute equation (1) in (2), we have
$2x + 3(10 - 4x) = 0$
$2x + 30 - 12x = 0$
$-10x = -30$
$x = 3$
from equation (1) $y = 10 - 4(3) = -2$
$\therefore x = 3, \quad y = -2$

12C SIMULTANEOUS EQUATIONS : USING SOLVE CAPACITY

To solve a pair of simultaneous equations on CAS, make use of the following steps:

- Main
- Keyboard
- 2D (skip this step for latest CAS)
- Select the symbol
- The following will appear on your calculator

- Insert the 1st equation in the first box, the 2nd equation in the second box and x,y in the third box.
- Press EXE and the answer will appear as shown below.

$$\left. \begin{array}{l} x + y = 50 \\ 2x + 3y = 120 \end{array} \right| x, y$$
$$\{x = 30, y = 20\}$$

Note that most graphic calculators have similar functions to solve simultaneous equations.

EXAMPLE
Use the solve facility on your calculator to solve the simultaneous equations
$$x + y = 50$$
$$2x + 3y = 120$$

EXERCISE 12C
Use the solve facility on your calculator to solve the following simultaneous equations.

1. $y = 2x + 5$ *and* $y = 3x - 1$ $x = 6, y = 17$	**2.** $3x - y = 10$ *and* $2x + 5y = 1$ $x = 3, y = -1$
3. $y = x + 4$ *and* $y = 4x - 8$ $x = 4, y = 8$	**4.** $4x + 3y = 11$ *and* $5x - y = 9$ $x = 2, y = 1$
5. $x + y = 10$ *and* $2x - y = 8$ $x = 6, y = 4$	**6.** $x + y = 9$ *and* $2x - y = 6$ $x = 5, y = 4$

CHAPTER 12 : SIMULTANEOUS LINEAR EQUATIONS: SOLUTIONS

12D USING GRAPHICAL FACILITY ON CALCULATOR

The next couple of examples will demonstrate how we can make use of technology to solve a pair of simultaneous equations by plotting both graphs and finding the coordinates of their point of intersection. Being straight lines obviously there will be only one point of intersection thereby only one value of x and one value of y. This method is slightly harder than the previous section as it requires some mathematical skills of re-arranging and making y the subject of the formulae each time.

Use the following steps on your calculator:

**MENU → GRAPH &TABLE →
INSERT BOTH EQUATIONS →
TICK THE BOX →
CLICK ON THE GRAPH ICON
(1ST ON TOP LEFT)**

EXAMPLES

Use the graphical facility on your calculator to solve the following pairs of simultaneous equations.

(a) $y = 2x + 5$
$x + y = 11$ ....(2)

Note that for equation (2), we have to make y the subject of the formula in order to be able to insert $11 - x$ in the required box.

As we can see from the graph, the point of intersection is (2,9).

Hence $x = 2$ and $y = 9$.

(b) $y = 3x - 1$
$2x + 3y = 19$ ....(2)

Again, we have to re-arrange equation (2) making y the subject of the formula as shown below.

$2x + 3y = 19$
$3y = 19 - 2x$
$\therefore y = \frac{1}{3}(19 - 2x)$

The point of intersection being (2,5).

Hence $x = 2$ and $y = 5$.

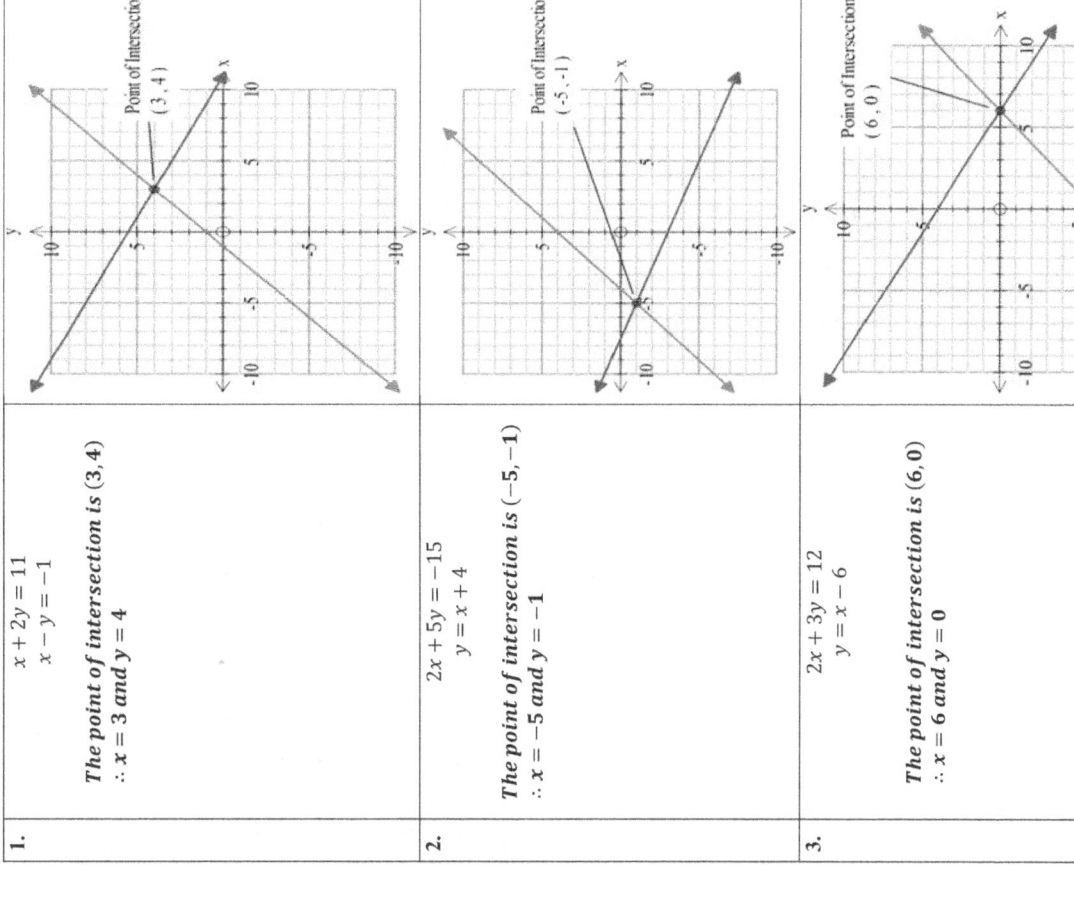

EXERCISE 12D

Using the graphical facility on your calculator solve the following simultaneous equations.

1.	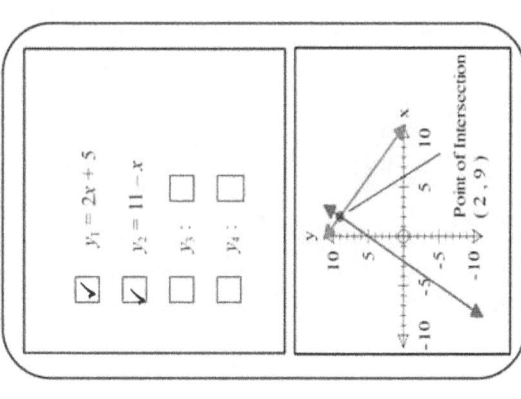	$x + 2y = 11$ $x - y = -1$ ***The point of intersection is (3,4)*** $\therefore x = 3 \text{ and } y = 4$
2.	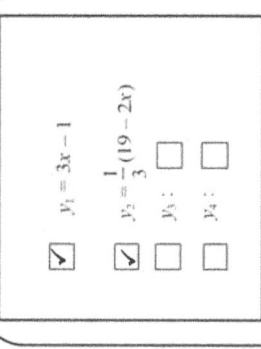	$2x + 5y = -15$ $y = x + 4$ ***The point of intersection is (-5,-1)*** $\therefore x = -5 \text{ and } y = -1$
3.	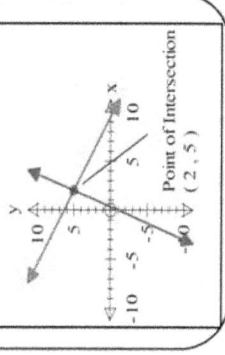	$2x + 3y = 12$ $y = x - 6$ ***The point of intersection is (6,0)*** $\therefore x = 6 \text{ and } y = 0$

CHAPTER 12 : SIMULTANEOUS LINEAR EQUATIONS: SOLUTIONS

12E THE GRAPHICAL METHOD STEP BY STEP APPROACH

Simultaneous equations can also be solved graphically. We have to graph both lines on the same set of axes and the solution is given by the coordinates of the point of intersection of the two lines.

It is definitely not the best method to solve a pair of simultaneous equations as it is often hard to graph the lines accurately. Furthermore, reading the point of intersection off the graph is sometimes difficult when the solutions are not whole numbers.

Nevertheless, the graphical method remains a very useful tool for solving simultaneous equations.

CLASS ACTIVITY

State the solution in each of the following

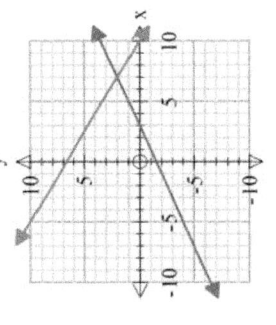

The graphs intersect at the point (1,3). Hence the solution is $x = 1$ and $y = 3$.

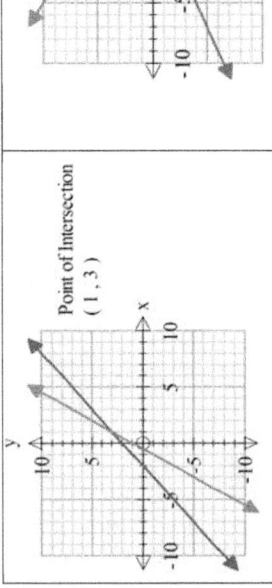

The solution is $x = 7$ and $y = 2$

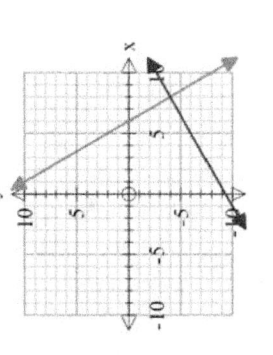

The solution is $x = 4$ and $y = 1$

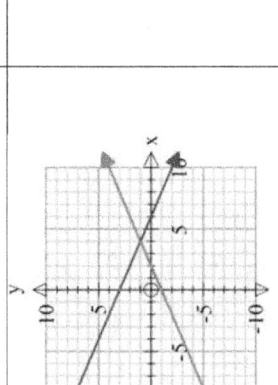

The solution is $x = 8$ and $y = -4$

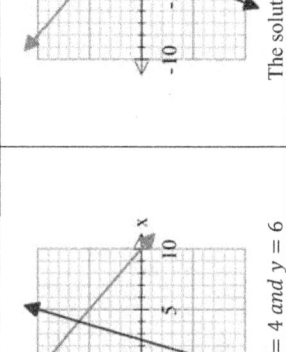

The solution is $x = 4$ and $y = 6$

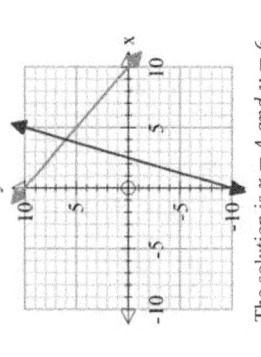

The solution is $x = -1$ and $y = 3$

EXAMPLES

Solve these simultaneous equations by using the graphical method.

1.
$$2x + y = 10$$
$$x + y = 7$$

SOLUTION

For example, to draw the line $2x + y = 10$ pick two easy numbers to plot. One when $x = 0$ and one where $y = 0$.

When $x = 0$ in the equation $2x + y = 10$
This means $y = 10$
So one point on the line is (0, 10)
When $y = 0$
$2x = 10$ so $x = 5$
So another point on the line is (5, 0)
Similarly, the line $x + y = 7$ crosses the axes at (0,7) and (7,0).
The graphs intersect at the point (3,4)
$\therefore x = 3, y = 4$

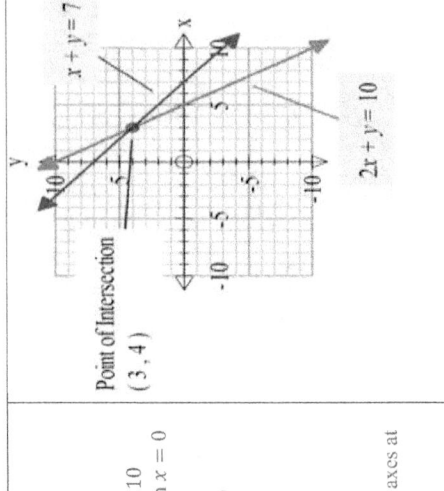

Point of Intersection (3, 4)

2.
$$y = x + 3$$
$$2x + y = 9$$

SOLUTION

To draw the line $y = x + 3$, we can construct a table of values as shown

x	0	1	2
y	3	4	5

We can then plot the pairs of values to draw the line $y = x + 3$.
To draw the line $2x + y = 9$, we can use the intercept method as shown in example 1 above.
When $x = 0, y = 9$
So one point on the line is (0, 9)
When $y = 0$
$2x = 9$ so $x = 4.5$
So another point on the line is (4.5, 0)
The graphs intersect at the point (2,5)
$\therefore x = 2, y = 5$

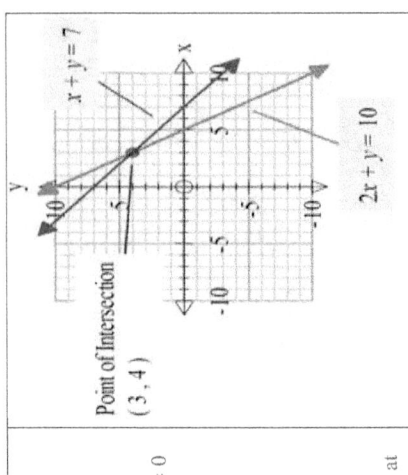

Point of Intersection (2, 5)

CHAPTER 12 : SIMULTANEOUS LINEAR EQUATIONS: SOLUTIONS

EXERCISE 12E

Solve these simultaneous equations by using the graphical method. Check your answer using CAS.

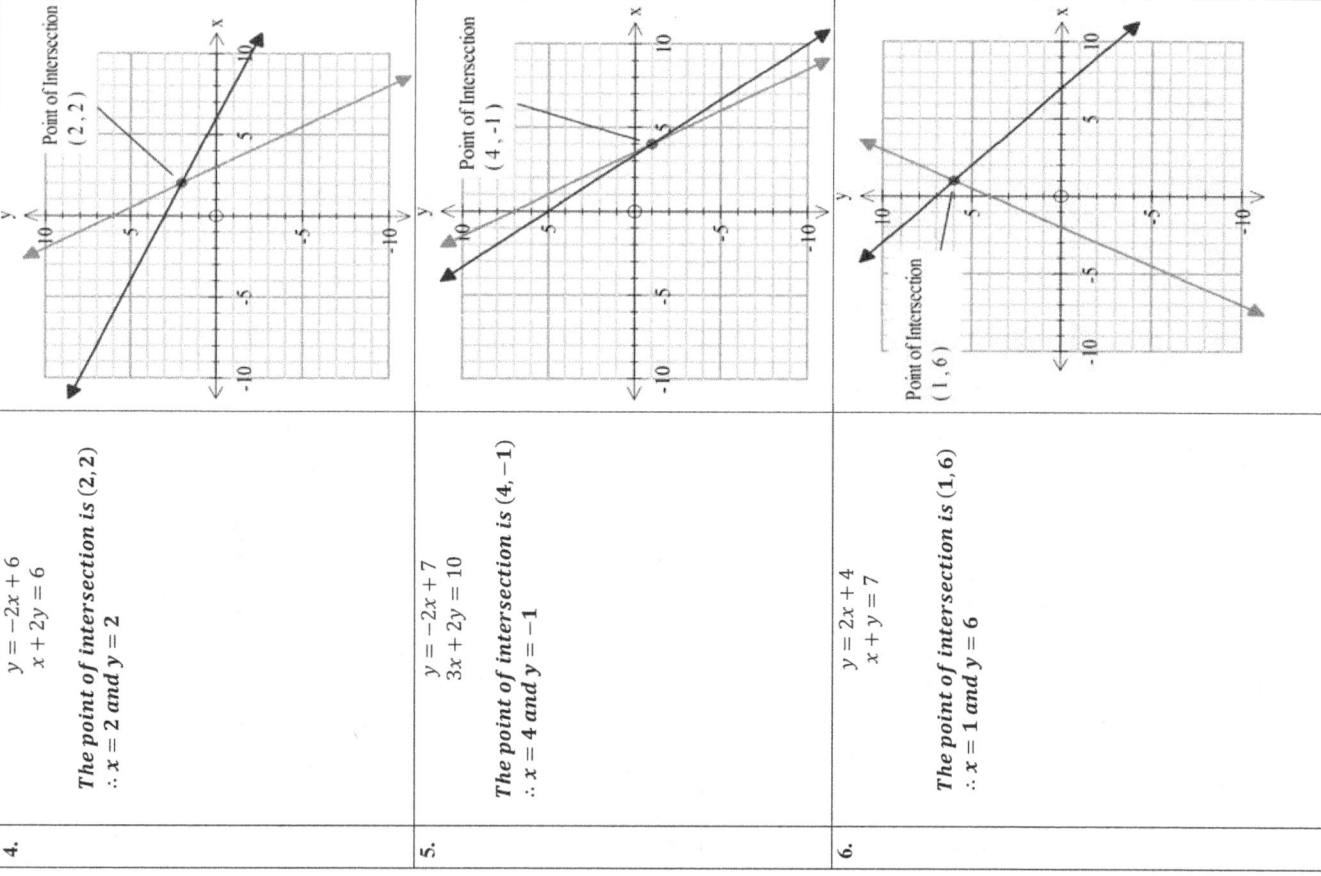

1. $x - 2y = 10$
    $y = x - 5$

    The point of intersection is $(0, -5)$
    $\therefore x = 0 \text{ and } y = -5$

2. $2x - y = 6$
    $y = x - 4$

    The point of intersection is $(2, -2)$
    $\therefore x = 2 \text{ and } y = -2$

3. $4x + y = 8$
    $y = x - 7$

    The point of intersection is $(3, -4)$
    $\therefore x = 3 \text{ and } y = -4$

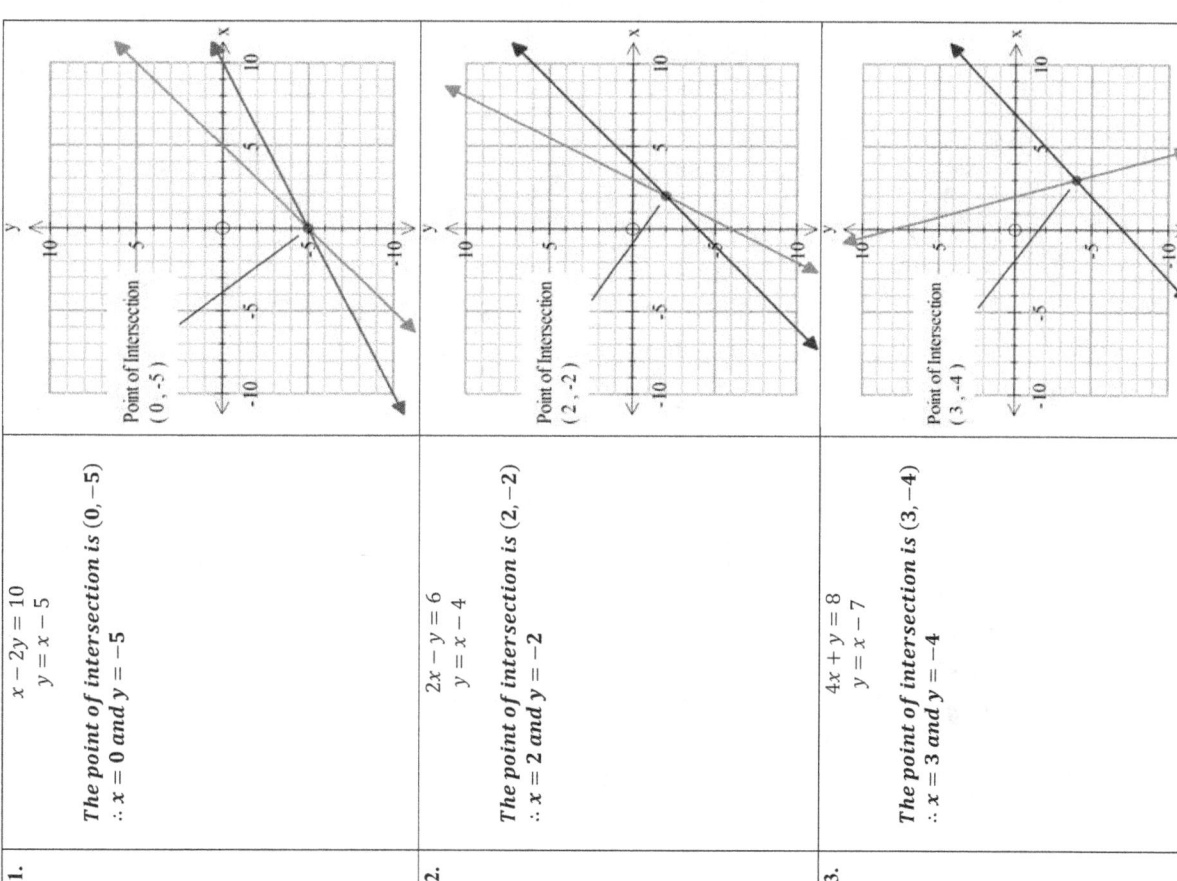

4. $y = -2x + 6$
    $x + 2y = 6$

    The point of intersection is $(2, 2)$
    $\therefore x = 2 \text{ and } y = 2$

5. $y = -2x + 7$
    $3x + 2y = 10$

    The point of intersection is $(4, -1)$
    $\therefore x = 4 \text{ and } y = -1$

6. $y = 2x + 4$
    $x + y = 7$

    The point of intersection is $(1, 6)$
    $\therefore x = 1 \text{ and } y = 6$

12F APPLICATIONS

EXAMPLES

1. The admission fee at the Royal Show is $3.50 for children and $5 for adults. On a certain day, 2200 people enter the fair and $8750 is collected. Two simultaneous equations can be written from this information.
One of the equation is $3.5x + 5y = 8750$

(a) Explain clearly what the term $5y$ represents in this situation.

Answer : money raised from adults admission fees

(b) Write down the second equation.

Answer : $x + y = 2200$

(c) Solve the pair of equations to determine the number of children and the number of adults who attended the show.

$$\left. \begin{array}{l} 3.5x + 5y = 8750 \\ x + y = 2200 \end{array} \right| x,y$$

$$\{x = 1500, y = 700\}$$

Therefore 1500 children and 700 adults attended the show.

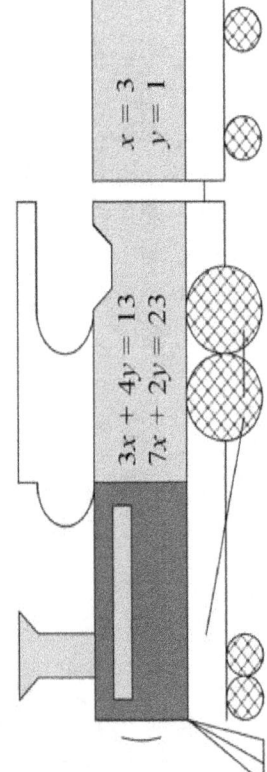

$$\begin{array}{l} 3x + 4y = 13 \\ 7x + 2y = 23 \end{array} \qquad \begin{array}{l} x = 3 \\ y = 1 \end{array}$$

2. Ten years ago, Alex was 12 times as old as Tim and in ten years' time, Alex will be twice as old as Tim. Find their present ages.

Solution
Let Alex's present age be x.
Let Tim's present age be y.
Ten years ago, Alex was $(x - 10)$ years old and Tim was $(y - 10)$years old.
Since Alex was 12 times as old as Tim, we have
$$x - 10 = 12(y - 10)$$
$$x - 10 = 12y - 120$$
$$x = 12y - 110 \quad \ldots\ldots equation\ (1)$$
Also in ten years' time
Alex will be $(x + 10)$ years old and Tim was $(y + 10)$years old.
As Alex will be twice as old as Tim,
$$x + 10 = 2(y + 10)$$
$$x + 10 = 2y + 20$$
$$x = 2y + 10 \quad \ldots. equation\ (2)$$
Solving equations (1) and (2) on CAS, we have

$$\left. \begin{array}{l} x = 12y - 110 \\ x = 2y + 10 \end{array} \right| x,y$$

$$\{x = 34, y = 12\}$$

Hence Alex is 34 years old and Tim is 12 years old currently.

EXERCISE 12F

1. Two numbers x and y are such that the sum of 2 numbers is 23 and their difference is 3. Write a pair of equations and use any appropriate method to find the 2 numbers.

$$\left. \begin{array}{l} x + y = 23 \\ x - y = 3 \end{array} \right| x,y$$

$$\{x = 13, y = 10\}$$

The two numbers are 10 and 13.

2. Seven footballs (x) and three soccer balls (y) cost a total of $314. Four footballs and five soccer balls cost a total of $255. Write down a pair of simultaneous equations involving x and y. Hence find the cost of each football and each soccer ball.

$$\left. \begin{array}{l} 7x + 3y = 314 \\ 4x + 5y = 255 \end{array} \right| x,y$$

$$\{x = 35, y = 23\}$$

Each football costs $35 and each soccer ball costs $23.

3. The length of a rectangle (y) is 5cm longer than its width (x). The perimeter of the rectangle is 54 cm.

(a) Write two equations that connect width (x) and length (y).

$$y = x + 5 \quad \ldots\ldots \textbf{ equation 1}$$
$$\textbf{Perimeter} = \textbf{54 cm}$$
$$2x + 2y = 54 \ (\div 2)$$
$$x + y = 27 \quad \ldots\ldots \textbf{ equation 2}$$

(b) Solve the equations using substitution and so state the *length* and *width* of the rectangle.

Substitute equation (1) in eqution (2), we have
$$x + x + 5 = 27$$
$$2x = 22$$
$$x = 11$$
$$y = 16$$
$$\therefore \textit{length} = \textbf{16cm and width} = \textbf{11 cm}$$

4. A car travels for x hours at 60 km/h and then travels for y hours at 80 km/h. If it has travelled for 8 hours and covered a total distance of 540 km, find x and y.

$$\left. \begin{array}{l} x + y = 8 \\ 60x + 80y = 540 \end{array} \right| x,y$$

$$\{x = 5, y = 3\}$$

5. A rectangle has a perimeter of 52 cm while the difference between the length and the width is 5 cm. Find the length and the width.

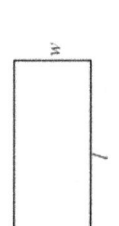

$l = w + 5$ .... equation 1
Perimeter = 52 cm
$2l + 2w = 52$
$l + w = 26$ .... equation 2 ($\div 2$)

$\left. \begin{array}{r} l = w + 5 \\ l + w = 26 \end{array} \right| l, w$

$\{l = 15.5, w = 10.5\}$

There are 18 50-cent coins and 7 20-cent coins.

6. A piggy bank contains 40 coins, all of them are either 5-cents coins or 20-cent coins. If the value of the coins in the piggy bank is $6.50, find the number of each kind of coin.

Let x be the number of 5-cent coins.
Let y be the number of 20-cent coins.

$\left. \begin{array}{r} x + y = 40 \\ 5x + 20y = 650 \end{array} \right| x, y$

$\{x = 10, y = 30\}$

There are 10 5-cent coins and 30 20-cent coins.

7. Alisha has 25 coins in her purse, consisting of 50-cent coins and 20-cent coins, which total $10.40. How many of each does she have?

Let x be the number of 50-cent coins.
Let y be the number of 20-cent coins.

$\left. \begin{array}{r} x + y = 25 \\ 50x + 20y = 1040 \end{array} \right| x, y$

$\{x = 18, y = 7\}$

8. At the Perth arena 1000 tickets were sold during a concert. Adult tickets cost $8.50, children's ticket cost $4.50, and a total of $7300 was collected. How many tickets of each kind were sold?

Let x be the number of adult tickets.
Let y be the number of children.

$\left. \begin{array}{r} x + y = 1000 \\ 8.50x + 4.50y = 7300 \end{array} \right| x, y$

$\{x = 700, y = 300\}$

There are 700 adults and 300 children.

9. Anna is x years old and Bob is y years old. Last year, Bob was 6 times as old as Anna.

(i) Form an equation in x and y and show that it simplifies to $y = 6x - 5$.

last year Anna was $x - 1$ years old and Bob was $y - 1$ years old.

$y - 1 = 6(x - 1)$
$y - 1 = 6x - 6$
$y = 6x - 5$

(ii) In 19 years' time, Bob will be twice as old as Anna.
Form another equation in x and y and show that it simplifies to $y = 2x + 19$.

$y + 19 = 2(x + 19)$
$y + 19 = 2x + 38$
$y = 2x + 19$

(iii) Hence find the present ages of Anna and Bob.

$\left. \begin{array}{r} y = 6x - 5 \\ y = 2x + 19 \end{array} \right| x, y$

$\{x = 6, y = 31\}$

Anna is 6 years old and Bob is 31 years old currently.

10. Mc Café charges $2.50 for a cup of tea and $3.25 for a cup of coffee. One morning the café sold 80 cups of tea and coffee altogether, and charged $245 in total. Two simultaneous equations can be written from this information.
One of the equation is
$2.5x + 3.25y = 245$.

(a) Explain clearly what the term $2.5x$ represents in this situation.

money raised from the sales of tea

(b) Write down the second equation.

$x + y = 80$

(c) Solve the pair of equations to determine the number of cups of each type of drink sold.

$\left. \begin{array}{r} x + y = 80 \\ 2.5x + 3.25y = 245 \end{array} \right| x, y$

$\{x = 20, y = 60\}$

They sold 20 cups of tea and 60 cups of coffee.

CHAPTER 13

THE NORMAL DISTRIBUTION

A continuous random variable X has a normal distribution if it has a p.d.f

$$f(x) = \frac{1}{\sigma\sqrt{2\pi}} e^{-\frac{(x-\mu)^2}{2\sigma^2}}$$

The graph of f(x) is given below.

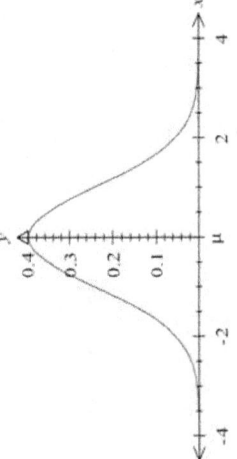

From this graph we can deduce some of the important properties of the normal distribution.

- The distribution is symmetrical about the mean μ.
- The mean, mode and median coincide and are equal due to symmetry.
- The domain of the function is $-\infty < x < \infty$
- The horizontal axis is an asymptote as $x \to -\infty$ and $x \to +\infty$.
- Area under the curve is 1.

From the p.d.f above, we can see that the probability distribution of X depends only on μ and σ. Hence instead of remembering the formula it is sufficient to refer to the random variable X as having a normal distribution by using the notation

$$X \sim N(\mu, \sigma^2)$$

CLASS ACTIVITY

Complete the following table.

	Mean μ	Standard deviation σ		Mean μ	Standard deviation σ
$X \sim N(50, 5^2)$	50	5	$X \sim N(65, 7^2)$	65	7
$X \sim N(42, 3^2)$	42	3	$X \sim N(35, 10)$	35	$\sqrt{10}$
$X \sim N(20, 6^2)$	20	6	$X \sim N(100, 9^2)$	100	3
$X \sim N(30, 25)$	30	5	$X \sim N(80, 10^2)$	80	10

13A   USING TECHNOLOGY

To solve problems in normal distribution, the use of technology is very important as in old days we use to make use of tables and it was really time consuming. Make use of the following steps on your calculator to determine probabilities in normal distribution questions.

- Menu
- Statistics
- Calc
- Distribution
- Normal CD
- Next
- Insert the values of μ, σ, lower and upper to obtain the answer.

EXAMPLES

1. If $X \sim N(50, 5^2)$, determine $P(X > 54)$.

SOLUTION

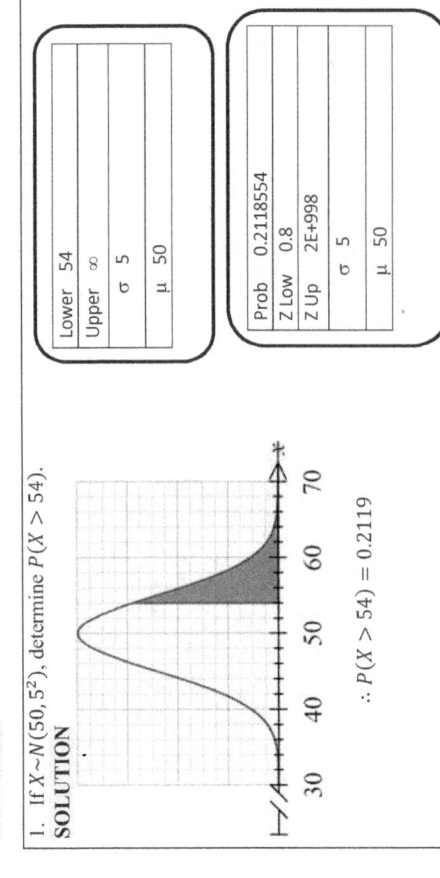

$$\therefore P(X > 54) = 0.2119$$

2. The length of rods in a large batch is normally distributed with mean 120 mm and standard deviation 1.5 mm. What percentage of rods would you expect to measure between 116 mm and 124 mm?

SOLUTION

$$P(116 < X < 124) = 0.9923$$

Hence 99.23% of the rods measure between 116 mm and 124 mm.

CHAPTER 13: THE NORMAL DISTRIBUTION : SOLUTIONS

EXERCISE 13A

Use technology to answer the following questions.

1. If $X \sim N(56, 10^2)$, determine $P(X < 66)$. **0.8413**	2. If $X \sim N(28, 3^2)$, determine $P(X > 36)$. **0.0038**
3. If $X \sim N(100, 25)$, determine $P(X > 100)$ **0.5**	4. If $X \sim N(60, 4^2)$, find $P(56 < X < 64)$. **0.6827**
5. If $X \sim N(-5, 9)$, determine $P(X > 0)$. **0.0478**	6. If $X \sim N(12, 2^2)$, determine $P(9 < X < 13)$. **0.6247**
7. The height of girls at a particular age follows a normal distribution with mean 130 cm and standard deviation 3 cm. Find the probability that a girl picked up at random from this age group has a height (a) Less than 134 cm **0.9088** (b) Between 131 cm and 133 cm. **0.2108**	8. A certain type of vegetable has a mass which is normally distributed with mean 2 kg and standard deviation 0.25 kg. In a lorry load of 600 of these vegetables, estimate how many will have a mass greater than 2.1 kg. $P(X > 2.1) = 0.3446$ $600 \times 0.3446 \approx 208$
9. The number of hours of the life of a torch battery is normally distributed with mean 120 hours and a standard deviation of 16 hours. Find the probability that a troch battery has a life of (a) More than 140 hours **0.1056** (b) Between 110 and 128 hours. **0.4255**	10. The number of marks of 1000 candidates in an examination is normally distributed with a mean of 58 marks and a standard deviation of 10 marks. Given that the pass mark is 50 marks, estimate the number of candidates who pass the examination. $P(X \geq 50) = 0.7881$ $No.\ of\ passes = 1000 \times 0.7881$ ≈ 788
11. A company packing spices knows that the weight of 500 packets form a normal distribution with a mean weight of 16 grams and a standard deviation of 0.2 gram. How many of these 500 packets are expected to weigh less than 15.6 grams? $P(X < 15.6) = 0.0228$ $No.\ of\ packets = 500 \times 0.0228 \approx 11$	12. The masses of tablets of chocolates produced by a certain machine are found to be normally distributed with a mean of 140g and a standard deviation of 5g. Estimate the number of tablets in a batch of 200 whose masses are greater than 145g. $P(X > 145) = 0.1587$ $200 \times 0.1587 \approx 32\ tablets$

13B   THE 68%, 95% AND THE 99.7% RULE

For continuous random variables, taking a large number of measurements and analysing the results usually give rise to a normal curve as shown below.

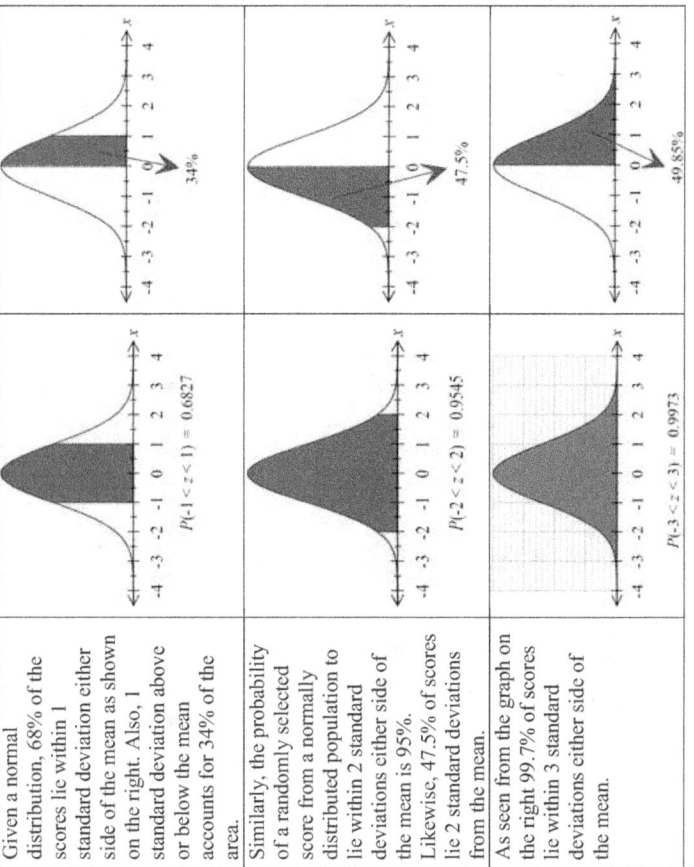

Given a normal distribution, 68% of the scores lie within 1 standard deviation either side of the mean as shown on the right. Also, 1 standard deviation above or below the mean accounts for 34% of the area.

Similarly, the probability of a randomly selected score from a normally distributed population to lie within 2 standard deviations either side of the mean is 95%.
Likewise, 47.5% of scores lie 2 standard deviations from the mean.

As seen from the graph on the right 99.7% of scores lie within 3 standard deviations either side of the mean.

EXAMPLES

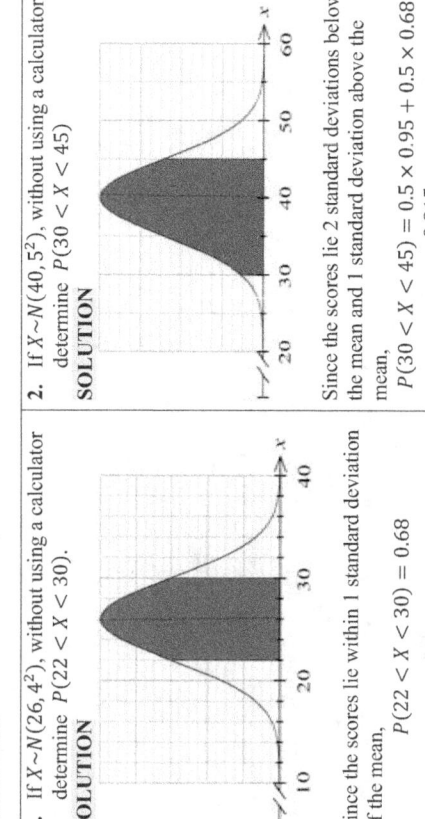

1. If $X \sim N(26, 4^2)$, without using a calculator determine $P(22 < X < 30)$.

SOLUTION

Since the scores lie within 1 standard deviation of the mean,

$P(22 < X < 30) = 0.68$

2. If $X \sim N(40, 5^2)$, without using a calculator determine $P(30 < X < 45)$

SOLUTION

Since the scores lie 2 standard deviations below the mean and 1 standard deviation above the mean,

$P(30 < X < 45) = 0.5 \times 0.95 + 0.5 \times 0.68$
$\phantom{P(30 < X < 45)} = 0.815$

CHAPTER 13: THE NORMAL DISTRIBUTION : SOLUTIONS

EXERCISE 13B

1. If $X \sim N(30, 5^2)$, without using a calculator determine $P(25 < X < 35)$

    0.68

2. If $X \sim N(40, 3^2)$, without using a calculator determine $P(34 < X < 46)$

    0.95

3. If $X \sim N(18, 2^2)$, without using a calculator determine $P(16 < X < 20)$

    0.68

4. If $X \sim N(50, 4^2)$, without using a calculator determine $P(X > 50)$

    0.5

5. If $X \sim N(30, 2^2)$, without using a calculator determine $P(24 < X < 36)$

    0.997

6. If $X \sim N(60, 4^2)$, without using a calculator determine $P(56 < X < 68)$

    $0.34 + 0.475 = \mathbf{0.815}$

7. If $X \sim N(40, 5^2)$, without using a calculator determine $P(X < 45)$

    $0.34 + 0.5 = \mathbf{0.84}$

8. If $X \sim N(20, 5^2)$, without using a calculator determine $P(X > 15)$

    $0.34 + 0.5 = \mathbf{0.84}$

9. If $X \sim N(25, 6^2)$, without using a calculator determine $P(X < 31)$

    $0.34 + 0.5 = \mathbf{0.84}$

10. If $X \sim N(20, 5^2)$, without using a calculator determine $P(X > 10)$

    $0.475 + 0.5 = \mathbf{0.975}$

11. After extensive testing, it was found that the lifetimes of Power bulbs had a mean of 2500 hours and a standard deviation of 100 hours. Assuming that the lifetime of a bulb is modelled by a normal distribution, find
    (a) the probability that a Power bulb will have a lifetime between 2400 and 2600 hours.

    0.68

    (b) The probability that a Power bulb has a lifetime exceeding 2700 hours.

    $0.5 - 0475 = \mathbf{0.025}$

12. The random variable X has a normal distribution with mean 6 and standard deviation 1.5. Calculate
    (a) $P(X < 6)$

    0.5

    (b) $P(4.5 < X < 9)$

    $0.34 + 0.475 = \mathbf{0.815}$

13. The weights of oranges in a supermarket shipment are normally distributed with a mean of 160 g and a standard deviation of 20 g. The distribution is such that 68%, 95% and 99.7% of the oranges have weights within one, two and three standard deviations from the mean respectively.

    Determine the probability that a randomly chosen orange from the supermarket

    (i) weighs between 120 g and 200 g.

    0.95

    (ii) weighs more than 160 g.

    0.5

    (iii) weighs exactly 100 g.

    0

    (iv) weighs between 140 g and 160 g.

    0.34

14. At a hardware store, the lengths of a large number of wooden rods marked as 2 m long, were actually normally distributed with a mean of 202 cm and a standard deviation of 3 cm.

    (i) State the median length of the wooden rods.

    202 cm

    (ii) Find the probability that the length of a randomly chosen plank is between 199 cm and 205 cm.

    0.68

    (iii) Find the probability that the length of a randomly chosen plank is less than 1.99 m.

    $0.5 - 0.34 = \mathbf{0.16}$

CHAPTER 13: THE NORMAL DISTRIBUTION : SOLUTIONS

13C INVERSE NORMAL DISTRIBUTION

In this section, we are going to make use of the capabilities of our calculators to use inverse normal distribution to determine unknown quantities.

Make use of the following steps on your calculator to achieve your goal.

- Menu
- Statistics
- Calc
- Inv. Distribution
- Inverse Normal CD
- Next
- Choose the correct tail setting and insert the values of μ, σ and the given probability to obtain the answer.
- Next

EXAMPLES

1. Determine the value of k below.

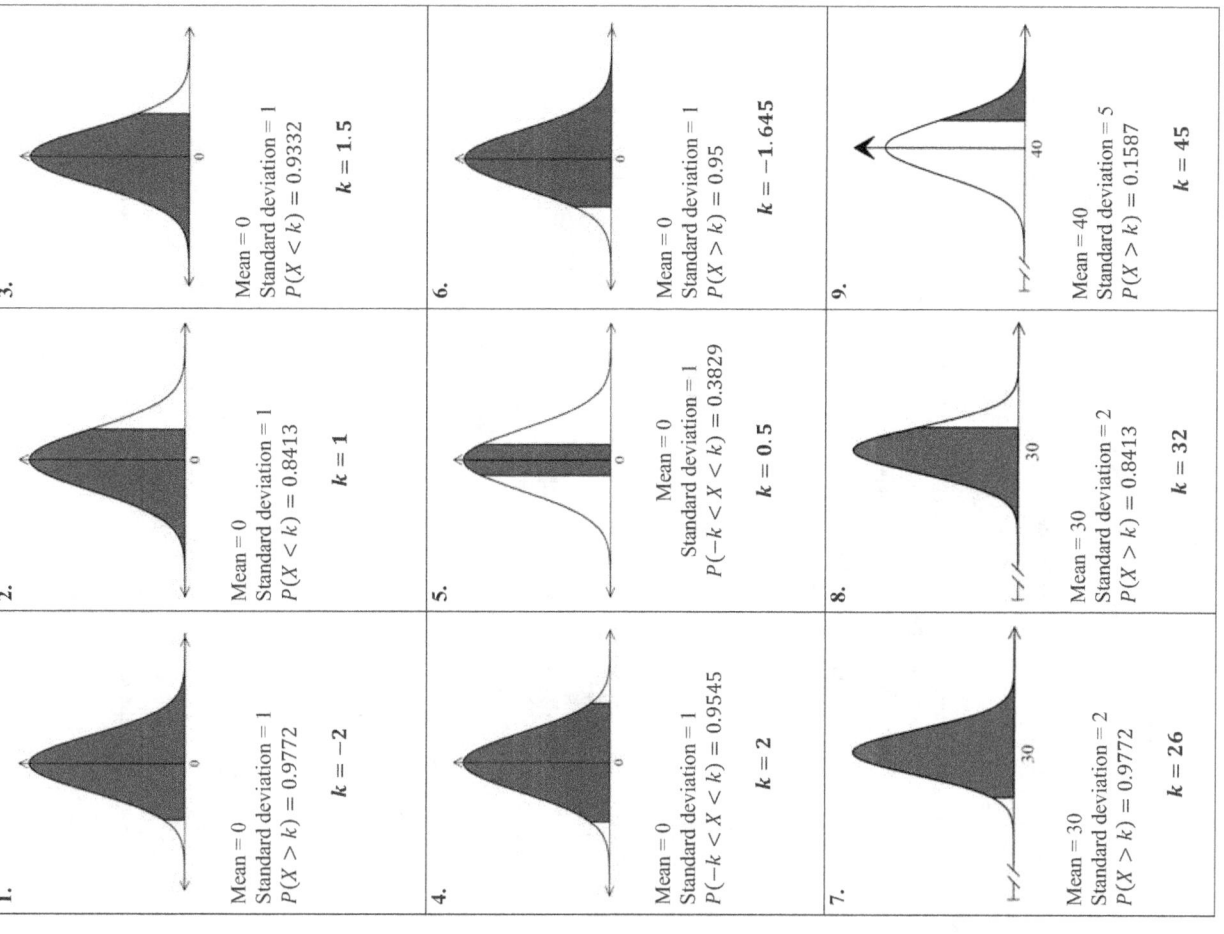

Mean = 0
Standard deviation = 1
$P(X > k) = 0.1587$

SOLUTION
Hence $k = 1$

2. The masses of apples sold at a fruit and vegetable shop are normally distributed with a mean mass 150 g and standard deviation 10 g.
Determine (a) the mass exceeded by 5% of the apples.
(b) the range of the masses of the central 50% of the apples (i.e IQR).

SOLUTION
(a) Mean = 150
Standard deviation = 10
$P(X > k) = 0.05$
Hence $k = 166$ g

(b) range = 157 − 143 = 14 g.

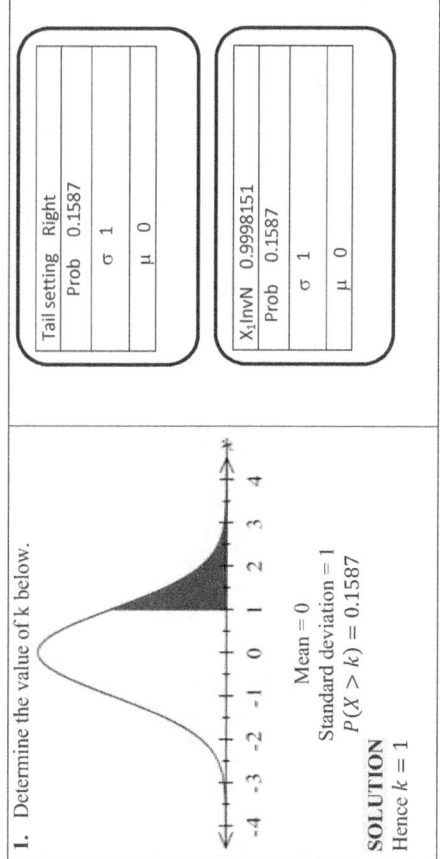

MATHEMATICS APPLICATIONS UNIT 2

EXERCISE 13C

Determine the value of k in each of the following.

1.

Mean = 0
Standard deviation = 1
$P(X > k) = 0.9772$

$k = -2$

2.

Mean = 0
Standard deviation = 1
$P(X < k) = 0.8413$

$k = 1$

3.

Mean = 0
Standard deviation = 1
$P(X < k) = 0.9332$

$k = 1.5$

4.

Mean = 0
Standard deviation = 1
$P(-k < X < k) = 0.9545$

$k = 2$

5.

Mean = 0
Standard deviation = 1
$P(-k < X < k) = 0.3829$

$k = 0.5$

6.

Mean = 0
Standard deviation = 1
$P(X > k) = 0.95$

$k = -1.645$

7.

Mean = 30
Standard deviation = 2
$P(X > k) = 0.9772$

$k = 26$

8.

Mean = 30
Standard deviation = 2
$P(X > k) = 0.8413$

$k = 32$

9.

Mean = 40
Standard deviation = 5
$P(X > k) = 0.1587$

$k = 45$

CHAPTER 13: THE NORMAL DISTRIBUTION : SOLUTIONS

10. The life span of a species of insects is modelled by a normal distribution with a mean of 360 hours and a standard deviation of 20 hours. Determine the life span exceeded by 7% of the insects.

$P(X > k) = 0.07$

$k = 389.5 \text{ hours}$

11. The weight of a consignment of sacks of sugar is normally distributed with a mean of 30 kg and a standard deviation of 2 kg. Determine to the nearest kg, the weight below which 15% of the sacks fall.

$P(X < k) = 0.15$

$k = 27.9 \approx 28 \text{ kg}$

12. The marks for a mathematics examination at a school are normally distributed with a mean of 59% and a standard deviation of 12%.

(a) State the median examination score.

59%

(b) Determine the interquartile range of the examination scores.

$Q_3 = 67.1$, $Q_1 = 50.9$

$IQR = 16.2$

(c) The top 10% students were awarded an A grade. Determine the minimum cut off for an A grade.

$P(X > k) = 0.1$

$k = 74.4$

13. The masses of cabbages sold at a supermarket are normally distributed with a mean mass 850 g and standard deviation 50 g.

(a) If a cabbage is chosen at random, determine the probability that its mass exceeds 900g.

0.1587

(b) Determine the mass exceeded by 15% of the cabbages correct to three (3) significant figures.

$P(X > k) = 0.15$

$k = 901.8 \approx 902 \text{ g } (3 \text{ s.f})$

14. Top Jewellery Ltd purchased fresh water pearls for the production of its necklaces. The diameters of the pearls were found to be normally distributed, with a mean of 1.2 cm and a standard deviation of 0.15 cm.

(a) What proportion of the pearls will have a diameter exceeding 1.05 cm?

$P(X > 1.05) = 0.8413$

84%

(b) Below what size will the diameter of 5% of the pearls fall?

$P(X < k) = 0.05$

$k = 0.95 \text{ cm}$

15. The marks for a Mathematics Applications Unit 2 examination at a High School are normally distributed with a mean of 61% and a standard deviation of 8%.

(a) State the modal examination score.

61%

(b) Determine the interquartile range of the examination scores.

$Q_3 = 66.4$, $Q_1 = 55.6$

$IQR = 10.8$

(c) The bottom 10% students were awarded a D grade. Determine the maximum cut off for a D grade.

$P(X < k) = 0.1$

$k = 50.7$

MATHEMATICS APPLICATIONS UNIT 2

13D QUANTILES OR PERCENTILES

Consider the diagram on the right. 90% of the scores lie below 50. We say 50 is the 90th percentile.

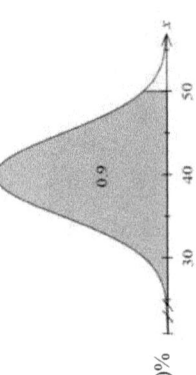

In general, the kth percentile of a set of values divides them so that k % of the values lie below and (100 − k)% of the values lie above.

- The 25th percentile is referred to as the lower quartile Q_1.
- The 50th percentile is referred to as the median Q_2.
- The 75th percentile is referred to as the upper quartile Q_3.

If we are given the percentage or statistical probability of being at or below a certain x-value, to find the percentile, we have to find the x-value that corresponds to it.

EXAMPLES

1. If $X \sim N(40, 5^2)$, determine the 60th percentile.
   SOLUTION

   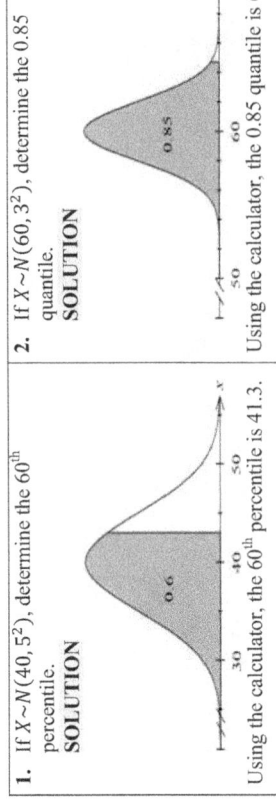

   Using the calculator, the 60th percentile is 41.3.

2. If $X \sim N(60, 3^2)$, determine the 0.85 quantile.
   SOLUTION

   Using the calculator, the 0.85 quantile is 63.1.

EXERCISE 13D

1. If $X \sim N(50, 4^2)$, determine the 65th percentile.

   51.5

2. If $X \sim N(60, 5^2)$, determine the 0.42 quantile.

   59.0

3. If $X \sim N(35, 3^2)$, determine the 78th percentile.

   37.3

4. If $X \sim N(20, 2^2)$, determine the 0.55 quantile.

   20.3

5. If $X \sim N(6,16)$, determine the 52nd percentile.

   6.2

6. If $X \sim N(10,25)$, determine the 33rd percentile.

   7.8

7. If $X \sim N(36, 3^2)$, determine the 88th percentile.

   39.5

8. If $X \sim N(12, 2^2)$, determine the 0.5 quantile.

   12

CHAPTER 13: THE NORMAL DISTRIBUTION : SOLUTIONS

13E THE Z-SCORE OR STANDARD SCORE

A normal distribution that is standardized has a mean of 0 and a standard deviation of 1. It is also referred to as the standard normal distribution. If we know the population mean μ and the standard deviation σ of a set of normally distributed scores, we can standardize each "raw" score, x, by converting it into a z score by using the following formula on each individual score:

$$Z = \frac{x - \mu}{\sigma} = \frac{raw\ score - mean}{standard\ deviation}$$

A z score reflects how many standard deviations above or below the population mean a raw score is.

EXAMPLE 1
Jack sat two different tests in Semester One. His results in each test, the class average as well as the standard deviation of each test are given in the table below. Express each as a standard score and comment on your result.

	Jack's score	Class Average (mean)	Standard deviation
Test 1	34	36	4
Test 2	45	40	5

SOLUTION
In Test 1, Jack's z-score is $Z = \frac{34-36}{4} = -0.5$

This indicates that Jack scored half a standard deviation below the mean.

In Test 2, Jack's z-score is $Z = \frac{45-40}{5} = 1$

This indicates that Jack scored one standard deviation above the mean.

EXAMPLE 2
The average score on a mathematics test was 68 marks with an standard deviation of 6 marks. If Adrian's z-score is -1.5, how many marks did he score?

SOLUTION

$$Z = \frac{x - 68}{6} = -1.5$$

Solve on CAS, $x = 59$. Adrian scored 59 marks.

EXAMPLE 3
In a normally distributed data set, the mean is 96 and the z-score for a raw value of 104 is 2. Find the value of the standard deviation.

SOLUTION

$$Z = \frac{104 - 96}{\sigma} = 2$$

Solve on CAS, $\sigma = 4$.

EXERCISE 13E

1. A set of data has a normal distribution with a mean of 20 and standard deviation 4. Find the z-scores of the measurements 16 and 28.

   $z = \frac{16 - 20}{4} = -1$

   $z = \frac{28 - 20}{4} = 2$

2. A set of data has a normal distribution with a mean of 500 and standard deviation 50. Find the z-scores of the measurements 450 and 525.

   $z = \frac{450 - 500}{50} = -1$

   $z = \frac{525 - 500}{50} = 0.5$

3. Cars currently sold by Cheap Cars Ltd have an average of 145 horsepower with a standard deviation of 25 horsepower. What is the z-score for a car with 180 horsepower?

   $z = \frac{180 - 145}{25} = 1.4$

4. Aisha's score on an Investigation Test was 72%. The class average was 75 and the standard deviation was 8%. What was her z-score?

   $z = \frac{72 - 75}{8} = -0.375$

5. Suppose a data set is normally distributed with a mean of 120 and a standard deviation of 10.

   (a) What data value is 2 standard deviations above the mean?

   Solve $\left(\frac{x-120}{10} = 2, x\right)$

   $\therefore x = 140$

   (b) What data value is 1.5 standard deviations below the mean?

   Solve $\left(\frac{x-120}{10} = -1.5, x\right)$

   $\therefore x = 105$

6. In a normally distributed data set, find the value of the standard deviation if the following additional information is given.

   (a) The mean is 240 and the z-score for a data value of 225 is -0.5.

   Solve $\left(\frac{225-240}{x} = -0.5, x\right)$

   $\therefore \sigma = 30$

   (b) The mean is 24 and a z-score for the data value of 18 is -2.

   Solve $\left(\frac{18-24}{x} = -2, x\right)$

   $\therefore \sigma = 3$

7. Harold sat two different tests in Semester Two. His results in each test, the class average as well as the standard deviation of each test are given in the table below. Express each as a standard score and comment on which test Harold's performance was better.

	Harold's score	Class Average (mean)	Standard deviation
Test 1	42	50	8
Test 2	38	42	5

Test 1 : $z = \frac{42-50}{8} = -1$ Test 2 : $z = \frac{38-42}{5} = -0.8$

Harold's performance was better in Test 2 as his z-score was lower.